Morphology, ontogeny and phylogeny of the Phosphatocopina (Crustacea) from the Upper Cambrian "Orsten" of Sweden

ANDREAS MAAS, DIETER WALOSZEK & KLAUS J. MÜLLER

Maas, A., Waloszek, D. & Müller, K.J. 2003 00 00: Morphology, ontogeny and phylogeny of the Phosphatocopina (Crustacea) from the Upper Cambrian "Orsten" of Sweden. *Fossils and Strata*, No. 49, pp. 1–238. Germany. ISSN 0300-9491.

About 2500 specimens of different Phosphatocopina species from the Upper Cambrian "Orsten" of Sweden were investigated using scanning electron microscopy (SEM) with reference to their morphology, ontogeny, systematics and phylogeny. Major morphological aspects of the shields and partly of the ventral morphology are described and documented. Ontogeny was described and documented in detail for the best preserved species, *Hesslandona unisulcata* Müller, 1982. Characters of all species were compared and a character matrix was drawn up, forming the basis for a computer-aided phylogenetic analysis of the Phosphatocopina. According to this analysis, the Phosphatocopina can be reconstructed as a monophyletic taxon, characterised by a special all-embracing bivalved shield and small antennulae (both autapomorphies). A recently discovered species of Phosphatocopina from the Lower Cambrian of England is, according to the analysis, the sister taxon to all investigated phosphatocopines, for which the name Euphosphatocopina is proposed. The characteristic interdorsum of several species, a cuticular plate separating the two shield halves, turned out to be an ingroup character, not present in the ground pattern of the Phosphatocopina, and still in the ground pattern of Euphosphatocopina only a simply furrow is developed. The next evolutionary stage has triangular plates anteriorly and posteriorly, which eventually become elongated into the dorsal interdorsum. This character state characterises the ground pattern of the Hesslandonina, the sister group of *Falites fala* Müller, 1964. Our intensive investigation of the Phosphatocopina and the phylogenetic analysis allows a better determination of the ground patterns of the taxa Labrophora *sensu* Siveter *et al.* (in press) (= Eucrustacea + Phosphatocopina) and Eucrustacea *sensu* Walossek (1999). The stem species of Labrophora is characterised by a large set of autapomorphies compared with any of the stem lineage taxa of the Crustacea (serving as outgroup taxa), such as the labrum, the sternum with paragnaths, and the development of coxal portions on the antennae and the mandibles (and only on these limbs). Besides the monophyletic Phosphatocopina, its sister taxon Eucrustacea can also be characterised by autapomorphies of its stem species, namely the development of a specific oligomeric hatching stage, the orthonauplius, and the modification of the first post-mandibular limb into a "mouthpart", the so-called maxillula, both characters plesiomorphically missing in the Phosphatocopina. Phosphatocopina can, therefore, be excluded from being an ingroup taxon of the Eucrustacea, nor are they Ostracoda, into which they were originally placed due to their superficially similar bivalved shield. They are instead a monophyletic taxon, restricted to the early Palaeozoic, and the sister group of the Eucrustacea.

Key words: Systematics; taxonomy; Crustacea; Eucrustacea; Phosphatocopina; Euphosphatocopina new name; Hesslandonina; Dorsospinata new name; morphology; ontogeny; phylogeny.

A. Maas [andreas.maas@biologie.uni-ulm.de] & D. Waloszek [dieter.waloszek@biologie.uni-ulm.de], Section for Biosystematic Documentation, University of Ulm, Helmholtzstrasse 20, D-89081 Ulm, Germany

K.J. Müller, Palaeontological Institute, University of Bonn, Nussallee 8, D-53113 Bonn, Germany

Contents

Introduction

"Orsten"-type fossils

The so-called "Orsten" type of preservation (Müller & Walossek 1991a,b) is a special form of fossil preservation in which mainly arthropods of a small scale, about 100–1000 µm in size, are preserved by the impregnation of their cuticle with phosphate. Shortly afterwards, they became embedded in a limestone matrix by concretionary growth of calcareous, bitumen-containing nodules [see also Seilacher (2001) for this type of nodule]. Neither the source of the phosphate nor the process is completely understood. Specimens of more than 1000 µm in length are rarely, if at all, preserved and either only fragmented or empty valves of bivalve head shields. The most striking fact is that this type of preservation resulted in a three-dimensional preservation of the fossils, which show features that are usually not recognisable in fossils, such as limbs and surface structures like bristles, hairs, pores, and sensilla. Although first described from the Upper Cambrian, "Orsten" type of preservation is not restricted to Cambrian sediments. It is known from different times and from different locations (e.g. Bate 1972: Lower Cretaceous material from the Santana Formation in Brazil; Weitschat 1983a, b: Lower Triassic ostracodes from Spitsbergen). From the Cambrian rocks found in southern Sweden, a series of well-preserved crustaceans and other arthropods has been described by K.J. Müller and D. Waloßek [e.g. Müller 1983; Müller & Walossek 1985a, 1988; Walossek & Müller 1990, 1994; Walossek 1993; see Walossek & Müller (1998a, b) for more general information]. The Swedish "Orsten" material also contains a large number of minute bivalved possible crustaceans, the phosphatocopines, which have never, apart from two earlier papers by K.J. Müller (1979a, 1982a), been described in detail. To fill this gap and to describe what is in fact the largest assemblage of species and taxa of the crustacean and arthropod fauna in the "Orsten", are the goals and subject of this investigation.

History of phosphatocopine research

In 1964, Müller, Germany, first described bivalved arthropods from about 490 to 500 million year old Upper Cambrian "Orsten" calcitic nodules of southern Sweden (Müller 1964a). He produced the material by micropalaeontological etching methods using acetic acid [see Bowring & Erwin (1998) and Landing et al. (2000) for recent dating of Cambrian and the Cambrian/Ordovician boundary]. The phosphatic composition of the shells of these fossils led him to combine 13 of the newly described species plus one already known, into a new taxon Phosphatocopina and he interpreted them as a group of early Palaeozoic Ostracoda.

No additional record of Phosphatocopina from the same sites was reported until Müller (1979a) – after another field trip to Sweden and further etching of nodules – described and illustrated phosphatocopine specimens with preserved appendages and other ventral soft cuticular details. At this time, not much could be added to the question of the true relationships of Phosphatocopina and particularly to the question of whether Phosphatocopina are Ostracoda at all. A new collection of limestones for etching such small fossils was undertaken in the summer of 1979. Although rare in the sediments, phosphatocopine material has been found in various localities. In 1982, K.J. Müller described a new species of phosphatocopines with three-dimensionally preserved soft parts in some detail, which he named *Hesslandona unisulcata* (Müller 1982a).

In the following years, several new phosphatocopines were described from China and Australia, from where Jones & McKenzie (1980) misinterpreted the notably preserved inner lamella of a phosphatocopine as a set of limbs. However, all this phosphatocopine material was very poorly preserved and – apart from one exception described by Walossek et al. (1993) – never showed any soft part morphology (e.g. Fleming 1973; Gründel 1981; Tong 1987; Huo et al. 1991; Hinz-Schallreuter 1993a–c). Again, none of the new forms has been investigated under phylogenetic perspectives (but see Walossek et al. 1993), nor has a systematic review of phosphatocopines been published.

Prior assumptions on the systematics of Phosphatocopina

Müller (1964a) placed the newly established taxon Phosphatocopina within the exclusively Palaeozoic taxon Bradoriida aside the exclusively Cambrian Bradoriina within the crustacean group Ostracoda. This was adopted by most succeeding authors. As a consequence, Bradoriida were regarded as Cambrian Ostracoda and, therefore, as the oldest Crustacea.

Yet, this kind of systematic assignment of the Phosphatocopina resulted from traditional taxonomic classification, and relationships within the Phosphatocopina were – besides the establishment of new families (Hinz-Schallreuter 1993c, 1998) – not considered at all. Indeed, both groups, Phosphatocopina and Bradoriina, superficially resemble the Ostracoda in having a bivalved shield. Despite new morphological evidence based on soft part data, any investigations on the Phosphatocopina continued to be based merely on morphological characters of the shield. Although the unity of Phosphatocopina was never seriously doubted, no phylogenetic characterisations of Phosphatocopina as a monophyletic group can be found in the literature apart from Walossek (1999, 2002).

Soft parts and new perspectives about the phylogeny

The quite varied material from the Swedish "Orsten" with more than 100,000 specimens now at hand, extensively collected previously in the 1970s and early 1980s, permitted a detailed study not only of the head shields but also of the body and appendages. In fact, the soft parts provided not only a good basis for a re-examination of the relationships within the group using a phylogenetic–systematic approach, but also allowed comparison of morphological details of the Phosphatocopina with living crustacean groups.

Müller & Walossek (1991a) pointed to the possibility that Phosphatocopina are not ostracodes but form a distinct group of Crustacea. Recently, Walossek (1999) proposed that the Phosphatocopina could be the sister taxon of the Eucrustacea, the crown-group of Crustacea, which includes all crustaceans with Recent derivatives. He listed several obvious synapomorphies of the two taxa (see p. 193), leaving out a discussion pending this detailed investigation, already underway at the time.

Scope and aims

The scope of this paper is to present a more complete description of the morphology and ontogeny and to discuss the phylogeny of the Phosphatocopina using the data currently available. All phosphatocopine species available from the Upper Cambrian Swedish "Orsten" are described in detail in accordance with a phylogenetic analysis considering shield characters. Of the soft part morphology of the available phosphatocopines, an in-depth account of ventral details and the larval series had to concentrate on *Hesslandona unisulcata* Müller, 1982. This species is not only known from many well-preserved specimens, but also from a large set of different growth stages. Other species do not show as many well-preserved individuals belonging to a comparably high number of growth stages and their body morphology and ontogeny are only briefly added for comparisons. Their description has been postponed to the second part of the description of the Swedish "Orsten" Phosphatocopina.

The phylogenetic analysis of the relationships of Phosphatocopina using soft part characters turned out to be not possible in the originally planned way because the amount of data on soft parts is still not sufficient for this approach. Only soft part characters were considered that were preserved in much of the material due to better sclerotisation of the cuticle, like the labrum or

the sternum and some limbs or limb parts, especially the limb stems. These structures are often recognisable even when the fossil is only badly preserved. Nevertheless, although most of the species at hand are known exclusively from their shields, the consideration of soft part morphology characters is an important tool, not only to reconstruct the relationships of the Phosphatocopina within the Crustacea, but also to help reconstruct the phylogeny within the Phosphatocopina.

Most species from the literature could not be included in the phylogenetic analysis because they are insufficiently known or poorly described. Only two phosphatocopine species from other localities could therefore be considered according to literature data. The phylogenetic analysis of Phosphatocopina presented herein is seen as a broad base for the future assignment of taxa that are still poorly known.

The major aims of this paper are:

- to emend the knowledge of the morphology and larval development of Phosphatocopina, exemplified by *H. unisulcata*;
- to present a list of all records of phosphatocopine taxa dealt with herein in the literature;
- to evaluate if the Phosphatocopina are really monophyletic and to present reasons, i.e. autapomorphic characters, for this decision and a ground pattern of this taxon;
- to evaluate the sister-group relationship between Phosphatocopina and Eucrustacea, and to confirm or reject the monophyletic group Labrophora Siveter, Waloszek & Williams (2003) composed of Phosphatocopina and Eucrustacea as proposed by Walossek (1999) and Siveter *et al.* (2001) and if the latter is the case, to present a set of autapomorphies in the ground pattern of this taxon;
- to evaluate the phylogeny and evolution of the Crustacea in the light of the evidence provided by the Phosphatocopina.

Material and methods

Material

General information

The specimens investigated in this study were collected by Müller, from different outcrops in Västergötland and on the Isle of Öland, Sweden [Figs. 1–4; see Eichbaum (1979) for additional information on the Isle of Öland]. The senior author later investigated the complete phosphatocopine material for this study in Ulm. About 1,300 specimens were previously mounted on SEM stubs, the rest, about 50,000 specimens, were already isolated and stored in Franke cells. From this bulk, another 1,000

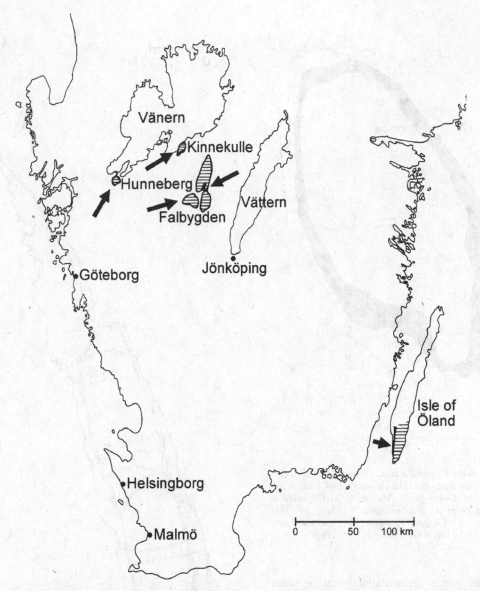

Fig. 1. General view of southern and central Sweden with outcrops of Upper Cambrian alum shale (hatched areas). Regions providing material for this study are marked by arrows. North to the top [after Müller & Hinz (1991)].

specimens were selected for more detailed investigations. All material is from limestone nodules (Fig. 5) intercalated in a black shale sequence, the "alum shale" succession, of Late Cambrian age.

Fossiliferous rocks were collected by Müller from Zones 1–6 of the Upper Cambrian, but only Zones 1, 2, and 5 contained phosphatocopines (Tables 1, 2). Most probably, the original integument of the animals had been impregnated by phosphate at, or immediately after, death and was embedded within calcareous concretions, forming nodules of a diameter of several centimetres up to more than a metre. These concretions are surrounded by black shales, the alum shales (Swedish: "alum skiffer"; Fig. 5). The limestone matrix prevented the fossils from subsequent compaction, which resulted in the three-dimensional preservation of the body. The nodules are

named *orsten*, which originates from a local Swedish name, possibly from "orne sten" = pig stone. The name refers to the early usage of such stones to cure pigs rather than a "stone smelling like a pig" in the sense of stinking stones because of the smell of rotten eggs, due to the high bitumen content when these nodules are broken or treated with acid.

Statistical data

The 2500 investigated specimens, which could clearly be identified as Phosphatocopina, range in size from about 0.1 to 2.4 mm in shield length. About 300 of them are smaller than 0.3 mm, 350 of them range from 0.3 to 0.5 mm, about 850 have a length between 0.5 and 1 mm, and 300 specimens are longer than 1 mm. From another 300 specimens, the length could not be measured suffi-

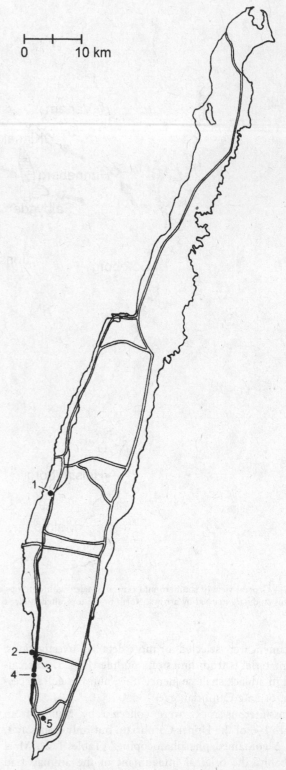

Fig. 2. Locality map of the Kinnekulle area, Västergötland, with out-crops of Upper Cambrian alum shale (black areas). 1 = Gössäter; 2 = Toreborg; 3 = Österplana; 4 = Haggården-Marieberg; 5 = Ödbogården; 6 = Brattefors; 7 = Sandtorp; 8 = Stubbegården; 9 = Gum; 10 = Eke-backa; 11 = Backeborg; 12 = Klippan; 13 = Kakeled; 14 = Pusabäcken; 15 = Trolmen. North to the top [after Müller & Hinz (1991)].

ciently. About 200 pieces are isolated limbs or other body parts. About two thirds of the material is early larval stages, while one third belongs to later larval instars. Size is not a good reference for staging and assignment to taxa. Some species either hatch at later instars than others, they may be smaller at particular stages than other species, or the youngest forms are absent in the material, possibly because they did not live in the same life zone. It remains unclear whether a specimen is fully grown, so it is not known if there are adults within the material at all. For the different species, the numbers of specimens assignable to them varies strongly. Remarkably, all species are restricted to a particular stratigraphic zone (Tables 1, 2).

In the case of *Hesslandona unisulcata* Müller, 1982, from which the ontogeny is described in detail, 120 specimens out of 168 specimens plus isolated legs could be assigned to a particular growth stage. The rest are unassignable fragments. For *H. unisulcata*, the isolated limbs could in almost all cases be assigned to a particular growth stage.

Fig. 3. Locality map of the Isle of Öland with outcrops of Late Cambrian alum shale (black circles). 1 = Eriksöre; 2 = Degerhamn; 3 = S. Möckleby; 4 = Mörbylilla; 5 = Grönhögen. North to the top [after Müller & Hinz (1991)].

Stratigraphy

The Late Cambrian is subdivided into six biozones, ran-ging from the oldest *Agnostus pisiformis* Zone (Zone 1)

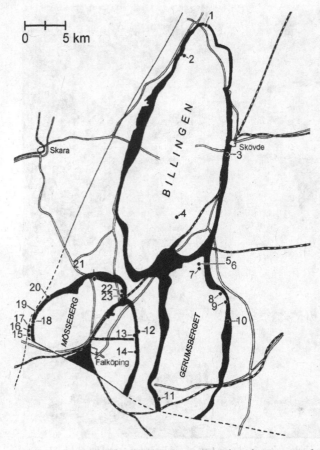

Fig. 4. Locality map of Falbygden, Västergötland, with outcrops of Upper Cambrian alum shale (black areas). 1 = St. Stolan; 2 = Karlsfors; 3 = Karlsro; 4 = Randstadsverket; 5 = Nya Dala; 6 = Stenstorp-Dala; 7 = Stenåsen; 8 = Smedsgården–Stutagården; 9 = Ekedalen; 10 = Ödegården; 11 = Milltorp; 12, 13 = Uddagården; 14 = Djupadalen; 15 = Nästegården; 16 = Ekeberget; 17 = Ledsgården; 18 = Skår; 19 = Kleva; 20 = St. Backor; 21 = Gudhem; 22 = Rörsberga; 23 = Tomten. North to the top [after Müller & Hinz (1991)].

to the youngest *Acerocare ècorne* Zone (Zone 6d) (Westergård 1947) (Table 2). The stratigraphy of the Cambrian southern Sweden area has been described in detail by Westergård (1922, 1947, 1953). The area has been thoroughly investigated (see, e.g. Martinsson 1974; Andersson *et al.* 1985; Berg-Madsen 1985a; Bergström & Gee 1985; Clarkson *et al.* 1998a; Terfelt 2000), with works even from the 19th century (e.g. Wallerius 1895). Wiman (1905) gives numerous references relating to early investigations. Assignment of samples containing phosphatocopines to time zones of the Upper Cambrian is supported by trilobites, representing markers for specific subzones (Westergård 1922; Henningsmoen 1957). Phosphatocopine material is restricted to Biozones 1, 2 and 5 (Table 2). Zones 3, 4 and 6 did not yield any fossil crustaceans at all; at least Zones 3 and 4 yield, e.g. conodonts (Müller & Hinz 1991).

Methods

Isolation of specimens (not done for this work)

Because the material for this study had already been processed in Bonn, Germany, no further preparation was needed. The procedure of processing "Orsten" nodules and the preparation of fossils in productive samples have been described in detail by Müller (1964b, 1985) and Müller & Walossek (1985b). Because this was not part of this work, it will be described only briefly (Fig. 6). The "Orsten" nodules are located within surrounding alum shales (Fig. 6). After collection (Fig. 6, see also Fig. 5) they are cracked to pieces of about the size of a walnut. The fragments are laid on a set of sieves of different mesh size within diluted acetic acid of approximately 10% to dissolve the limestone (Fig. 6). After 14 days the residues are washed and dried. The residues contain the undissolved phosphatic material. Suitable

Table 1. Species of Phosphatocopina from the Late Cambrian Swedish "Orsten" with notes on the numbers of individuals investigated in this study.

Species	Zone	Number of individuals	Page
Hesslandona unisulcata Müller, 1982	1	168	15
Hesslandona necopina Müller, 1964	1	286	58
Hesslandona kinnekullensis Müller, 1964	2	78	67
Hesslandona trituberculata (Lochman & Hu, 1960)	2	14	75
Hesslandona ventrospinata Gründel, 1981	1	105	75
Hesslandona suecica n. sp.	1	280	83
Hesslandona angustata n. sp.	1	31	89
Hesslandona curvispina n. sp.	1	80	91
Hesslandona toreborgensis n. sp.	2	88	104
Trapezilites minimus (Kummerow, 1931)	2	26	106
Waldoria rotundata Gründel, 1981	1	160	113
Veldotron bratteforsa (Müller, 1964)	2	11	121
Falites fala Müller, 1964	5	34	123
Vestrogothia spinata Müller, 1964	5	77	139
Unassignable, mostly larval, individuals	1, 2, 5	570	–

Fig. 5. An Upper Cambrian calcitic "Orsten" nodule embedded within alum shale, taken in the quarry of Uddagården (see Figs. 3:12, 13) near Falköping, Västergötland, Sweden, by D. Waloszek in 1988.

specimens are sorted using a stereo microscope (Fig. 6) and are then stored in Franke cells and labelled according to the sample in which they were found.

Preparation for SEM

About half of the material for this study was already mounted on SEM stubs. Another half was still in the Franke cells and was treated as follows. Selected material from the Franke cells was glued to SEM stubs (Fig. 6:5) using a special kind of wax. The wax is hard at room temperature but melts when warmed with a small flame. When it was soft and sticky, up to eight specimens were arranged in a circle on it, with one distinctive gap to make each specimen easily accessible. Thereafter, the stubs were coated with a 20 nm layer of gold/platinum. The specimens taken for this study were observed mainly under SEM (Zeiss DSM 920) at the "Zentrale Einrichtung Elektronenmikroskopie" of the University of Ulm (Fig. 6:6). Older images were previously photographed in Bonn; some have been documented at the University of Copenhagen. The SEM in Ulm – like the one in Copenhagen – supplies digital images (Fig. 6:7). These digital images were further processed for their use in tables using Adobe Photoshop™ 5.05 and lettered and assembled using Deneba Canvas™ 3.5.5.

Measurements

For ontogenetic studies and for morphometric data, measurements of every single specimen were taken from the SEM images; rounded to the nearest 5 or 10 μm in most cases. Only those values yielding sufficient quantity were documented, such as the greatest length, the greatest height of the shields, the length of the straight dorsal area, the length of the anterior and posterior spines, and the anterior and posterior cardinal angle (Fig. 7A). In addition, the anterior, median and posterior width of the doublure were measured (Fig. 7B) and also the width of the interdorsum (Fig. 7C). These measurements were necessary because traditional taxonomy was always done on the basis of the valves. A parallel system of measuring parts of the ventral soft part morphology was used, thus acquiring additional information for each individual. For the legs, the length of the post-mandibular limbs, the medio-lateral extension of the first endopodal portion and the length of the exopod were documented only for *H. unisulcata* (Fig. 10) because this was the only species for which enough data could be accumulated.

Terminology

The terms used to describe the morphology of the Phosphatocopina and their abbreviations (Table 4) have

Table 2. Stratigraphy of the Late Cambrian of Scandinavia [based on Westergård (1947, 1953) and Henningsmoen (1957); see also Gründel & Buchholz (1981)] considering Phosphatocopina and other "Orsten" components. The youngest zone is above and the oldest zone is below. For phosphatocopine species see Table 1. References for other major Orsten components are Müller (1983), Müller & Walossek (1986b), Walossek & Müller (1990), and Walossek & Szaniawski (1991).

Div.	Step	Subzones	Ph	Other major "Orsten" components
6	d	*Acerocare ecorne*	−	
	c	*Westergardia* sp.	−	
	b	*Peltura costata*	−	
6	a	*Peltura transiens*	−	
	f	*Peltura paradoxa*	−	
	e	*Parabolina lobata*	+	*Dala peilertae,*
	d	*Ctenopyge linnarssoni + Ctenopyge bisulcata*	+	*Bredocaris admirabilis,*
	c	*Ctenopyge tumida + Ctenopyge affinis*	+	*Cambrocaris baltica*
	b	*Ctenopyge flagellifera + Ctenopyge similis*	+	type-A larva
5	a	*Leptoplastus neglectus + Ctenopyge postcurrens*	+	
	e	*Leptoplastus stenotus*	−	
	d	*Leptoplastus angustatus*	−	
	c	*Leptoplastus ovatus + L. crassicorne*	−	
	b	*Leptoplastus raphidophorus*	−	
4	a	*Leptoplastus paucisegmentatus*	−	
	b	*Parabolina spinulosa*	−	
3	a	*Protopeltura aciculata*	−	
	f	*Olenus rotundatus + Olenus scanicus*	−	
	e	*Olenus dentatus*	−	
	d	*Olenus attenuatus*	−	
	c	*Olenus wahlenbergi*	−	
	b	*Olenus truncatus*	+	
2	a	*Olenus gibbosus*	+	*Walossekia quinquespinosa, Oelandocaris oelandica*
1		*Agnostus pisiformis*	+	*Agnostus pisiformis, Walossekia quinquespinosa, Skara minuta, Skara anulata, Rehbachiella kinnekullensis, Martinssonia elongata, Henningsmoenicaris scutula, Cambropachycope clarksoni, Goticaris longispinosa,* type-A larva

Div. = division; Ph = Phosphatocopina; + = present; − = absent.

been traditionally derived from ostracode terminology, because phosphatocopines were originally described as ostracodes due to the superficially similar shield (cf. e.g. Jaanusson 1957; Scott 1961; Williams & Siveter 1998). Ostracode terminology, however, differs in many aspects from general crustacean terminology. Therefore, a more generalised terminology has been applied, which facilitates comparability but also acknowledges some special features of Phosphatocopina. For shield terminology, a feature missing in most other crustaceans, we usually use crustacean terminology supplemented by ostracode terminology in shield description (Fig. 8).

We use the term doublure instead of duplicature for the free inner rim of the valves (cf. Fig. 7B). The inner lamella is the little sclerotised part of the exoskeleton between the lateral extensions of the shield (Fig. 7B). The interdorsum is the continuous bar dorsally between the shield valves of some phosphatocopine species (cf. Fig. 7C).

We do not apply the term "carapace" for the cephalo-thoracic shield of Phosphatocopina, as in the literature (e.g. Hinz-Schallreuter 1998; Siveter *et al.* 2001; Williams & Siveter 1998) and frequently for the shields of various crustacean and even arthropod groups. Walossek (1993)

suggested restricting the term "carapace" to the situation in Crustacea when thoracomeres I–VII are included in the shield. This is the case in Euphausiacea and Decapoda, where the first seven thoracomeres (not eight as often claimed in the literature, e.g. Newman & Knight 1984; Gruner 1993) are incorporated in the shield (cf. Maas & Waloszek 2001b). We agree with Walossek (1993) in using the term carapace only for this special condition. All other crustacean shields can be recognised as either simple head shields retained from the ground pattern of Eucrustacea, including segments of antennula and three post-antennular limbs, or cephalothoracic shields with a different number of thoracomeres integrated. The term domicilium, characterising the inner space of the valves in ostracode terminology is adopted herein (Fig. 9) because it clearly describes the morphological situation.

The description and terminology of the limbs follows that of Walossek as layed down in various papers (e.g. Walossek 1993) (Fig. 10). We use the terms antennula, antenna and mandible for the first three pairs of appendages because they are similarly specialised as in eucrustacean taxa. We only count the post-mandibular appendages from the anterior to the posterior because

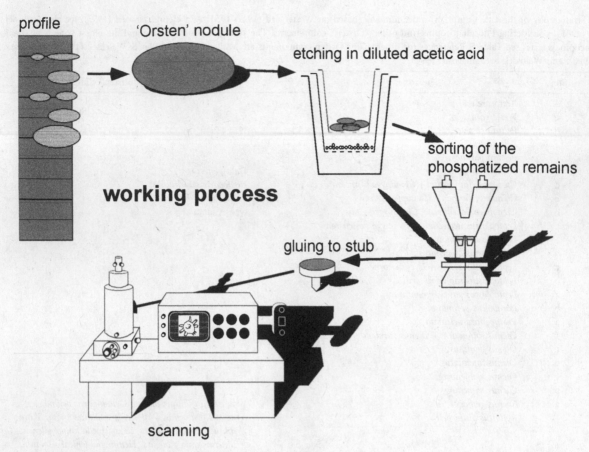

profile

'Orsten' nodule

etching in diluted acetic acid

working process

sorting of the
phosphatized remains

gluing to stub

scanning

Fig. 6. Schematic view of the working process of "Orsten" material. For an explanation see the text.

they are subequal; and a use of the eucrustacean terms maxillula or first maxilla and maxilla or second maxilla would be misleading. Ostracode terminology distinguishes between lobes, nodes and bulbs on the surface of valves. We do not follow this system because labelling a structure as a lobe and another one as a bulb would imply that they have nothing in common. Such an a priori decision could be misleading in a phylogenetic analysis. Therefore, we call any node- or bulb-like structure a lobe in order to retain a neutral position and have numbered the lobes as a combination of that proposed by Gründel (1981, fig. 2) and Williams & Siveter (1998, text-fig. 4e) (Fig. 11). The ostracode terminology uses the term tecnomorph for valves belonging to young and male adults; adult females are called heteromorphs (Hartmann 1966). This terminology is not applied herein as there is no evidence for a clear distinction between sexes in the material studied. It is not known whether there are adults in the material at all.

The signs used in the synonymy lists are supposed to avoid the search for unimportant literature or correspond to the degree of certainty of assignment of a citation to a respective species. They are based on Richter

(1948) and Rabien (1954) and were summarised by Matthews (1973) and Frenzel (2000). They are used herein with a little deviation (see Table 3). Struve (1966) has made some supplementary proposals. They are very useful in brachiopod taxonomy. The phosphatocopine terminology is, however, less complicated due to the low number of species, so his proposals are not applied herein.

Systematics

When first established, the Phosphatocopina were classified as members of the crustacean taxon Ostracoda included in a taxon Bradoriida (= Archaeocopida) housing all Cambrian ostracodes. Many authors are still following this traditional, typologically based system (e.g. Hinz-Schallreuter 1998; McKenzie *et al.* 1999). In 1991, Müller & Walossek (1991a) doubted the assignment of the Phosphatocopina to the Ostracoda, a suggestion followed by other authors (Siveter & Williams 1997; Williams & Siveter 1998, Martin & Davis 2001). Subsequently Walossek & Müller (1998b) supposed

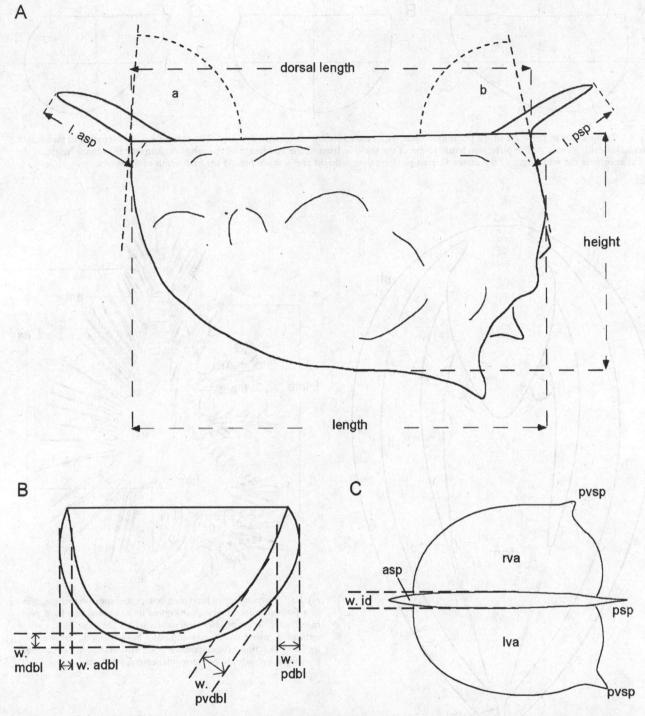

Fig. 7. Terminology and measurements of different structures in phosphatocopine shells. A: Shield measurements including the length and the height of the valves, the length of the anterior (asp) and posterior spines (psp) and the anterior (a) and posterior cardinal angle (b), exemplified by *Hesslandona ventrospinata* Gründel *in* Gründel & Buchholz, 1981. B: Measurement of the width (w.) of the anterior (adbl), median (mdbl) and posterior (pdbl) part of the doublure, exemplified by *Veldotron kutscheri* Gründel, 1981. C: Measurement of the width (w.) of the interdorsum (id), exemplified by *Hesslandona ventrospinata* with unequal right (rva) and left (lva) valves and postero-ventral spines (pvsp).

that the Phosphatocopina are the sister taxon to the Eucrustacea (see also Walossek 1999). Of fundamental importance for this paper is the competitive assignment of Phosphatocopina within Crustacea. In traditional systems the Ostracoda are considered as a monophyletic group of Crustacea, including Phosphatocopina:

A   B   C

Fig. 8. Terminology of the shape of the phosphatocopine shell exemplified by schematised drawings. The anterior left, mid-length and mid-height are marked with a cross. A: Pre-plete: maximum height of the shield in front of the mid-length of the valves. B: Amplete: maximum height of the shield at or near the mid-length of the valves. C: Post-plete: maximum height of the shield behind the mid-length of the valves.

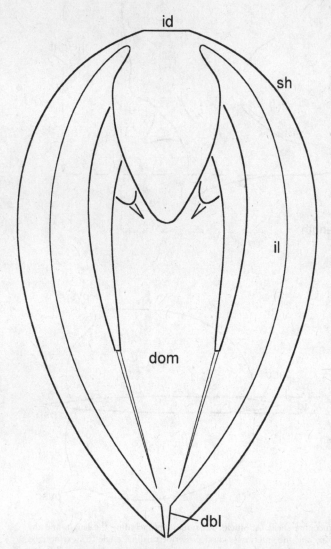

Fig. 9. Terminology of the closed valves of a phosphatocopine. dbl = doublure; dom = domocilium; id = interdorsum; sh = shield. Setae only partly considered.

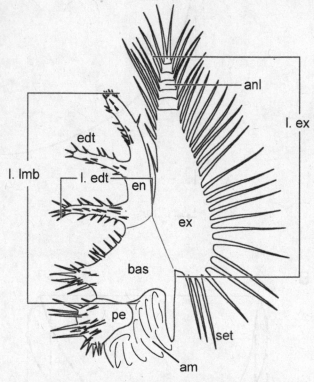

Fig. 10. Terminology of limb morphology and measurement, applied to a post-mandibular phosphatocopine limb, exemplified by a post-mandibular limb of *Hesslandona unisulcata* Müller, 1982. am = arthrodial membrane; anl = annulus; bas = basipod; edt = enditic median projection of endopodal articles; en = endopod; ex = exopod; l. = length; lmb = limb; pe = proximal endite; set = seta.

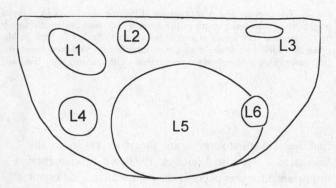

Fig. 11. Terminology of the lobes (L; intentionally drawn as circles or ellipses to show their extent) of the phosphatocopine shield. Lateral view of the left valve based on species of *Cyclotron* Rushton, 1969. Anterior left. L_1–L_6 = lobes 1–6 [after Gründel (1981), but he used H instead of L (for the German word "Höcker")].

Table 3. List of signs used in the synonymy lists of each species, introduced by Richter (1948) if not differently indicated. A combination of several signs in a single citation is possible. Only signs that are used in this work are cited.

Sign	Explanation
Year in Roman	Citation contributes to knowledge of species
Year in italic	Name of the species only mentioned, e.g. in a table or list, without description or illustration
.	Certainty in assignment of individuals to species
?	Assignment of individuals to species uncertain
(?)	Assignment of individuals to species probable but not provable (Rabien 1954)
	Certainty in assignment of individuals to species not clear but likely
*	First valid description
non	Not belonging to the respective species (Frenzel 2000)
p	[pars], only part of the specimens belong to the respective species (Rabien 1954)
v	[vidimus], specimens observed by A. Maas
(v)	Specimens partly observed by A. Maas (newly introduced herein)
cop.	[copy], "cop. *Author*" in citation means that the author used an image that was already illustrated by another *author*

Crustacea Brünnich, 1772[1]

  . (higher taxonomic units not named, see below)

  .

Ostracoda Latreille, 1806
  †Archaeocopida Sylvester-Bradley, 1961
    †Bradoriina Raymond, 1935
    †Phosphatocopina K.J. Müller, 1964
    †Leperditicopida Scott *in* Moore, 1961
  Myodocopa Sars, 1866
    Myodocopida Sars, 1866
    Halocyprida Dana, 1853
  Podocopa G.W. Müller, 1894
    Platycopida Sars, 1866
    Podocopida Sars, 1866

We do not name higher taxonomic units between Crustacea and Ostracoda as it is not important for this consideration. We also do not consider the suggestions concerning the "classification" and phylogeny of Archaeocopida (=Bradoriida) and Leperditicopida nor the idea of Abdomina and Lipabdomina within bradoriids by Shu (1990a) (see also Hinz-Schallreuter 1993c). This discussion is beyond the scope of this paper and does not touch the Phosphatocopina. The system for Recent Ostracoda is adopted from Martin & Davis (2001). In the new suggestion of Walossek & Müller (1998b), the

[1]Concerning the authorship of Crustacea, mostly Pennant (1777) is credited. Holthuis (1991) noted that Brünnich (1772) actually introduced the name.

Phosphatocopina are reconstructed as the sister taxon to the Eucrustacea, while the Ostracoda form a monophyletic group within the entomostracan taxon Maxillopoda:

Crustacea Brünnich, 1772
  "Derivatives of the eucrustacean stem lineage" (five fossil species, see below)
  Labrophora Siveter, Waloszek & Williams
    †Phosphatocopina K.J. Müller, 1964
  Eucrustacea Kingsley, 1894
    Malacostraca [Aristoteles] Latreille, 1806
    Entomostraca O.F. Müller, 1785
      Cephalocarida Sanders, 1955
      N.N.
        Branchiopoda Latreille *in* Cuvier, 1817
        Maxillopoda Dahl, 1956
          "copepodan lineage" (no formal name yet)
          "thecostracan lineage" (no formal name yet)
          Branchiura Thorell, 1864 (affinities unclear)
          Ostracoda Latreille, 1806 (affinities unclear)

Some authors regard the Ostracoda as Maxillopoda (e.g. Grygier 1983); others do not (e.g. Hartmann 1963). The reasons for the decisions are, however, barely or not clearly expressed. The Remipedia Yager, 1981 are herein regarded as maxillopods with affinities to Copepoda because both taxa share some morphological features, and molecular data also point in this direction (Spears, pers. comm. 2001). The "Derivatives of the eucrustacean stem lineage" are a number of species that may not form a monophylum. However, synapomorphies are missing. The species are *Martinssonia elongata* Müller & Walossek, 1986, *Henningsmoenicaris scutula* (Walossek & Müller, 1990), *Cambrocaris baltica* Walossek & Szaniawski, 1991, and the sister taxa *Cambropachycope clarksoni* Walossek & Müller, 1990 and *Goticaris longispinosa* Walossek & Müller, 1990 (see Walossek & Müller 1990, 1998a, b; Walossek & Szaniawski 1991; Walossek 1999).

Phylogenetic analysis

Suitable characters based on our own investigations of the species plus some literature data were encoded in a data matrix (Appendix A.3, see Appendix A.2 for a list of all coded characters) using MacClade (Maddison & Maddison 1992) and were analysed in PAUP (Swofford 1990) with unordered and equally weighed status. As the outgroup taxon for the analysis, *Agnostus pisiformis* [Linné, 1757] Wahlenberg, 1822 (see Müller & Walossek 1987) from Zone 1 of the Upper Cambrian "Orsten" of Sweden was chosen as representing a member of the Euarthropoda but being unassignable yet to any main taxon, i.e. †Trilobita, Chelicerata, Crustacea,

Table 4. List of terms used and their abbreviations. (A) Sorted by term. (B) Sorted by abbreviation.

(A)		(B)	
Term	Abbreviation	Abbreviation	Term
annulus	anl	adbl	anterior part of doublure
antenna, "second antenna"	ant	acsp	antero-central spine
antennula, "first antenna"	atl	am	arthrodial membrane
anterior part of doublure	adbl	an	anus
anterior plate	apl	anl	annulus
anterior spine	asp	ant	antenna
antero-central spine	acsp	apl	anterior plate
anus	an	asp	anterior spine
arthrodial membrane	am	atl	antennula
basipod	bas	bas	basipod
central spine	csp	cox	coxa
coxa	cox	csp	central spine
domicilium	dom	dbl	doublure
doublure	dbl	dom	domicilium
enditic protrusion	edt	edt	enditic protrusion
endopod	en	en	endopod
exopod	ex	ex	exopod
inner lamella	il	id	interdorsum
interdorsum	id	il	inner lamella
latero-caudal spine	lcsp	l.	length
lobe	L	L_{1-6}	lobe 1–6
length	l.	lcsp	latero-caudal spine
limb	lmb	lf	lateral flap (mouth)
limb stem	lst	lmb	limb
lateral flap (mouth)	lf	lst	limb stem
left valve	lva	lva	left valve
mandible	mdb	mdb	mandible
median (ventral) part of doublure	mdbl	mdbl	median (ventral) part of doublure
membrane between limb portions	mem	mem	membrane between limb portions
mouth (opening)	mo	mo	mouth (opening)
posterior part of doublure	pdbl	pdbl	posterior part of doublure
posterior plate	ppl	pe	proximal endite
posterior spine	psp	ppl	posterior plate
postero-ventral part of doublure	pvdbl	pvdbl	postero-part of doublure
postero-ventral spine	pvsp	psp	posterior spine
proximal endite	pe	pvsp	postero-ventral spine
right valve	rva	rva	right valve
seta(e)	set	saf	supra-anal flap
setula(e)	stl	set	seta(e)
shield	sh	sh	shield
supra-anal flap	saf	stl	setula(e)
upper flap (mouth)	uf	uf	upper flap (mouth)
ventro-caudal spine	vcsp	vcsp	ventro-caudal spine
width	w.	w.	width

Atelocerata = Tracheata (see Appendix A.1 for main PAUP settings).

The assumption of a sister-group relationship of Eucrustacea and Phosphatocopina and a sister-group relationship of these two taxa together facing the stem lineage derivatives of Eucrustacea should be evaluated by the phylogenetic analysis. This is only possible if all representative taxa are considered as ingroup taxa along with the phosphatocopines. Apart from the phosphatocopines, all fossil taxa are marked by a "†" in the description of the results and the discussion). The species chosen are:

- †*Cyclotron lapworthi* (Groom, 1902) (see Williams *et al.* 1994b) from the Upper Cambrian of England as a further representative of Phosphatocopina;
- †*Martinssonia elongata* Müller & Walossek, 1986 (see Müller & Walossek 1986a) from Zone 1 of the Upper Cambrian "Orsten" of Sweden as one of the "derivatives of the eucrustacean stem lineage";
- †*Rehbachiella kinnekullensis* Müller, 1983 (see Walossek 1993) from Zone 1 of the Upper Cambrian Swedish "Orsten" as an example of entomostracan eucrustaceans (Branchiopoda);
- *Euphausia superba* Dana, 1852 (see Fraser 1936; Maas

& Waloszek 2001b) from marine Antarctic waters as an example of malacostracan eucrustaceans (Eumalacostraca).

Ostracode taxa are not considered in the phylogenetic analysis. This is because most of the characters dealing with the shield of phosphatocopines had to be coded for ostracode shields as well. This would lead to a distorted analysis and ostracodes would show up near phosphatocopines due to the bivalved shield. If the ostracode shields were coded as different from the phosphatocopine shield they would show up together with *Rehbachiella kinnekullensis* in one branch of the resulting tree.

Results

In the following section, all investigated species of Phosphatocopina from the Late Cambrian of southern Sweden are described in detail with reference to the character list used for the computer-aided phylogeny analysis (Appendix A). The reference to the character list is given as "(chX:Y)" within the following description of features; X refers to the character number and Y is the coded character state (see Appendix A for a list of all characters coded and the data matrix). All records of the respective species we were able to find are listed in the list of synonymies given for each species. The records are listed using a method created by Richter (1948) to make qualifying comments on each single citation (see Table 3, p. 13).

Hesslandona Müller, 1964

*   1964a *Hesslandona* n. g. – Müller, p. 21.
.   1965 *Hesslandona* Müller – Adamczak, pp. 29, 33.
.   1972 *Hesslandona* Mueller – Taylor & Rushton, p. 13.
.   1974 *Hesslandona* – Kozur, p. 823.
.   1975 *Hesslandona* – Müller, p. 177.
.   1978 *Hesslandona* Müller, 1964 – Rushton, p. 279.
.   1979 *Hesslandona* Müller, 1964 – Bednarczyk, p. 218.
.   1982a *Hesslandona* Müller, 1964 – Müller, p. 279ff.
.   1983 *Hesslandona* – Briggs, pp. 9–11.
.   1983 *Hesslandona* – Müller, p. 94.
.   1986 *Hesslandona* – Huo *et al.*, pp. 23–25.
.   1986 *Hesslandona* – Schram, p. 415.
.   *1986a* *Hesslandona* Mueller, 1964 – Kempf, p. 400.
.   *1986b* *Hesslandona* Mueller, 1964 – Kempf, p. 677.
.   *1987* *Hesslandona* Mueller, 1964 – Kempf, p. 436.
.   1990 *Hesslandona* Müller, 1964 – Abushik *et al.*, p. 41.
.   1990 *Hesslandona* Müller, 1964 – Melnikova & Mambetov, p. 57.
.   1990a *Hesslandona* – Shu, pp. 66, 76.
.   1991 *Hesslandona* Müller, 1964 – Melnikova & Mambetov, p. 56.
.   1992a *Hesslandona* – Hinz, p. 15.
.   1993 *Hesslandona* Müller, 1964 – Hinz, p. 12, fig. 2b3.
.   1993b *Hesslandona* Müller, 1964 – Hinz-Schallreuter, pp. 333, 343, 344, 347.
.   1993c *Hesslandona* Müller, 1964 – Hinz-Schallreuter, pp. 386, 395, 396, 402.
.   1994 *Hesslandona* Müller – Hinz-Schallreuter, p. 13.
.   1998 *Hesslandona* – Cohen *et al.*, pp. 251, 253.
.   1998 *Hesslandona* Müller, 1964 – Hinz-Schallreuter, pp. 106–108, 110, 115, 116, 132; text-fig. 1.
.   1998 *Hesslandona* Müller, 1964 – Williams & Siveter, p. 30.
.   1998 *Hesslandona* – Ziegler, p. 223.
.   2001 *Hesslandona* – Chen *et al.*, fig. 4.
.   *2001* *Hesslandona* – Martin & Davis, p. 10.

Derivation of name. – In honour of Ivar Hessland.

Type species. – *Hesslandona necopina* Müller, 1964 by original designation (Müller 1964a).

Original diagnosis. – (Müller 1964a, translated). "Outer and inner lamella with inclusions of apatite. Valves of equal size, surface mostly without sculptures. They may show three lobes, which are missing in juveniles. The morphology of the hinge is generally different from that of all other Ostracoda. In the dorsal midline, a flattened interdorsum is developed that may bear spines at the anterior and posterior end. Interdorsum bordered on both sides by adont hinges, at which the articulation of the respective valve worked. Thus, here are two 'hinges' present. The doublure is rather small. Sexual dimorphism was not observed."

Emended diagnosis. – Valves of equal size, close tightly. Maximum length of valves on the dorsal rim or between dorsal rim and midline. Surface without lobes, with one prominent lobe antero-dorsally or with three lobes in a line close to the dorsal rim, two anteriorly, another one in the last third of the total length. Interdorsum with small loop-like thickenings anteriorly and posteriorly or anterior and posterior end drawn out into more or less long spines. Doublure rather small or relatively wide.

Species referred to taxon. – The taxon *Hesslandona* possibly encompasses 11 species, of which nine are considered in this work. The synonymy list is specified in the respective descriptions:

Hesslandona angustata n. sp. (p. 89);
Hesslandona curvispina n. sp. (p. 93);
Hesslandona kinnekullensis Müller, 1964 (p. 67);
Hesslandona necopina Müller, 1964 (type species, p. 58);

Hesslandona suecica n. sp. (p. 85);

Hesslandona toreborgensis n. sp. (p. 104);

Hesslandona trituberculata (Lochmann & Hu, 1960) Rushton, 1978 (p. 71);

Hesslandona unisulcata Müller, 1982 (p. 16);

Hesslandona ventrospinata Gründel *in* Gründel & Buchholz, 1981 (p. 77);

Another two species, which are not investigated herein, probably belong to *Hesslandona* (synonymy not evaluated herein):

Hesslandona abdominalis Hinz-Schallreuter, 1998

1985b Hesslandonid ostracode (new species ?) – Berg-Madsen, p. 140, figs. 5A–D.

1993c *Hesslandona reichi* ssp. n. A – Hinz-Schallreuter, p. 399.

1998 *Hesslandona abdominalis* n. sp. – Hinz-Schallreuter, pp. 103, 112, 115, 116, 118, 122, 124, 126, pl. 1, figs. 1, 2; pl. 9, fig. 3; pl. 10, fig. 1; table 4.

Hesslandona reichi Hinz-Schallreuter, 1993

1993c *Hesslandona reichi* n. sp. – Hinz-Schallreuter, pp. 392, 396, 399, figs. 6.1–6.3.

1998 *Hesslandona reichi* Hinz-Schallreuter, 1993 – Hinz-Schallreuter, pp. 115, 116.

Remarks. – *Naviformella antiquata* Zhao & Xiao *in* Xiao & Zhao (1986) and *Xiaoyangbaella nudata* Zhao & Xiao *in* Xiao & Zhao (1986) (see Appendix B) from the Lower Cambrian of the Aksu/Wushi region of Xinjiang, China, were referred to *Hesslandona* by Zhao (1989b). Species of Hesslandonina including *Hesslandona* have a bivalved shield with an interdorsum. Both Chinese species, however, have a univalved shield without dorsal structures. Therefore, an inclusion of the species of Zhao & Xiao in *Hesslandona* can be ruled out. They are possibly synonyms of *Dabashanella hemicyclica* Huo, Shu & Fu, 1983 (Siveter, pers. comm. 2002).

Hesslandona unisulcata Müller, 1982

.   1974 *Falites fala* – Martinsson, p. 208 (= *Hesslandona unisulcata*).

p 1978 *Falites fala* Müller – Rushton, pp. 276, 277 [*partim:* specimens BGS BDA 1820 (pl. 26, fig. 12), BDA 1844, BDA 1855, BDA 1863]; non BGS BDA 1824 (= *Waldoria rotundata*); text-fig. 2. (= *H. unisulcata*).

v* 1982a *Hesslandona unisulcata* sp. nov. – Müller, p. 279, pl. 1–8; text-figs. 1–5.

v. 1982c *Hesslandona unisulcata* – Müller, fig. 2.

.   1983 *Hesslandona unisulcata* Müller, 1982 – McKenzie *et al.*, figs. 2, 3.

v. 1983 *Hesslandona unisulcata* Müller, 1982 – Reyment, fig. 2 (cop. Müller 1982a, pl. 6, fig. 1b).

v. 1985a *Hesslandona unisulcata* Müller, 1982 – Müller & Walossek, p. 161, fig. 2a.

.   1986 *Hesslandona unisulcata* – Schram, fig. 33-10(A).

.   1986a *Hesslandona unisulcata* Mueller, 1982 – Kempf, p. 400.

.   1986b *Hesslandona unisulcata* Mueller, 1982 – Kempf, p. 625.

.   1987 *Hesslandona unisulcata* Mueller, 1982 – Kempf, p. 730.

.   1990 *Hesslandona unisulcata* Mueller, p. 275

.   1991a *Hesslandona unisulcata* Müller, 1982 – Müller & Walossek, figs. 1, 2.

.   1993c *Hesslandona unisulcata* Müller, 1982 – Hinz-Schallreuter, p. 400.

.   1993c *Falites unisulcatus* (Müller, 1982) – Hinz-Schallreuter, pp. 392, 400, 402, fig. 7.1.

.   1993 *Hesslandona unisulcata* Müller, 1982 – Whatley *et al.*, p. 345

.   1995 *Hesslandona unisulcata* Müller, 1982 – Müller *et al.*, fig. 4F.

.   1996 *Hesslandona unisulcata* Müller, 1982 – Hou *et al.*, fig. 9b, c (UB 1628, 658).

.   1998 *Hesslandona sulcata* Müller – Boxshall, p. 162 (sic!).

.   1998 *Falites unisulcatus* – Hinz-Schallreuter, p. 114.

v. 1998a isolated post-mandibular limb of a phosphatocopine – Walossek & Müller, fig. 12.3b.

v. 1998a *Hesslandona unisulcata* Müller, 1982 – Walossek & Müller, fig. 12.3a.

.   1998 *Hesslandona unisulcata* Müller, 1982 – Williams & Siveter, p. 28.

.   1998 *Hesslandona* – Ziegler, fig. 234.1 (cop. Müller 1982a, fig. 5).

.   1999 *Hesslandona unisulcata* Müller, 1982 – Whatley *et al.*, p. 345.

.   2001 "einer der ältesten bekannten Krebse aus der Gruppe Phosphatocopina" – Schmidt-Rhaesa & Bartolomaeus, fig. 9A.

.   2001 *Hesslandona unisulcata* – Siveter *et al.*, p. 481.

Derivation of name. – Named after the prominent anterior lobe [erroneously called a "sulcus" by Müller (1982a), meaning depression].

Holotype. – A closed shield illustrated by Müller (1982a) in his pl. 5, figs. 5a, b (UB 674), length 910 μm, height 620 μm (Pl. 1A, C–E).

Remarks. – Originally described by Müller (1982a) as a new species of hesslandonid Phosphatocopina from the Upper Cambrian of Sweden (Zone 1), this species was transferred to the vestrogothiid taxon *Falites* by Hinz-Schallreuter (1993c). She argued that within the Vestrogothiidae, traditionally the second group of Phosphatocopina besides the Hesslandonidae, there are species of *Vestrogothia* from the Middle Cambrian that have a broad interdorsum and species from the Upper Cambrian with an extremely narrow interdorsum. Accordingly, she disregarded the presence or absence of an interdorsum as a generic characteristic. As will be demonstrated in chapter "Hesslandonina" pp. 170–171, this is simply not true. The interdorsum is a reliable character, which developed only once in phosphatocopine evolution and never becomes reduced secondarily. Consequently, any described *Vestrogothia* species having an interdorsum must be a hesslandonid. Notably, the taxon *Falites*, to which Hinz-Schallreuter (1993c) assigned *H. unisulcata*, has no interdorsum, but *H. unisulcata* does. In fact, Rushton (1978) had found some phosphatocopine specimens, which he incorrectly determined as *Falites fala*, but which are specimens of *H. unisulcata* (Williams & Siveter 1998). Furthermore, Martinsson (1974) accepted the incorrect information given by Müller (1964a) to state that the *Agnostus pisiformis* Zone (Zone 1) also contains *Falites fala*, although it occurs exclusively in Zone 5 of the Upper Cambrian as currently recognised. Martinsson (1974) therefore correctly listed *Falites fala* for Zone 5, but his mentioning of *Falites fala* for Zone 1 erroneously refers to *H. unisulcata* (see also Hinz-Schallreuter 1996b). The isolated limb of the Upper Cambrian Swedish material dealt with herein and illustrated by Walossek & Müller (1998a) is now known to belong to *H. unisulcata*.

Type locality. – Gum, Kinnekulle, Västergötland, Sweden (see Fig. 2).

Type horizon. – Upper Cambrian, *Agnostus pisiformis* Zone (Zone 1).

Material examined. – One hundred and sixty-eight specimens of different stages and from different areas, *Agnostus pisiformis* Zone (Zone 1) (Table 5).

Dimensions. – Smallest specimen: valves about 230 μm long and about 170 μm high. Largest specimen: valves about 1600 μm long and about 1050 μm high.

Additional material. – Rushton (1978) had found four isolated valves of *H. unisulcata* from the Outwoods Formation, *Agnostus pisiformis* Zone, Nuneaton District, Warwickshire, UK, but which he erroneously assigned to *Falites fala* Müller, 1964 (see remarks above). Hinz-Schallreuter (1993c) noted a single further specimen from a drift bolder near Stoltera, Mecklenburg-Vorpommern, Germany, dated to the uppermost Middle Cambrian (Table 6).

Distribution. – From the uppermost Middle Cambrian to the first zone (*Agnostus pisiformis* Zone) of the Late Cambrian, northern central Europe.

Original diagnosis. – "Species of *Hesslandona* with prominent anterior sulcus. Dorsal bar without spines but with loop-like thickenings at ends" (Müller 1982a).

Emended diagnosis. – Species of *Hesslandona* with a prominent lobe in the anterior region of the valves. Lobe obliquely elongate, long axis pointing postero-ventrally. Valves slightly pre-plete. Anterior and posterior ends of dorsal margin of valves slightly lifted. Interdorsum with shallow swellings at anterior and posterior ends. Doublure wide, maximum width postero-ventrally, at least 1/10 the length of the valve.

Description. – The material provided different growth stages of *Hesslandona unisulcata* Müller, 1982. Preservation ranges from isolated valves to shields with the whole body still present. Assignment of individuals to ontogenetic stages is based on morphometric data (length, height and shape of the shield, cf. Table 16) as well as the developmental state of limbs and other soft parts, such as eyes, labrum, sternum, and hind body.

Adult material is probably not at hand. The description of *H. unisulcata* is based on data of the largest individuals which show the respective details, therefore no size data are given.

Shield (Fig. 12)

The shield is bivalved (ch1:2) with a long and straight dorsal rim (ch2:2) (Pl. 1A, B; Fig. 12A, B). The maximum length of the shield is slightly dorsal to the midline (ch3:3). The right and left valves are of equal size and symmetrical shape. The maximum height of the valves is anterior to the midline (pre-plete) (ch4:1), about 1.4 times longer than high. The free anterior part of the shield margin starts from the dorsal rim, curves antero-ventrally (antero-dorsal angle nearly 70 degrees) and equally postero-ventrally towards the ventral maximum of the valve (ch6:2). The ventral part of the margin is gently curved (ch7:2). More posteriorly, the margin swings gently upwards more rapidly than anteriorly (ch5:2, angle 60 degrees) to recurve even slightly anteriorly to meet the posterior end of the dorsal rim (ch8:2). The maximum width is slightly dorsal to the midline. The margins are straight throughout, without

Plates

Anterior is in most cases to the left, sometimes up. Single Arabic numbers refer to respective post-mandibular pairs of limbs or segments. A combination of Arabic numbers with limb parts such as "bas" (basipod), refers to the limb parts of the respective post-mandibular limb (e.g. 1pe = proximal endite of the first post-mandibular limb); a combination of "en" (endopod) with Arabic numbers refers to portions of the endopod, counted from the proximal one towards the distal. Lower case letters are used to distinguish between more than one arrow on a single image. Abbreviations are listed alphabetically in Table 4B.

PLATE 1

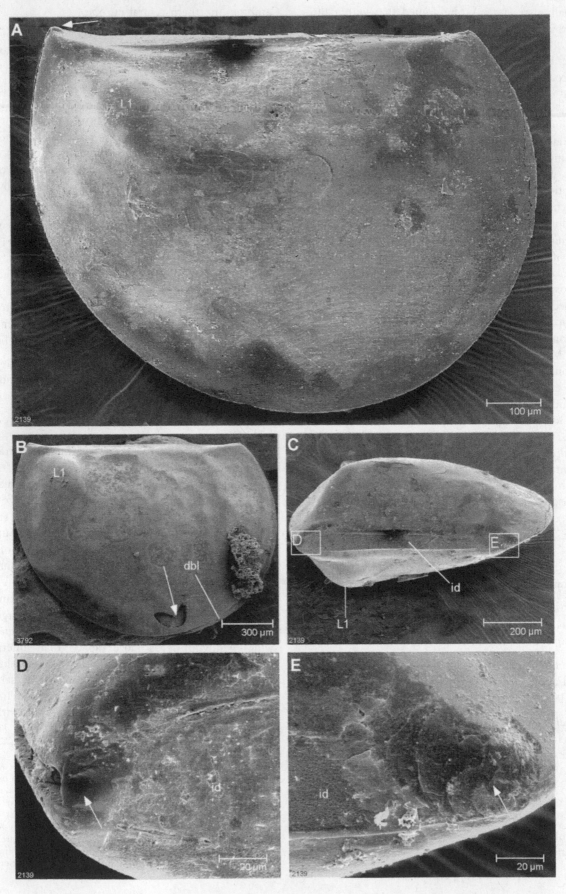

outgrowths (ch19:0, ch20:0), and the shield valves close tightly along the whole margin (ch9:1). The surface of the valves is smooth, but with one large lobe, L_1 (ch10:1) (Pl. 1A, B; cf. Fig. 9) on the anterior area closer to the interdorsum than to the midline of the valves. The lobe is ellipsoidal, with its long axis expanding in the antero-dorsal/postero-ventral direction (Pl. 1A). Other lobes or spines are not present (chs11–18:0, ch21:0).

Interdorsum

The interdorsum is continuous from the anterior to the posterior end of the shield (ch22:3), being laterally bordered by narrow furrows on both sides. From the anterior, the interdorsum widens first, narrows again and remains of nearly the same width until it tapers to its posterior end (Pl. 1C). The maximum width is about 1/14 the length of the valves (ch23:1). The interdorsum is almost flat (ch24:1) with no ornamentation in the middle part (ch25:1) but with small, hump-like thickenings at the anterior (ch26:1, ch27:1, ch28:0) and posterior ends (Pl. 1D, E) (ch29:1, ch30:1, ch31:0, ch32:0, ch33:0).

Doublure

A doublure is present (ch34:1) along the inner margin of the valves, being narrowest anteriorly (ch35:1), ventrally and posteriorly wider, and postero-ventrally slightly wider (ch36:4) (Pl. 2A, B). The maximum width of the doublure is about one sixth the length of the valves (ch37:2). Approximating the dorsal rim, the doublure narrows rapidly, fading out into a membranous area. Several specimens show a more or less irregular row of numerous pits of about 6 μm in diameter near the outer margin of the doublure (ch38:2; Pl. 2C) enclosing about 5 μm long and about 3 μm wide bottle-shaped structures (ch39:2, ch40:2) (Pl. 2C, black arrows), possibly representing sensory structures. The appearance varies, however. Specimens may have only small nodes instead; others may have pits and nodes, while most of the specimens do not show such structures on the doublure at all. It is therefore difficult to state whether this is a real character or simply preservational. More pores of about 1 μm in diameter occur on the outer margin of the ventral part of the doublure (Pl. 2C, white arrows), maybe serving as holes for sensilla (ch41:2) (Pl. 2C, circle). Rhomboid-like structures may occur on the complete surface of the doublure (Pl. 2C, double arrow). No other structures such as stripes are present (ch42:1).

Inner lamella

The inner lamella (Pl. 2B) expands along the whole doublure and extends medially to the dorso-lateral side of the body. The antero-dorsal and postero-dorsal area of the inner lamella extends into the membranous parts of the doublure. It is concave, fitting to the concave inner surface of the valves, leaving only a small internal space between it and the outer dorsal cuticle. The inner lamella frequently shows a wrinkled texture, probably due to its lack of sclerotisation. There are no structures on the inner lamella.

Body

The body proper, which is completely enveloped by the bivalve shield, comprises at least eight segments. Trunk segmentation of the preserved body portion is retained in the insertions of the limbs, isolated sternal plates and depressions of the inner-segmental area of the trunk. Minimally, the first six segments are dorsally fused to the shield (ch44:6, cf. Pl. 2D), being a cephalothoracic shield including two limb-bearing trunk segments. The area of fusion of the body proper to the shield is very narrow, approximately corresponding to the width of the interdorsum. Starting a few micrometres posterior to the dorsal anterior membranous area of the doublures (ch43:1) (Pl. 2B, D), the body proper expands along the inner dorsal length of the shield. At about five sixths of its whole length, the body extends free from the shield into the domicilium. The body proper is, in ventral aspect, oval with a blunt anterior and an elongated posterior end. The maximum width of the body proper is located at the second post-antennular appendages, the mandibles. The cross-section of the body proper anterior to the mandibular part is almost trapezoidal with the long axis dorso-ventrally oriented. The cross-section of the body proper at the mandibular segment is more or less half-oval with the long axis in antero-dorsal aspect, the height of the body proper in this position is about one third the height of the valves. Posterior to it, the body proper changes its cross-section to half-circular and becomes circular from the free segment of the fourth pair of post-mandibular limbs onwards. The height of the body decreases gradually towards the posterior to

---

Plate 1.   *Hesslandona unisulcata* Müller, 1982

A: Holotype [UB 674 (Müller 1982a, pl. 6, fig. 5a, b)]. Image flipped horizontally. A specimen representing possible growth stage VIII, about 910 μm in length. A lateral view of the left valve. Note the large lobe L_1 antero-dorsally on the valve (L1), the arrow points to the loop-like thickening of the anterior-most part of the interdorsum (cf. Pl. 1D).

B: UB W 141. A specimen representing a late growth stage, about 1210 μm in length. A lateral view of the left valve. A part of the shield is broken such that a small window to the inner surface of the doublure is present (arrow). The right valve in the back is somewhat displaced against the left one, displaying the outer rim of its doublure (dbl).

C: Holotype. Dorsal view. Rectangles mark areas magnified in Pl. 1D, E.

D: Close-up of area marked in Pl. 1C, showing the antero-dorsal area of the shield. The arrow points to the loop-like thickening of the anterior-most part of the interdorsum (cf. Pl. 1A).

E: Close-up of area marked in Pl. 1C, showing the postero-dorsal area of the shield. The loop-like thickening of the posterior-most part of the interdorsum (id) is inconspicuous due to the coarse preservation (arrow).

Table 5. Sample productivity of examined specimens of *Hesslandona unisulcata* Müller, 1982.

Sample	Zone	Found at	Number of specimens
5952	–	Between Stenstorp and Dala	1
6364	1	St. Stolan, Falbygden–Billingen	2
6365	1	St. Stolan, Falbygden–Billingen	2
6367	1	St. Stolan, Falbygden–Billingen	1
6404	1–2	West Kestad, between Haggården and Marieberg	2
6408	1	Gum (Kinnekulle), Falbygden–Billingen	4
6409	1	Gum (Kinnekulle), Falbygden–Billingen	15 (holotype included)
6410	1	Gum (Kinnekulle), Falbygden–Billingen	1
6411	1	Gum (Kinnekulle), Falbygden–Billingen	2
6414	1	Gum (Kinnekulle), Falbygden–Billingen	24
6416	1	Gum (Kinnekulle), Falbygden–Billingen	4
6417	1	Gum (Kinnekulle), Falbygden–Billingen	24
6722	1	Trolmen (Kinnekulle), Falbygden–Billingen,	2
6730	1	NNE Backeborg (Kinnekulle), Falbygden–Billingen	1
6732	1	NNE Backeborg (Kinnekulle), Falbygden–Billingen	1
6734	1	NNE Backeborg (Kinnekulle), Falbygden–Billingen	1
6736	1	NNE Backeborg (Kinnekulle), Falbygden–Billingen	1
6739	1	Blomberg (Kinnekulle), Falbygden–Billingen	3
6741	1	Northwest Blomberg (Kinnekulle), Falbygden–Billingen	2
6743	1	Between Blomberg and Kakeled (Kinnekulle), Falbygden–Billingen	2
6748	1	Gum (Kinnekulle), Falbygden–Billingen	3
6749	1	Gum (Kinnekulle), Falbygden–Billingen	1
6750	1	Gum (Kinnekulle), Falbygden–Billingen	13
6755	1	Gum (Kinnekulle), Falbygden–Billingen	2
6758	1	Gum (Kinnekulle), Falbygden–Billingen	2
6759	1	Gum (Kinnekulle), Falbygden–Billingen	2
6760	1	Gum (Kinnekulle), Falbygden–Billingen	11
6761	1	Gum (Kinnekulle), Falbygden–Billingen	5
6762	1	Gum (Kinnekulle), Falbygden–Billingen	6
6763	1	Gum (Kinnekulle), Falbygden–Billingen	2
6764	1	Gum (Kinnekulle), Falbygden–Billingen	13
6765	1	Gum (Kinnekulle), Falbygden–Billingen	1
6768	1	Gum (Kinnekulle), Falbygden–Billingen	1
6771	1	Gum (Kinnekulle), Falbygden–Billingen	2
6772	1	Gum (Kinnekulle), Falbygden–Billingen	2
6773	1	Gum (Kinnekulle), Falbygden–Billingen	2
6774	1	Gum (Kinnekulle), Falbygden–Billingen	1
6783	1	Gum (Kinnekulle), Falbygden–Billingen	2
6785	1	Gum (Kinnekulle), Falbygden–Billingen	1

–=unknown zone.

Table 6. Complete list of finds of *Hesslandona unisulcata* Müller, 1982, reported by different authors, with localities and horizons.

Locality	Horizon	Reference
Nuneaton District, Warwickshire, UK	Upper Cambrian, Zone 1	Rushton 1978
Västergötland, southern Sweden	Upper Cambrian, Zone 1	Müller 1982a and herein
Stoltera, northeastern Germany	Uppermost Middle Cambrian	Hinz-Schallreuter 1993c

about one eighth the height of the valves at its posterior end. Details of the part of the body proper being free from the shield are uncertain.

The anterior part of the body proper is the hypostome/labrum complex consisting of the hypostome, the labrum and the median eye. The hypostome forms the anterior sclerotised ventral surface, which is somewhat rhomboid and becomes gradually higher towards the posterior. Within its antero-lateral edge, the antennulae arise from a peduncle-like protrusion on a distinct lateral slope. The antennae insert postero-lateral of the hypostome in a spindle-shaped insertion area that is located on a lateral slope. The posterior end of the hypostome is drawn out into a lobe-like protrusion, the labrum. Posterior to the hypostome, the ventral surface between the first to third pair of post-antennular limbs is a sclerotic plate, the sternum. It is slightly domed and bears one pair of humps anteriorly, the paragnaths, between the second pair of post-antennular limbs, the mandibles. Posterior to the sternum, the ventral body surface is marked by at least two single sternal plates occurring between the second and third pairs of

A

B

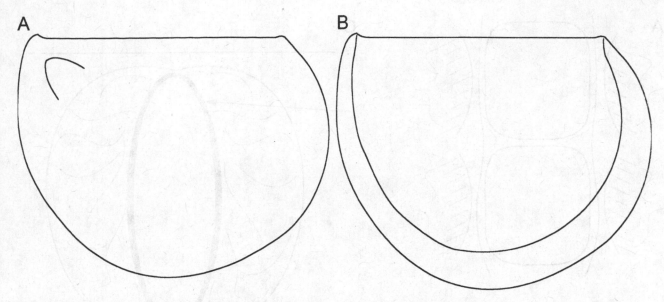

Fig. 12. Schematic drawing of the shield of *Hesslandona unisulcata* from outside lateral (A) and from inside lateral (B).

post-mandibular limbs. The plates are rectangular in antero-posterior aspect with rounded edges; being slightly domed (Fig. 13A) and becoming progressively shorter posteriorly. The post-mandibular limbs insert in a more nor less regular row, corresponding to the sternal plates, the distances between the limbs become progressively shorter towards the posterior, so that they almost touch each other. The anterior-most post-mandibular limbs insert on a lateral slope corresponding to the trapezoidal cross-section of the body. More towards the posterior the insertion areas are located more and more ventrally on the body proper corresponding to the gradual change in the cross-section of the body to the posterior (cf. Fig. 27). The greater insertion areas comprise a sclerotised oval ring flanked by two crescent-shaped membranous fields (Fig. 13B). The lateral walls of the "post-maxillulary" segments are slightly bulged; the inner-segmental areas are slightly depressed (Pl. 12B). ("Post-maxillulary" = posterior to the first post-mandibular limb, the eucrustacean maxillula). The body ends freely from the shield in a more or less conical portion of oval to circular cross-section. The number of segments involved in the hind body is not determinable. There is no limb-less part of the body, i.e. an abdomen, present. The free part of the hind body curves ventrally. The last segment, which is dorsally fused to the shield, is already curved ventrally; realised by a flexible connecting membrane.

A pair of spatula-like, paddle-shaped, flat elements, the furcal rami, inserts terminally on the body (ch47:2) (cf. Fig. 60). Their insertion areas are long-ellipsoid in dorsal to ventro-lateral direction, forming an angle of 90 degrees between one another. The furcal rami are adorned with at least seven setae on their inner and outer lateral margins plus a terminal seta (ch48:2,

ch49:4) (Pl. 12D for possible growth stage VII). The furca is not known from stages earlier than possible growth stage III and later than possible growth stage VII.

Hypostome, labrum and eyes
The hypostome is a rhomboid case with blunt anterior and posterior ends. Its long axis is in the antero-posterior direction, its maximum width is located anteriorly, its minimum width is located posteriorly (Pl. 2C, D). Its maximum height is located posteriorly and is slightly smaller than its maximum width. The hypostome protrudes from near the antero-dorsal rim of the shield to the insertion area of the antennae and posterior to the sternum (Fig. 14A). It is strongly sclerotised. Its antero-lateral edges are slightly extended: the so-called anterior wings. The antero-ventral and latero-ventral margin of the hypostome is slightly excavated, leaving space for a small sclerotised area extending into the inner lamella (Pl. 2E, arrow; Fig. 14B). The postero-lateral edges are slightly extended but not as distinctly as the antero-lateral edges, forming the so-called posterior wings. In some specimens available, the antero-ventral part of the hypostome shows one large circular depression with a diameter of almost the width of the hypostome (Pl. 2E). In most specimens, the structure comprises one pair of oval depressions antero-ventral on the hypostome. Each depression is slightly less than half as wide as the hypostome. In some specimens there is a third depression posterior to the other ones. It is ellipse shape in right–left extension (Pl. 3A). The cuticle in these depressed areas seems to be poorly sclerotised or membranous (Pl. 2E). In a few specimens, the membranous part is expanded to large bulged blisters (Pl. 3B, C); presumably representing the original *in situ* shape of this organ which is interpreted as the median eye. The diameter of the blister is almost as large as the width of the hypostome. A pore

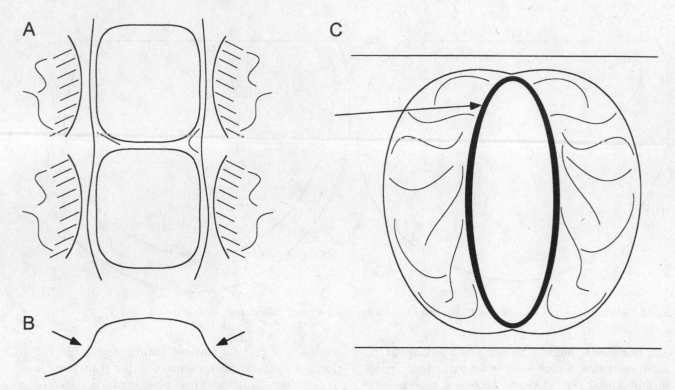

Fig. 13. A: Schematic drawing of two sclerotic plates of the second and third post-mandibular segments from ventral. B: One of these plates as an antero-posterior cross-section, the greater insertion areas are marked as grey lines (arrows). C: Schematic drawing of the greater insertion area of a post-mandibular limb from lateral, showing the two membranous fields and the median oval sclerotised (dark shaded) ring (arrow), leaving space for muscles and nerves (light shaded).

with a small bristle-like protrusion, inserting anteriorly and directing posteriorly, is located medio-distally on the unpaired anterior holes or blister of the median eye (Pl. 3C, D). At the rear of the hypostome, behind the median eye and without being distinctly demarcated off from the hypostome, a more or less conical protrusion with a blunt tip occurs: the labrum (ch61:1). The labrum is almost as large as the hypostome (Pl. 3A, B). It is triangular in antero-posterior aspect with a basal area as wide as the hypostome and tapering only slightly towards the distal end. By about halfway along its length, it narrows with a slight concave curvature of its lateral margins to a more or less pointed tip (Fig. 14A). The labrum is conical in lateral aspect with a slight posterior elevation (Fig. 14C), running straight from proximal to distal in the middle of the posterior surface. The elevation starts from the insertion area and fades out before reaching the tip. The tip of the labrum is directed ventrally so its posterior surface is perpendicular to the ventral body surface (Pl. 3E). Pores are arranged regularly in the midline at the posterior side of the labrum, flanked by sensilla and groups of papilliform structures or hairs at the antero-lateral side (Pl. 3F, G; Fig. 15).

Mouth

The mouth is located proximally on the posterior side of the labrum. It is a not more than 50 μm wide cavity

opening between three flaps, an upper flap and a pair of lateral flaps with curved free margins (Pl. 4A; Fig. 15; cf. Pl. 25C, D). The lateral flaps are slightly adorned with rows of setulae subterminally and terminally (Pl. 4B). The ventral sternal surface seems to run into the opening, probably representing the antenna sternite. Therefore, it is possible that the oesophageal opening is somewhat recessed into the cavity.

Sternum

The post-oral ventral surface of the segments of the second to fourth post-antennular pair of limbs is combined in a single sternitic unit, the sternum (Fig. 14B). It expands from behind the labrum and the mouth to the posterior end of the first post-mandibular segment (ch62:1) (Pls. 2D, 3A, F). It is tongue-shaped and slopes posteriorly (Fig. 14B), its widest part being anteriorly and the narrowest part posteriorly. The sternum has a strongly sclerotised cuticle. In the median line of the anterior part, the sternum is slightly depressed along its length, forming a groove. This groove separates a pair of humps, the paragnaths (ch63:1). The paragnaths are adorned with numerous small hairs in several distinctive laterally bent parallel rows directed from the anterior to the posterior (Pls. 3F, 4B). The posterior part of the sternum is more or less slightly domed and slopes towards the lateral sides. The posterior end is slightly

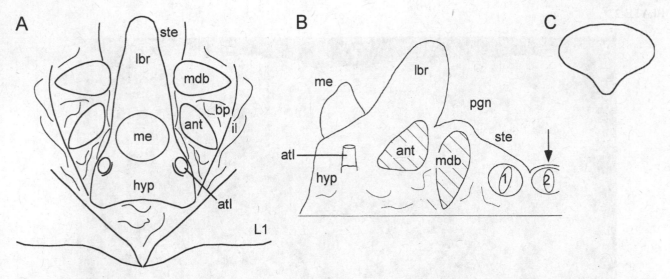

Fig. 14. The hypostome/labrum complex. A: Schematic drawing of the hypostome/labrum complex from the anterior. B: Schematic drawing of the hypostome/labrum complex from antero-ventral. C: Schematic drawing of a cross-section through the labrum at its mid-length, anterior above. See Table 4 for abbreviations.

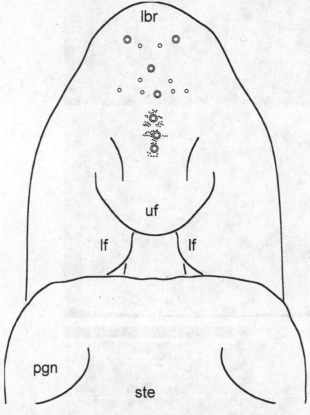

Fig. 15. Schematic drawing of the labrum from the posterior, showing the mouth cavity (pale grey) and the covering flaps (uf, lf) plus the distribution pattern of sensilla (double rings), pores (rings) and papilliform structures (black circles).

sloping and continues into a narrow membrane, which separates the sternum from the sternal plate belonging to the segment of the second pair of post-mandibular limbs. The posterior part of the sternum is smooth.

Antennulae

The antennulae (Pls. 6B, 7B, 9A, 11A) insert in distinct circular openings on both sides of the hypostome antero-proximally about double the diameter of their insertion area behind the anterior margin of the hypostome. The insertion area is located on a peduncle-like structure (Pl. 7B) protruding from the steep lateral wall of the hypostome. The antennulae are smaller than 1/10 the length of the valves (ch46:1). They are rod-shaped, slightly conical and consist of eight irregularly arranged but weakly defined annuli (ch45:1). The basal five annuli are seta-less. The third last annulus bears one seta medio-distally, the seta having about half the length of the antennula. The penultimate annulus bears one seta medio-distally, the seta being about approximately two thirds the length of the antennula. Three bristles insert on the terminal annulus. The medio-distally and latero-distally inserting ones are about as long as the antennula. The third seta inserts distally between the others. It is distinctly thinner but as long as the other setae.

Antennae

The antennae (Pls. 6A, 9A, 10A; Fig. 57A) insert laterally to the posterior part of the hypostome and labrum and antero-laterally to the sternum in the middle of the first third of the length of the shield on the anterior margin of the prominent lobe on the valves. The insertion area is located on a laterally directed slope. It is triangular with a laterally pointed tip and directed at a 45 degree angle of the body axis from near the postero-lateral edge of the hypostome antero-laterally. The antennae show right/left symmetry and consist of a single limb stem and two rami in all stages. This limb stem is a fusion product of the coxa and basipod (see p. 154; ch50:2,

PLATE 2

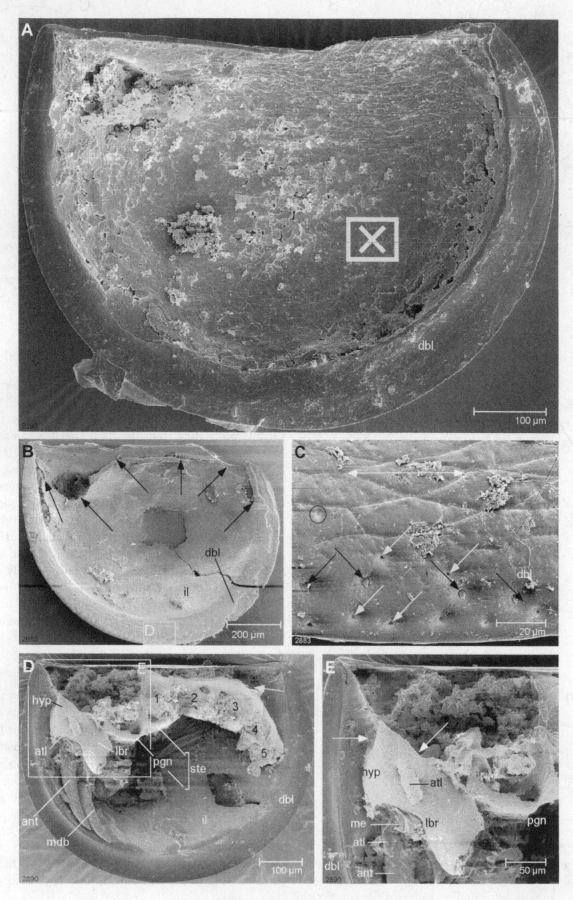

ch51:2, ch52:2), a two-divided endopod (ch53:5) and a multi-annulated exopod. The prominent limb stem is – in posterior view – subtrapezoidal with a pointed lateral tip directed antero-laterally. Laterally, the limb stem slopes antero-laterally in a slight convex curvature. The antero-proximal margin is more or less straight. The postero-proximal margin is concave. The limb stem is medially drawn out into an oblique, antero-posteriorly compressed median gnathobase, which is curved from postero-distal down to antero-proximal around the flanks of the labrum. The gnatho-base bears a row of asymmetrical spines in three distinc-tive groups proximally, medially and distally at its median edge. The arthrodial membrane of the limb stem is very extensive on the median side. The endopod arises medio-distally from the apex of the limb stem distal to the gnathobase and consists of two portions. The prox-imal portion is spindle-shaped and medio-distally drawn out into a short thorn-like spine, guided by two spinules distally and proximally on the anterior and one proxim-ally inserting spinule on the posterior side. The terminal segment is also drop-shaped and drawn out medio-distally into a spine, flanked by one spinule laterally on either side. The position of the endopod corresponds to the latero-distal excavation of the labrum. The exopod

---

Plate 2. *Hesslandona unisulcata* Müller, 1982

A: UB 683 (Müller 1982a, pl. 8, fig. 4a, b). A specimen representing possible growth stage VI, about 730 μm in length. A lateral view inside the right valve. The inner lamella is torn off so that the inner side of the valve is exposed (⊠).

B: UB W 142. A specimen representing a late growth stage, about 980 μm in length. A lateral view inside the right valve. The inner lamella (il) is preserved, the body is torn off but its original expansion is retained at the dorsal margin of the inner lamella (arrows). Rectangles mark areas magnified in Pl. 2C.

C: Close-up of area marked in Pl. 2B, displaying the outer rim of the ventral doublure (dbl). Bottle-like structures arise at the bottom of cone-like pits (black arrows), pores are located irregularly in between (white arrows), one pore shows a small pin (circle) which might be interpreted as the remains of a sensillum. The surface displays rhombus-like structures (double arrow) which might have originally been cells.

D: UB W 143. A specimen representing possible growth stage VI, about 750 μm in length. A lateral view inside the right valve with a partly preserved body. The distal part of the antennula (atl) is broken off, as is the distal part of the labrum (lbr). The right antenna and mandible are represented by their exopods only (ant, mdb). Only five post-mandibular segments (1–5) are preserved, the hind body is miss-ing. The shield of this growth stage is fused to the body up to the third post-mandibular segment (arrow). A rectangle marks the area magnified in Pl. 2E.

E: Close-up of area marked in Pl. 2D, displaying the anterior part of the body. The antennula of the right side is completely preserved (white atl). The median eye (me) is represented by a single circular depression with remains of the membranous surface. Arrows mark the excavated anterior and lateral margins of the hypostome.

arises laterally some distance from the outer slope of the limb stem. The distance to the insertion area of the endopod is about the diameter of the exopod. The exopod consists of more than 30 annuli, which are incomplete sclerotic rings with a more or less membran-ous area medially (Fig. 17A–C). This area flanks a ped-uncle-like structure, a socket that extends into a seta. The first five to eight annuli are seta-less, the distal ones bear a single long seta of at least half the length of the exopod, and the terminal annulus bears two setae. The proximal setae insert medially on the exopod, but towards the tip of the exopod, the membranous insertion area of the setae changes progressively to a more anterior position. Few small seta-less incomplete annuli occur irregularly between the seta-bearing annuli along the exopod. They have no membranous peduncle-like areas. [Such an arrangement is known from eucrustacean exopods as well, e.g. †*Bredocaris admirabilis* Müller, 1983, see Müller & Walossek (1988), or †*Rehbachiella kinnekullensis* Müller, 1983, see Walossek (1993), and the rhizocephalan cirripede *Briarosaccus tenellus* Boschma, 1970, see Walossek *et al.* (1996)]. The setae of the exopod are adorned with fine secondary hairs at least 10 μm in length.

Mandibles

The insertion areas of the mandibles are lateral to the anterior part of the sternum at the end of the first third of the length of the shield at the posterior edge to the prominent lobe L_1 on the valves and slant down from the body proper laterally. They are spindle-shaped with a pointed curve laterally. The mandibles show right/left symmetry (Fig. 18). They are directed at a 90 degree angle against the body axis from the antero-lateral edge of the sternum to far on the lateral side of the body. They consist of a prominent limb stem and two rami. The limb stem is subtrapezoidal in anterior view with a drop-shaped cross-section. The lateral tip is pointed, reaching far to the lateral side, more strongly than in the antenna, in a concave curvature to almost the dorsal end of the body. The anterior and posterior margins are excavated, the posterior one being more strongly excav-ated than the anterior one. Medially, the limb stem is drawn out into an oblique, proximo-distally compressed gnathobase. Its originally distal surface is slightly tilted anteriorly. The gnathobase bears a row of short asym-metrically arranged spines along the whole length of its median edge, a set of two setae is located antero-medially; another two setae are located postero-medially (cf. Fig. 66A, B). The gnathobase is curved around the postero-lateral margin of the labrum. The arthrodial membrane of the limb stem is very extensive on the median and posterior side, larger than the same mem-brane of the antenna, less so anteriorly. From the medio-distal area of the limb stem, a three-segmented ramus

PLATE 3

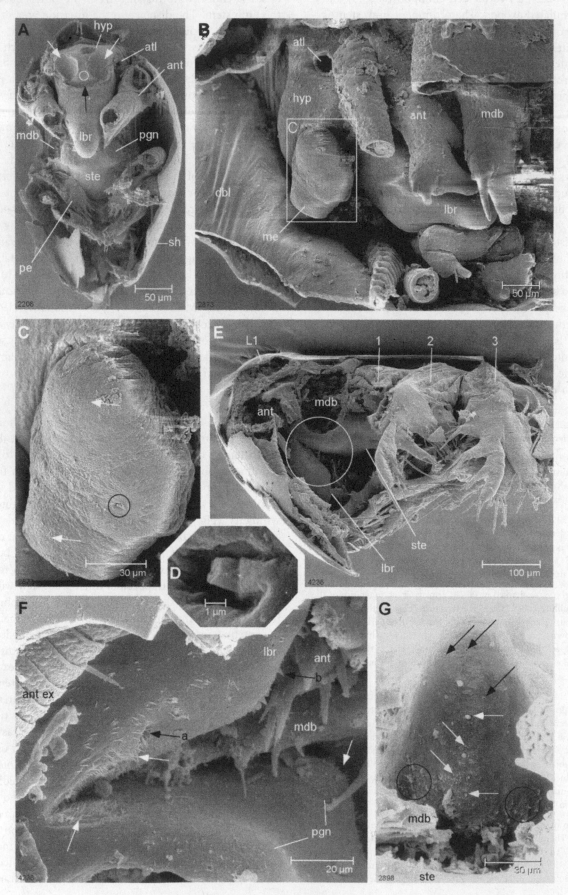

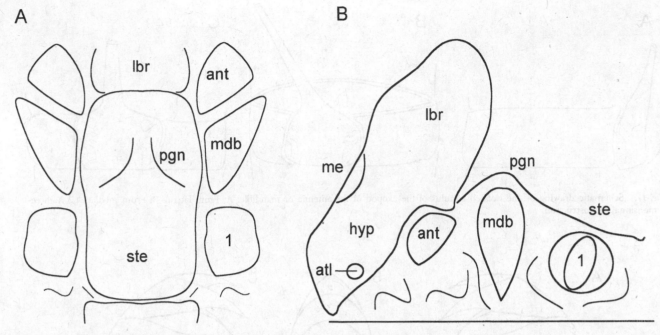

Fig. 16. Schematic drawing of the sternum (ste) from ventral.

arises. As will be shown later (pp. 154–155), this structure consists of a largely reduced basipod and the two-divided endopod, the remaining part of the limb

---

Plate 3.   *Hesslandona unisulcata* Müller, 1982

A: UB 659 (Müller 1982a, pl. 2, fig. 2a,b). A specimen representing possible growth stage III, about 420 μm in length. Antero-ventral view. The median eye consists of two oval depressions (white arrows) plus a third long-ellipse depression (black arrow) distal to them.

B: UB 776 (Müller 1985, pl. 1, fig. 5). A specimen representing a late growth stage, about 1300 μm in length. An antero-lateral view of the head. The median eye (me) is inflated. It is magnified in Pl. 3C.

C: Close-up of the median eye in Pl. 3B, displaying two globes (arrows). The circle marks an area shown magnified in Pl. 3D in a slightly different view.

D: Close-up of the median eye as marked in Pl. 3C. An antero-ventral view, displaying the sensillum with a diameter of between 1 and 2 μm extending into the mouth of a pore.

E: UB W 145. Image flipped horizontally. A specimen representing possible growth stage V, about 630 μm in length (see also Pl. 9D). A latero-ventral view, yielding a view inside the labrum–sternum complex from the lateral side because the antenna and mandible of the right side are missing, being represented only by their insertion scars (ant, mdb). The shield is largely broken and missing. The circle marks the area shown magnified in Pl. 3F in a slightly different view.

F: A postero-lateral view of the transition area of the labrum (lbr) and sternum (ste), showing numerous fine hairs on both structures (white arrows). The black arrow "a" points to a pore, the black arrow "b" points to a putative sensillum (cf. Pl. 3G).

G: UB W 146. A posterior view of the labrum, displaying pores (black arrows), possible sensilla (white horizontal arrows, cf. Pl. 3F), groups of papilliform structures (white sloping arrows), and groups of hairs (circles).

stem being the coxa (ch54:2, ch55:2, ch56:2, ch57:5). The basipodal part is spindle-shaped in antero-posterior aspect with its blunt tip medially, proximo-distally compressed and is drawn out medially into a large enditic spine, flanked by two smaller spines on each side. They are equally long and located proximal and distal to the large enditic spine, pointing medially. The basipodal endite bears additional small tooth-like setulae.

The endopod arises more laterally of the basipodal piece on the apex of the limb stem distal to the gnatho-base and overhangs the basipod endite medially. The proximal endopodal portion is spindle-shaped in antero-posterior aspect with its pointed tip medially. It is medio-distally drawn out into a spine, guided by one proximally inserting and one distally inserting spinule on the anterior side and one proximally inserting spinule on the posterior side. It is distinctly larger than the basipodal endite. The terminal segment arises latero-terminally on the first one. It is comparably drop-like and drawn out medio-distally into a spine, flanked by one spinule laterally on either side. The endopod together with the basipodal endite form a row of three medially directed enditic protrusions. The exopod is equal to the exopod of the antenna (see above).

Post-mandibular limbs ("maxillula", "maxilla" and "thoracopods")

The post-mandibular limbs are equal to each other in general shape and size (ch58:1). The insertion area of the post-mandibular limbs progressively changes its location from the lateral edge of the body proper lateral to the sternites of the respective segment anteriorly to the

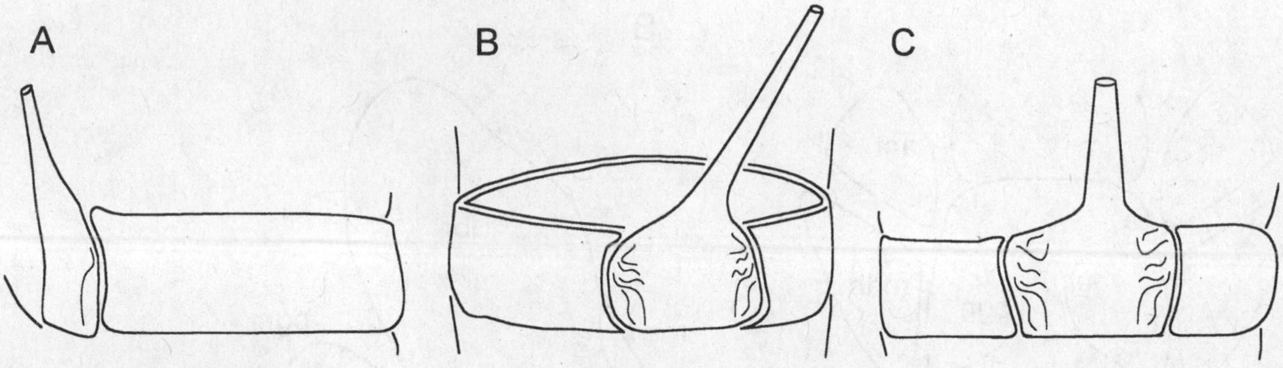

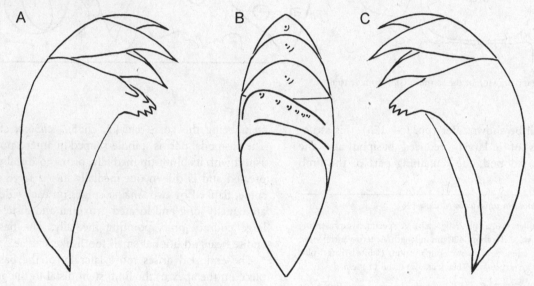

Fig. 17. Schematic drawing of one isolated annulus of the exopod of the antenna or mandible. A: From lateral. B: From median. C: A three-dimensional reconstruction.

Fig. 18. Schematic and simplified drawing of the mandible of either growth stage (cf. Fig. 67A, B). Exopods omitted. A: Left mandible from the posterior. B: Right mandible from median. C: Right mandible from the anterior.

ventral surface of the body proper posteriorly. The insertion areas form a sclerotised oval ring flanked by two crescent-shaped membranous fields (Fig. 13C). The limbs consist of the basipod, the proximal endite, and two branches, endopod and exopod (ch59:1). The limbs are directed anteriorly at an angle of about 60–80 degrees against the body axis. They are subequal but the sizes of the limbs differ slightly. The third one is the largest, while the preceding and succeeding ones become gradually smaller. The limbs are antero-posteriorly compressed, laterally more strongly than medially. The proximal endite inserts medio-proximally to the basipod. It is drop-like with the blunt end medially and bordered by a small membrane proximally and distally. It is irregularly adorned with several setae around its blunt median edge. The basipod is subtrapezoidal with a concave latero-distal margin (Pl. 9C), where the exopod inserts;

it is laterally extended into a concave spatula-like sclerotic bar with a pointed tip, such that the lateral margin of the basipod is an elongate s-shape (Pl. 9C). The antero-proximal margin is almost straight; the postero-proximal margin is excavated (Pl. 9C). The median edge of the basipod is drawn out into an enditic horn-shaped tube. Its cross-section changes from proximo-distally oval to circular from proximal to distal. The medio-distal edge slopes significantly less than the medio-proximal margin. Several setae are located at the blunt tip and spinose setae insert irregularly on the medio-distal slope of the horn. The endopod arises medio-distally from the basipod, the joint membrane being very small. Its lateral margin curves medially from proximal to distal (Pl. 9B, D). The endopod consists of three portions (ch60:4) with prominent medially elongated segments, serving as endites. The first two portions are of almost the same size and shape. They are medially extended to a rod-shaped tube, which is irregularly armed with setae on its whole length and its slightly pointed tip. The second portion arises latero-distally from the first; the third portion arises latero-distally from the second. The third portion is medially curved with fewer and less long setae directed medially. The insertion area of the exopod is oval in lateral direction. It is concave and slightly sloping disto-laterally towards the lateral side of the basipod. The cuticle of the basipod extends, without showing a distinct joint membrane, into the cuticle of the exopod. The exopod is longer than the endopod. It is a flat and drop-shaped plate, which is inserted proximo-laterally at the basipod. The maximum width of the exopod is located off the tip of the basipod. The exopodal plate is armed with regular marginal setation, ranging on the inner side from half-way to the top until the lateral end of the exopod so that the lateral-most four setae are directed proximally. The setae of the exopod are always adorned with fine secondary hairs. The largest limb available shows a distal annulation of the exopod.

Comparisons

Hesslandona unisulcata is very similar to *Trapezilites minimus*. The valves of both species have one prominent lobe L_1, which contrasts with all other species of *Hesslandona* having either no such lobe or more than one. A single lobe is also present in *Falites fala*, but this species lacks the interdorsum. *Hesslandona unisulcata* differs from *Trapezilites minimus* in the outline of the shield which is somewhat shifted towards the posterior with the free posterior margin curving more strongly back than the free anterior margin, while the shield of *Trapezilites minimus* is almost antero-posteriorly symmetrical. The ratio of length to height in *H. unisulcata* is about 1.4 throughout ontogeny, in *Trapezilites minimus* it is about 1.5 throughout ontogeny.

Ontogeny

The ontogeny of *H. unisulcata* could be studied using material from several larval stages, enough to describe some major morphological aspects of differentiation during growth in this species (Fig. 22). The respective possible growth stages are described consecutively beginning with the first known stage. The first one is described in full detail, thereafter only changes in the morphology are noted. Major changes in lengths of different limb aspects are noted for each limb respectively (see Table 16; cf. Fig. 9).

Possible growth stage I

Material. – This stage is represented by six specimens with the shield preserved, among these there is only one specimen with limbs, which are poorly preserved (Pl. 4D). The length of the valves ranges from 230 to 320 μm (Table 7). It cannot be ruled out that the two smallest specimens of 230 (Pl. 4E) and 240 μm in length (Pl. 4F) belong to an even earlier stage because of the significant gap to the other four specimens, but the lack of ventral soft parts prevents any further discussion. Because both specimens do not show any soft part details, there is no evidence for such an assumption. However, the shield of the next larger specimen is, at 290 μm in length, significantly larger.

Shield

The shield is bivalved with a long and straight dorsal rim (Pl. 4D, E). The maximum length of the shield is slightly dorsal to the midline. The dorsal rim of the shield is shorter than the length of the valves. The right and left valves are of equal size and symmetrical shape. The maximum height of the valves is anterior to the midline (pre-plete), about 1.35 times longer than high (Table 7). The free anterior part of the shield margin starts from the dorsal rim, curves antero-ventrally (antero-dorsal angle nearly 70 degrees) and equally postero-ventrally towards the ventral maximum of the valve. The ventral part of the margin is gently curved. More posteriorly, the margin swings gently upwards more rapidly than anteriorly (angle 60 degrees) to recurve even slightly anteriorly to meet the posterior end of the dorsal rim. The maximum width is slightly dorsal to the midline. The margins are straight throughout, without outgrowths, and the shield valves close tightly along the whole margin. The surface of the valves is smooth, but with one large lobe, L_1 (Pl. 5B) on the anterior area closer to the interdorsum than to the midline of the valves. The lobe is ellipsoidal, with its long axis expanding in the antero-dorsal/postero-ventral direction. Other lobes or spines are not present.

Interdorsum

The interdorsum is continuous from the anterior to the posterior end of the shield, being laterally bordered by narrow furrows on both sides. The interdorsum is nearly the same width throughout but tapers to its anterior and posterior ends. The maximum width is about 1/14 the length of the valves. The interdorsum is almost flat with no ornamentation throughout.

Doublure

A doublure is present along the inner margin of the valves, narrowest anteriorly, ventrally and posteriorly equally wider. The maximum width of the doublure postero-ventrally (Pl. 4E) is about 1/10 the length of the valves. Approximating the dorsal rim, the doublure narrows rapidly to merge into a membranous area. There are no structures on the doublure, i.e. pores or supposed sensilla.

PLATE 4

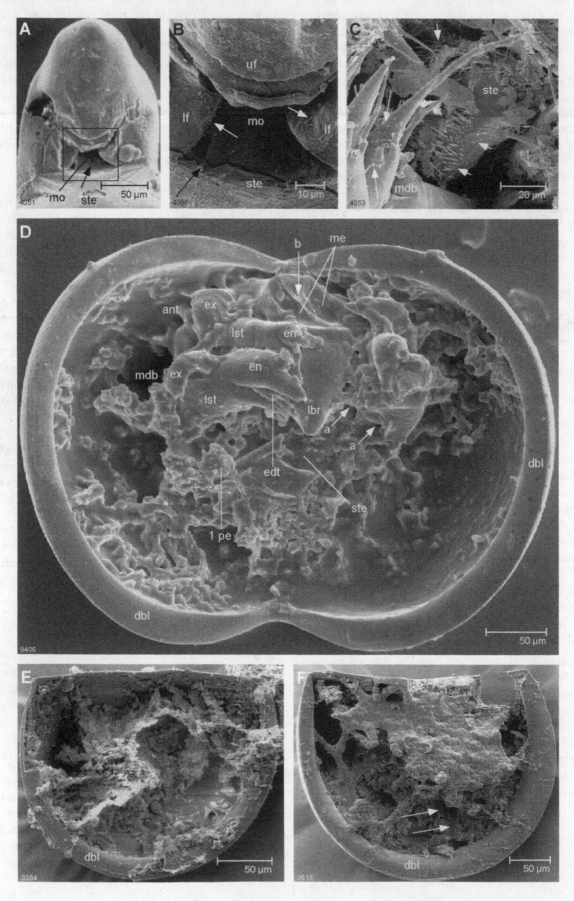

Table 7. List of specimens assigned to the first growth stage, ordered by the length and the height of the valves (see Fig. 7a). Right column: proportion of the mean length to the mean height by height set to 1. Note the large difference between the first two and the third specimens.

| | Specimen | | | | | | |
	1	2	3	4	5	6	Proportion
Length	230	240	290	315	315	315	1.35
Height	170	190	230	205	230	240	1

Inner lamella

The inner lamella expands along the whole doublure and extends medially along the dorso-lateral side of the body. It embraces the body postero-dorsally. Antero-dorsally and postero-dorsally the inner lamella extends into the membranous parts of the doublure. Besides a frequently present texture, there are no structures present on the inner lamella.

Body

The body proper is preserved in one specimen only (Pls. 4C, 5A, B), consisting of at least four segments, dorsally fused to the shield with all four segments

Plate 4. Undetermined phosphatocopine from the Upper Cambrian Zone 1 of Gum, Kinnekulle, southern Sweden. Specimen about 1500 μm in shield length

A: UB W 147. The mouth from the posterior. The width of the mouth opening (mo) is about 30 μm. The anterior sternal plate protrudes into the mouth opening (black arrow). The rectangle marks the area of the image magnified in Pl. 4B.

B: Close-up as indicated in Pl. 4A, displaying the mouth opening with upper and lateral flaps (uf, lf). Fine hairs are located medially on the lateral flaps (lf, white arrows) and right lateral border of the plate-like protrusion of the sternum (ste) into the mouth opening.

Hesslandona unisulcata Müller, 1982

C: UB W 148. A specimen of possible growth stage III, about 460 μm in length. A latero-ventral view of the sternum (ste), displaying the dense setation of the sternum (short arrows). The long arrows point to the fine hairs developed on the mandibular (mdb) enditic protrusion.

D: UB W 149. The smallest specimen having soft parts assignable to *Hesslandona unisulcata*, representing possible growth stage I, about 290 μm in length. Ventral view. A larva with four pairs of limbs, antennulae are covered. The specimen was unintentionally coated with a layer of gold which was too thick. Arrows "a" point to the voluminous median and posterior arthrodial membrane of the mandible. Arrow "b" points to the membrane separating the two axial cups of the median eye (me).

E: UB W 150. The smallest specimen clearly assignable to *Hesslandona unisulcata*, representing possible growth stage I, about 230 μm in length (see also Pl. 5A–E; cf. Table 6). A lateral view inside the right valve. The surface of the inner lamella is covered with dirt.

F: UB W 151. A specimen representing possible growth stage I, about 315 μm in length. A lateral view inside the right valve. The surface is covered with dirt; probably including badly preserved parts of the body and its appendages (arrows).

(Fig. 27A; Table 16), the shield therefore being a head shield. Each segment bears one functional pair of limbs. The body proper is limited to a narrow area expanding between the dorsal anterior and posterior membranous areas of the doublures. It is drop-shaped in antero-posterior aspect and fades out in the second half of its length. Its lateral extension is distinctly broader than the width of the interdorsum. In ventral view, the maximum width of the body proper is located at the second post-antennular appendages, the mandibles. Its cross-section is almost trapezoidal with the long axis abaxially oriented. The height of the body proper in the segment of the mandible is about one quarter the height of the valves. The height of the body decreases gradually posteriorly to about one eighth the height of the valves at its posterior end. The body proper is dorsally fused with the shield on its whole length. The prominent anterior part of the body is made up of the hypostome/labrum complex. Posterior to it, the ventral surface of the body proper is a single sclerotic plate, the sternum, expanding from the mouth posteriorly just behind the insertions of the first post-mandibular limbs. A pair of humps, the paragnaths, marks the mandibular part of the sternum. Segmentation of the body is only recognisable through the insertions of the limbs. There is no distinctive hind body or trunk bud. The hind body fades out into the inner lamella postero-dorsally (cf. Fig. 27A). An anus could not be found.

Hypostome, labrum and eyes

The anterior-most portion of the body is a compound of two structures, the hypostome and the labrum (hypostome/labrum complex; Pl. 5A, B). The heavily sclerotised hypostome protrudes from near the antero-dorsal rim of the shield to the insertion area of the antennae and posterior to the sternum. It is rhomboid-like with its maximum width located anteriorly. Its maximum height is located anteriorly and is slightly larger than its maximum width. Its antero-lateral sides are slightly extended. The antero-ventral and latero-ventral margin of the hypostome is slightly excavated. On the anterior third of the hypostome, a pair of less-sclerotised areas, which are oval and elongated in antero-posterior direction, might represent a paired eye (Pl. 5A, B). The areas are half as wide as long, the maximum diameter is almost as large as the width into the labrum. The original shape of this structure is not preserved in this stage, but as described for the species. Behind the eyes the labrum continues without distinct demarcation from the hypostome. The labrum is as long as its basal width, 50 μm, slightly longer than the hypostome. Its lateral margins are straight and its tip is bluntly pointed. It is triangular in antero-posterior aspect with a basal area as wide as the hypostome, gradually tapering towards the distal end. In lateral aspect, the labrum is conical with a slight elevation in cross-section, running straight in the middle of the posterior surface. The elevation starts from the insertion area and fades out before reaching the tip. The tip of the labrum is directed ventrally so its posterior surface is at about a 90 degree angle to the ventral body surface. It is not known if there are pores, sensilla and fine hairs at the postero-lateral side of the labrum.

Mouth

The mouth is not known from specimens assigned to this growth stage.

Sternum and paragnaths

The whole post-oral ventral surface behind the labrum representing the segments of the second to fourth post-antennular pairs of limbs forms a single sternitic unit, the sternum (Pl. 5A). It has a strongly sclerotised cuticle and is tongue-shaped, with its widest part anteriorly and the narrowest part posteriorly. The median line of the sternum is slightly depressed anteriorly and forms a groove between paired humps, the paragnaths, left and right of it. It is not known if the paragnaths are adorned with hairs.

Antennulae

The antennulae from this instar are not preserved (Pl. 5B).

Antennae

The antennae insert laterally to the posterior part of the hypostome and labrum and antero-laterally to the sternum at the middle of the first third of the length of the shield in front of the prominent lobe L_1 on the valves. The insertion area is located on a slope. It is oval and directed almost antero-posteriorly along the lateral edge of the hypostome/labrum with a slight twist to postero-median (Pl. 4D). The antennae consist of a single limb stem and two rami. The prominent limb stem is about 75 μm long (Table 16). It is pyramid-shaped with a triangular basal area and a pointed lateral angle. The lateral margin is a slope, the proximal margins are excavated, the anterior one more strongly than the posterior, corresponding to the slope of the insertion area. The limb stem is medially drawn out into an oblique, antero-posteriorly compressed, spinose gnatho-base, which is slightly curved around the flanks of the labrum. The median edge of the limb stem is not recognisable in detail. The width of the limb stem including the gnathobase is about 75 μm (Table 14). The arthrodial membrane of the limb stem is very extensive on the median side. The endopod arises medio-distally from the apex of the limb stem distal to the gnathobase. It consists of two portions. The proximal portion is about 60 μm long. It is drop-shaped with the pointed end medially and medio-distally drawn out into one large enditic spine guided by at least one smaller spine prox-

imal to it and an even smaller one antero-distally of it (Pl. 5C). The terminal portion of the endopod is a dome-like protrusion. It is directed medio-distally and is drawn out into a large enditic spine (Pl. 5C). The exopod arises from the lateral slope of the limb stem (Pl. 4D). Its distance to the endopod is less than the diameter of the insertion area of the exopod. The exopod is about 50 μm long and consists of an unknown number of annuli, probably less than 10 in number. The annuli are incomplete sclerotic rings with a more or less membranous area medially (cf. Fig. 17A–C). This area forms a ped-uncle-like structure that extends into a seta. All known annuli, including the proximal ones, carry a seta. The distal-most annulus bears two setae. The proximal setae insert medially on the exopod, but towards the tip of the exopod, the membranous insertion area of the setae changes progressively to a more anterior position. It is not known if seta-less incomplete annuli are present. It is not known if the setae of the exopod are adorned with fine secondary hairs.

Mandibles

The insertion areas of the mandibles are lateral to the anterior part of the sternum and slant down from the body proper laterally. It is oval to drop-shaped with a pointed curve laterally. The mandibles show right/left symmetry and consist of the prominent limb stem and two rami. The limb stem is about 75 μm long (Table 16) and directed at an angle of about 60 degrees against the body axis anteriorly to the lateral edge of the hypostome/labrum complex (Pls. 4C, 5D). It is pyramid-shaped with a triangular basal area. The lateral edge is extended to a pointed angle and forms a steep slope. The proximal margins are excavated, the anterior one more strongly than the posterior, corresponding to the slope of the insertion area. The limb stem is drawn out medially into an antero-posteriorly compressed, spinose gnathobase. The gnathobase is curved antero-proximally around the postero-lateral margin of the labrum and bears a row of asymmetrical spines (Pl. 5C, D). The width of the limb stem including the gnathobase is about 70 μm (Table 14). The arthrodial membrane is very large on the median and posterior sides (Pl. 4D). An additional enditic element is located medio-distally on the limb stem. It is proximo-distally compressed and is drawn out medially into a large enditic spine, surrounded by two distinctly smaller spinules posteriorly and a third one distally (Pl. 5D). It bears tooth-like outgrowths posteriorly. The endopod arises from the apex of the limb stem right above the gnathobase and is directed medially above the additional enditic element. The endopod consists of two segments. The proximal segment is about 60 μm long. It is club-shaped and medio-distally drawn out into a spine, which is guided by two subequal spin-ules anteriorly and posteriorly. It is distinctly smaller than the enditic element. The second segment is located latero-terminally on the first one (Pl. 5D). It is directed medio-distally and is drawn out into a large enditic spine. The exopod arises from the lateral slope of the limb stem. Its distance to the endopod is less than the diameter of the insertion area of the exopod (Pl. 5E). The exopod consists of an unknown number of annuli, probably not more than 10 in number. The morphology of the annuli and the composition of the exopod are the same as in the antenna. It is not known if the setae of the exopod are adorned with fine secondary hairs.

First post-mandibular pair of limbs

The first post-mandibular limb is not known in detail but is clearly not an anlage or limb bud, because it has a proximal endite as a drop-shaped element medio-proximally at the basipod (Pls. 4D, 5A). The proximal endite has one more or less median spine and at least four additional distinctly smaller spinules in an irregular row around the median spine. More smaller spines and/or hairs insert irregularly on the blunt tip of the proximal endite (Pls. 4D, 5A). The exopod is seemingly a multi-annulated ramus with a similar substructure as the exopods of the antenna and the mandible, but with less, about seven, setae. Other structures of the first post-mandibular limb are not known.

Possible growth stage II

Material. – This growth stage is represented by a single specimen with a preserved head shield and parts of the soft parts (Pl. 5F). Length of valves 360 µm, height 260 µm. The proportion of length to height about 1.38 to 1. This stage is distinguished from the previous stage because (cf. Table 16):

- the valves are about 20% longer than in the previous stage;
- the body proper with five segments, is completely fused to the shield, being a head shield (previous stage: four segments, completely fused to the shield);
- the exopods of the antenna and the mandible have 15 annuli (previous stage: not more than 10);
- the second pair of post-mandibular limbs is an anlage (previous stage: not present).

Shield, interdorsum, doublure, inner lamella

There are no morphogenetic changes concerning the shield and its associated structures such as the interdorsum, doublure and inner lamella. All these structures are slightly larger without changes in the proportions, compared with the previous stage.

Body including the hypostome, labrum, eyes, sternum; excluding limbs

The body proper is longer and more slender compared with the previous stage, and is composed of at least five limb-bearing segments, dorsally fused to the shield. Four

pairs of limbs are fully developed, another pair of limbs occurs as unitipped, dome-like buds. Other structures such as the hypostome, eyes, and sternum are not significantly changed compared with the previous larval stage. The labrum is about 60 µm long and slightly wider than its basal width of 50 µm (Table 16).

Antennulae

The antennulae (Pl. 5F; Fig. 19) insert in distinct circular openings on both sides of the hypostome proximally about three times the diameter of their insertion area behind the anterior margin of the hypostome. The insertion area is located on a peduncle-like structure protruding from the steep lateral wall of the hypostome. The antennulae are about 50 µm long, about one seventh the length of the valves. They are rod-shaped, slightly conical and consist of six irregularly arranged but weakly defined annuli (Fig. 19). The basal four annuli are seta-less. The penultimate annulus bears one seta medio-distally and almost vertical; the seta has less than half the length of the antennula. The terminal annulus bears three bristles. The medio-distally and latero-distally inserting ones are almost as long as the antennula. The third seta inserts distally between the others. It is distinctly thinner but as long as the other setae (Fig. 19).

Antennae

The limb is larger than in the previous stage (Table 16), with no changes in shape and proportion. The limb stem is undivided, 90 µm long, its median edge is straight and bears a row of four teeth-like spines inserting one above each other (Pl. 5F). The medio-lateral extension of the limb stem including the gnathobase is about 80 µm. The endopod is unknown. The exopod is about 100 µm in length, consisting of 15 annuli (Table 16), the proximal without setae, the terminal with two, the remainder with one seta.

Mandibles

The proximal portion and endopod are insufficiently known. The limb stem is about 90 µm long, and the medio-lateral extension of the limb stem including the gnathobase is about 80 µm. The exopod (Pl. 5F) is about 100 µm long (Table 16) and consists of 15 annuli (Table 16).

First pair of post-mandibular limbs

The first pair of post-mandibular limbs (Pl. 5F) is too poorly preserved in the single specimen at hand to permit a detailed description and to determine any length data (cf. Table 16).

Second pair of post-mandibular limbs

The second pair of post-mandibular limbs (Pl. 5F) is too poorly preserved to permit any information (cf. Table 16).

PLATE 5

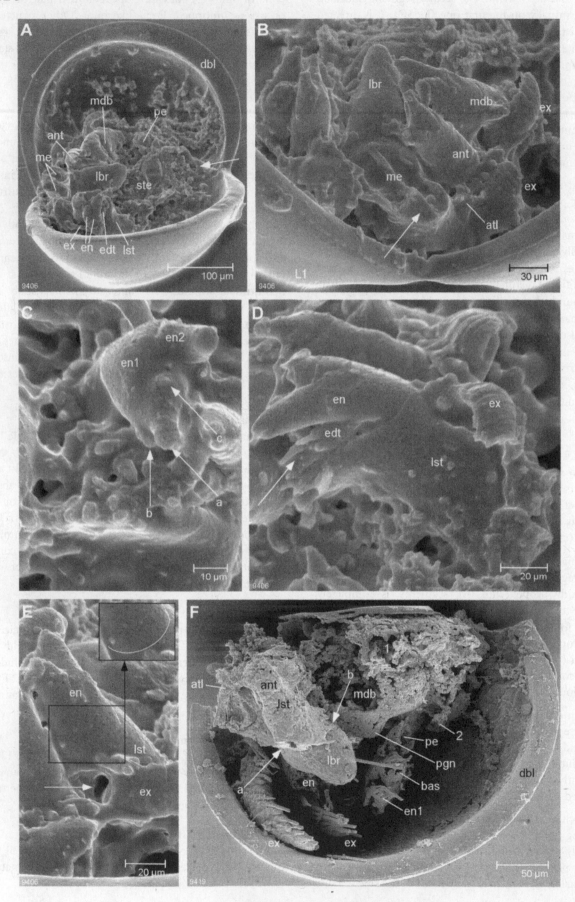

Hind body

The body is not preserved behind the segment of the second pair of post-mandibular limbs, but seemingly ruptured around the time of the death of the animal.

Possible growth stage III

Material. – Nine specimens, with the shield preserved, three of them with more or less well-preserved limbs (Pls. 6A, B, 7A; Table 16). The length of the valves ranges from 410 to 470 μm (Table 8). This stage is distinguished from the previous stage because (cf. Table 16):

- the valves are about 22% longer than in the previous stage;
- the body proper has seven segments, fused to the cephalothoracic shield with six segments (previous stage: five segments, completely fused to the head shield);
- there is an isolated sternitic plate, belonging to the segment of the second pair of post-mandibular limbs, posterior to the sternum (previous stage: segment without such a plate);

Plate 5. *Hesslandona unisulcata* Müller, 1982

A: Image flipped horizontally. The same specimen as in Pl. 4D. Latero-ventral view. The antenna and mandible of the left side cover the hypostome. The first post-mandibular limb is represented almost only by its proximal endite (pe). The hind body is not preserved (arrow).

B: The same specimen as in Pls. 4D and 5A. Anterior view. The body is somewhat distorted (arrows). Only the proximal portion of the right antennula (atl) is preserved, without displaying details due to the thick layer of gold.

C: The same specimen as in Pls. 4D, 5A, B. A median view of the antennal endopod. Note the large spine arising from the enditic protrusion (arrow "a") of the first endopodal segment (en1) and the guiding spines underneath (arrow "b") and antero-distally (arrow "c"), which is represented by a scar only.

D: Image flipped horizontally. The same specimen as in Pls. 4D, 5A–C. A posterior view of the mandible. Note the enditic protrusion (edt) squeezed between the limb stem (lst) and the endopod (end) with its median spine and guiding spinules (arrow).

E: The same specimen as in Pls. 4D, 5A–D. A lateral view of the mandible. The exopod (ex) is broken off (arrow). The insertion area of the endopod (end) on the limb stem (lst) is indistinct, probably due to the thick gold layer, traced in the inset.

F: UB W 152. Image flipped horizontally. The only specimen assigned to possible growth stage II, about 360 μm in length. A lateral view inside the left valve. The right antenna is only represented by its limb stem (lst), the endopod (arrow "a") and the exopod are torn off. The medio-distal part of the limb stem shows a row of four setae (arrow "b"). The right mandible is missing (arrow "c"); the left one is represented by its exopod (ex) and endopod (end) sticking out from behind the labrum (lbr). The first post-mandibular limb (1) is preserved only on the left side, but shows no details because of the coarse preservation of the whole specimen. The second pair of post-mandibular limbs (2) might be represented by a bud, which does not show any details.

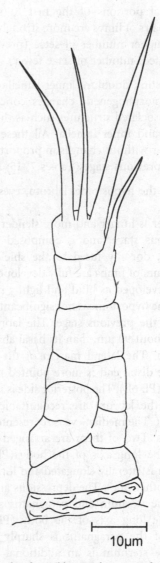

10μm

Fig. 19. Antennula of possible growth stage II from the posterior, showing the irregular annulation of the limb and the distal setation.

Table 8. List of specimens assigned to the third growth stage, ordered by the length and the height of the valves (see Fig. 6a). Right column: proportion of the mean length to the mean height by height set to 1.

| | Specimen | | | | | | | | | |
	1	2	3	4	5	6	7	8	9	Proportion
Length	410	415	415	430	440	440	450	450	470	1.33
Height	310	320	320	320	330	330	330	330	350	1

- the exopods of the antenna and the mandible have about 24 annuli (previous stage: exopods with 15 annuli);
- the second pair of post-mandibular limbs are completely developed (previous stage: as anlagen);
- the third pair of post-mandibular limbs are completely developed (previous stage: not present);
- the fourth pair of post-mandibular limbs is an anlage (previous stage: not present);

- the endopodal portions of the first to third post-mandibular pairs of limbs are more strongly elongated and with a higher number of setae (previous stage: not as elongated, number of setae fewer).

Shield, interdorsum, doublure, inner lamella

There are no morphogenetic changes concerning the shield and its associated structures such as the interdorsum, doublure and inner lamella. All these structures are slightly larger without changes in proportions, compared with the previous stage (Tables 7, 15; Fig. 11).

Body including the hypostome, labrum, eyes, sternum; excluding limbs

The body proper is longer and more slender compared with the previous stage, and is composed of at least seven segments, dorsally fused to the shield with six segments. Six pairs of limbs are fully developed, another pair of limbs developed as bifid and lightly setose buds (Pls. 6B, 8B). The hypostome is not significantly changed compared with the previous stage. The labrum is distinctly longer, about 90 μm, than its basal abaxial width of about 60 μm. The lateral margin of the labrum is constricted. The distal end is more pointed than in the preceding stage (Pl. 6B). The posterior side is not known. No setulae on the labrum are recognisable but were probably present. The median eye is represented by three oval depressions. Two of them are arranged parallel to each other in the long axis of the body (Pl. 6B). The third depression is laterally elongated and long-ellipsoid anterior to the other ones. The depressions are all about 50 μm long. The sternum and paragnaths are slightly elongated in the long axis of the body (Pl. 6B). The outer margin of the paragnaths is sharply truncated. Posterior to the sternum is an additional sclerotised rectangular and domed sternitic plate (Pls. 6B, 8B), belonging to the second post-mandibular segment. Lateral to it, the second pair of post-mandibular limbs inserts into an area consisting of a sclerotised oval ring flanked by two crescent-shaped membranous fields (cf. Fig. 12B). The subsequent body proper is oval in abaxial cross-section and becomes gradually smaller towards the posterior. The segments of the third and fourth post-mandibular pairs of limbs have no distinct ventral sternitic plate (Pl. 8B). The area is membranous and slightly narrower than the respective preceding one. The segment of the fourth pair of post-mandibular limbs is free from the shield and is slightly turned ventrally (Pl. 6B). A pair of flat spatula-shaped, soft and bud-like furcal rami of about 40 μm length occur at the terminal end of the body at an angle of 90 degrees to each other. They are not demarcated from the body by distinct joints. They are dorso-ventrally flattened and bear two setae proximally and subterminally on the lateral, two setae subterminally on the median side and two setae at its terminal end (Pl. 8C, F; Fig. 20).

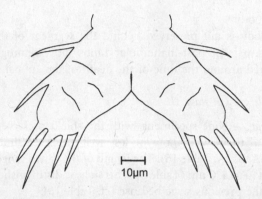

Fig. 20. Furcal rami of possible growth stage III. See Pl. 8F for details.

Antennulae

The antennulae (Pls. 6B, 7B) are not changed compared with the previous stage. They are about 50 μm long, as in the preceding stage, about one ninth the length of the valves.

Antennae

The antennae (Pl. 6A) are larger (Table 16) and more densely setose than in the preceding stage. The gnathobase of the limb stem bears a row of five short spines. The endopod is almost unchanged. The terminal segment is directed medio-distally, drawn out into a large enditic spine, guided by at least one seta latero-distally (Pl. 6A). The exopod (Pls. 6A, 7A) is about 130 μm long and consists of about 24 annuli. The distance between the insertion areas of the endopod and the exopod is as long as the diameter of the insertion area of the exopod.

Mandible

The mandibles (Pls. 6A, 7A) are larger (Table 16) and more densely setose than in the preceding stage. The limb stem and endopod are unchanged. The exopod is about 130 μm long and consists of about 24 annuli. The distance between the insertion areas of the endopod and the exopod is as long as half the diameter of the insertion area of the exopod.

First pair of post-mandibular limbs ("maxillulae")

The first pair of post-mandibular limbs inserts lateral to the posterior part of the sternum in an area of about 35 μm in diameter. It is about 140 μm long (Table 16). Because the first pair of appendages is poorly known from the preceding stages, it is described in detail here for the first time in ontogeny. The limbs consist of the basipod, the proximal endite, and two branches, the endopod and exopod (Pl. 7C; Fig. 24A). The proximal endite inserts medio-proximally into the basipod. It is drop-like with the blunt end medially and bordered by a small membrane proximally and distally. The tip of the proximal endite bears a row of six setae anteriorly and about three setae posteriorly. A seta more than twice as long, about 100 μm, arises distally on the median tip of the proximal endite (Pl. 7A; cf. Pl. 4C). The basipod is subtrapezoidal in profile with a laterally curved slope (cf. Pl. 7A). It is slightly compressed in antero-posterior aspect, laterally more strongly than medially. The median edge of the basipod is dome-like with a blunt tip, comparable with the median edge of the proximal endite. Its cross-section is circular. Setae are located at the blunt tip of the basipodal endite, forming an anterior row and a posterior row with more setae plus a group of one large and two smaller setae medially. The endopod arises medio-distally from the basipod. It is about 100 µm long and consists of three portions, slightly elongated in a medio-distal direction, serving as endites. The first two portions are of almost the same size and shape. They are tube-like with a blunt rounded median edge. The medio-lateral extension of the proximal portion including the enditic elongation is about 60 µm (Table 16). The enditic tip of the proximal portion has four setae, arranged more or less irregularly. The medio-distal enditic tip of the second endopodal podomere bears two setae, one at the anterior edge, one at the posterior edge. The tip of the third portion has one spine-like seta and three additional distinctly weaker setae arranged in a curved row proximal to the spine. A finer seta inserts on the termino-lateral edge of the third segment. The insertion area of the exopod is circular. It is slightly sloping from disto-laterally towards the lateral side of the basipod, distinctly separated from the basipod by a membrane. The exopod is 120 µm long, longer than the corresponding endopod. It consists of about 15 conical annuli of the same substructure as the annuli of the antenna and the mandible. The first five annuli are laterally fused and depressed and slightly drawn out to a narrow lateral edge (Pl. 7C; Fig. 24A) that bears five long irregularly setose setae laterally, seemingly belonging to the respective proximal five annuli.

Second and third pairs of post-mandibular limbs ("maxillae", "thoracopod I")

The second and third pairs of post-mandibular limbs are similar to the first pair and subequal. A seta inserting at the medio-distal part of the proximal endite, as in the first pair of post-mandibular limbs, is in fact present but not longer than 25 µm. The second and third pairs of post-mandibular limbs are about 140 µm long and insert at the postero-lateral edges of the respective sternite. The arthrodial membranes of the limbs are very extensive (Pl. 8A; Fig. 25A). The proximal endite bears a more or less irregular circle of setae around one additional distinctly stronger thorn-like seta in the middle of the blunt enditic tip. The basipod is subtrapezoidal and bears a long dome-like endite with six setae which are arranged in a circle around two additional spine-like setae in the middle of the enditic tip which is not as elongated as the enditic tip of the proximal endite. Three additional setae are located more proximally on the

posterior side (Pl. 7A). The endopod inserts distally on the basipod. It consists of three portions with large medially elongated segments, serving as endites. The first two portions are of almost the same size and shape. The medio-lateral extension of the proximal segment including the enditic elongation is about 60 µm (Table 16). The proximal portion bears one long spine-like seta at the distal part, a small seta is located posterior to it, another one inserts at the antero-distal part of the enditic tip, three additional setae are located postero-proximally at the enditic tip (Pl. 7A, arrow). The second portion is rod-shaped. The third portion of the endopod is tube-like and medially curved. Its median tip bears three bristles directed medially, arranged in a similar way as in the second podomere. A large spine-like seta inserts at the distal part of the podomere, another one is located antero-proximal to it and a third one inserts postero-proximally (Pl. 7E). The exopod arises from a proximo-lateral oval-shaped area. The insertion area is slightly sloping from disto-laterally towards the lateral side of the basipod, distinctly separated from the basipod by a membrane. The exopod is about 120 µm long (Table 16) and as long as the endopod. It is a plate compressed in antero-posterior profile and somewhat drop-shaped. Its maximum width is distal of the basipod. The exopodal plate shows no segmentation proximally, while its distal part is annulated with nine annuli, which are oval in profile, antero-posteriorly compressed. The more distal the annuli are located the more their profile changes from long-oval to circular. The exopod is armed with regular marginal setation on the inner side from halfway to the top; the lateral side is armed with five setae proximally. The setae of the exopod are always lightly adorned with irregularly arranged fine secondary setulae.

Fourth pair of post-mandibular limbs ("thoracopods II")

The fourth pair of post-mandibular limbs ("thoracopods II") inserts on the ventro-lateral edges of the posterior part of the sternal surface of the only segment which is not fused to the shield (Pl. 8B). It consists of an undivided limb stem and two rami (Pl. 8C). A proximal endite is not present. The limb stem is slightly oval in cross-section and is more or less dome-like with an oval-straight distal surface. It is about 20 µm long and 40 µm wide (Table 16). It is poorly sclerotised, obviously soft and bears no setae (Pl. 8C). The endopod inserts medio-distally on the limb stem. It is about 40 µm long (Table 16) and consists of one spindle-shaped portion bearing two terminal setae. The exopod inserts at the latero-distal edge of the limb stem. It is about 40 µm long and consists of three annuli, which are oval in cross-section. The proximal one is seta-less, the middle one bears one seta latero-distally and the terminal one bears two setae distally (Fig. 26A).

PLATE 6

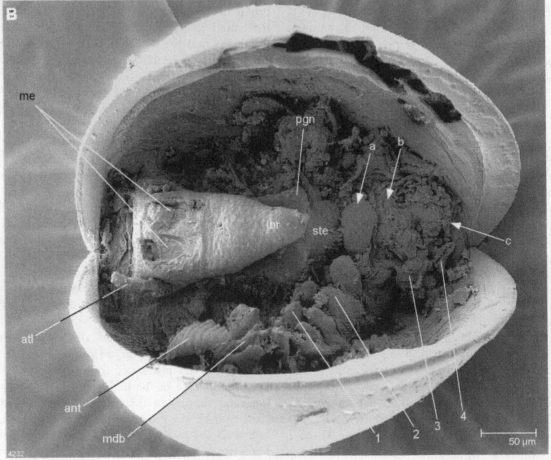

Possible growth stage IV

Material. – Nine specimens with the shield preserved, one of them with more or less well-preserved limbs (Pl. 9A), and two additional isolated limbs assignable to this stage by their details and size. Length of valves ranging from 510 to 560 μm (Table 9). This stage is distinguished from the previous stage because (cf. Table 16):

- the valves are about 22% longer than in the previous stage;
- there is an additional isolated sternitic plate, belonging to the segment of the third pair of post-mandibular limbs (previous stage: only isolated sternitic plate belonging to the second pair of post-mandibular limbs);
- the endopodal portions of the post-mandibular pairs of limbs are more strongly elongated and with more setae (previous stage: not as elongated, number of setae fewer);
- the exopods of the first to third post-mandibular pairs of limbs have 17 annuli (previous stage: 15 annuli);
- the proximal eight annuli of the exopods of the second and third post-mandibular pairs of limbs are laterally fused and bearing eight lateral setae (previous stage: proximal five annuli of the exopods of the post-mandibular pairs of limbs are laterally fused and bearing five lateral setae).

Shield, interdorsum, doublure, inner lamella

There are no morphogenetic changes concerning the shield and its associated structures such as the interdorsum, doublure and inner lamella. All these structures are slightly larger without changes in the proportions, compared with the previous stage (Tables 9, 15; Fig. 11).

Body

The body proper is longer and more slender compared with the previous stage, comprising at least seven segments as indicated by appendages, dorsally fused to the shield with six segments (Pl. 9D), two segments being free from the shield. Six pairs of limbs are fully developed, the morphology of the third pair of post-mandibular limbs is unknown, as well as whether there are even more limbs more posteriorly. There are no significant changes in the hypostome, including the median eye, compared with the preceding stage. The labrum is not distinctly different in proportion and shape. It is about 100 μm long and its basal width is about 80 μm. It has hairs, papilliform structures and pores posteriorly – as described for the species (Pls. 9D, 3G, cf. 4A). The surface of the sternum is adorned with numerous setulae arranged in distinct rows and areas, especially on its anterior part and the paragnaths (Pls. 9D, 3F). A second sternitic plate, located just posterior to the sternitic plate of the second post-mandibular limb and belonging to the segment of the third pair of post-mandibular limbs, is significantly separated from the preceding one by a membranous area on the ventral body surface. The succeeding ventral part of the body is smooth, without such plates (Pl. 9A). The plate is rectangular in long axis and somewhat roof-like domed medially. Lateral to the plate the third pair of post-mandibular limbs inserts (Pl. 9B) in an extensive membrane. The insertion area itself is almost vertical. The limbs are not significantly larger than in the preceding stage (Table 16) and similar in design. The hind body of this growth stage, including the fourth pair of post-mandibular limbs, is not known.

Antennulae

The antennulae (Pl. 9A) of this growth stage are only known from their insertion areas at the lateral flanks of the hypostome.

Antennae

The antennae (Pl. 9A) are not significantly different from the antennae in the preceding stage (Table 16).

Mandibles

Apart from the larger size (Table 16), the mandibles (Pl. 9A) are too poorly known to make detailed comparisons with the same limbs of the previous stage.

First pair of post-mandibular limbs ("maxillula")

The first pair of post-mandibular limbs (Pl. 9B, cf. 9C) is larger (Table 16) and more densely setose than in the preceding stage. It is about 150 μm long. The basipod is laterally extended into a curved spatulate sclerotic bar with a pointed tip which reaches far on to the lateral side of the body proper. The endopodal podomeres are more elongated medially, the medio-lateral extension of the first one is about 75 μm (Table 16). The exopod is about 160 μm long and consists of 17 annuli, the eight proximal annuli are fused laterally, adorned with one lateral seta each. The distal nine annuli are separated

---

Plate 6. *Hesslandona unisulcata* Müller, 1982; possible growth stage III

A: UB W 153. The specimen is about 450 μm in length. Latero-ventral view. Most of the left valve is lacking. The first two post-mandibular limbs (1, 2) of the left side are pressed on to each other, the first one almost totally covering the second one. The tips of the limbs of the right side are broken off (white arrows). Note the seta latero-distally on the terminal segment of the antennal endopod (black arrow) right behind the extending spine; both spine and seta are basally broken off. The ellipse marks the distance between the insertion areas of the endopod and exopod of the mandible (mdb).

B: UB W 154. Image flipped horizontally. The specimen is about 410 μm in length. A ventral view, the valves are partly closed, displaying the isolated sternitic plate (arrow "a") behind the sternum (ste). The subsequent sternal surface is membranous without sclerotised plate-like structures (arrow "b"). The hind body is lifted, not fused to the shield (arrow "c"). Note the pointed tip of the labrum (lbr).

PLATE 7

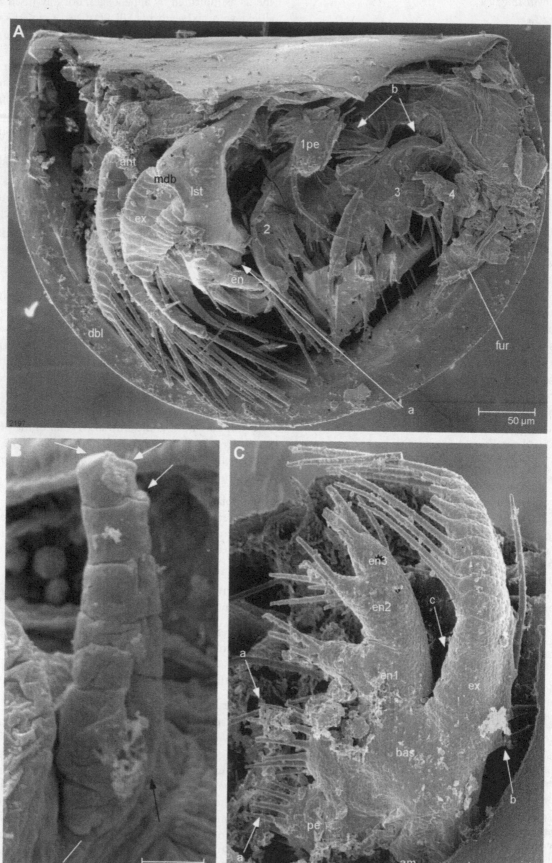

and bear one seta medially, the terminal annulus bears two setae.

Second and third pairs of post-mandibular limbs ("maxilla", "thoracopod I")

The second and third pairs of post-mandibular limbs (Pl. 9B) are not significantly changed compared with their shape in the preceding stage, both being about 150 μm long. As in the first pair of post-mandibular limbs, the basipod is laterally extended into a concave spatulate sclerotic bar with a pointed tip. The endopodal podomeres are more elongated medially, the medio-lateral extension of the first one is about 75 μm (Table 16). The third portion of the third pair of post-mandibular limbs is not known. The exopod of the second pair of post-mandibular limbs is about 160 μm long, consisting of 17 annuli. Its proximal eight annuli are laterally fused, eight lateral setae arising from this part of the ramus (Fig. 25B), the distal part has nine annuli which are medially setose, comparable with the exopod of the first pair of post-mandibular limbs. The exopod of the third pair of post-mandibular limbs is not known (Pl. 9C), its shape is probably that of the first and second pair of post-mandibular limbs.

Possible growth stage V

Material. – Seventeen specimens with the shield preserved, among these a few specimens with more or less well-preserved limbs (Pls. 9D, 10A), and one additional isolated limb assignable to this stage by its details and size. The length of the valves is between 580 and 650 μm (Table 10). This stage is distinguished from the previous stage because (cf. Table 16):

Plate 7. *Hesslandona unisulcata* Müller, 1982; possible growth stage III

A: UB 658 (Müller 1982a, pl. 1, fig. 1a–d); UB 1570 (Müller & Walossek 1991a, fig. 1). The specimen is about 470 μm in length. A lateral view, displaying both pairs of antennae and mandibles and the succeeding limbs (1–4) of the left side. Arrow "a" points to the median outgrowth squeezed somewhat between the limb stem (lst) and the endopod (en). The right first post-mandibular limb is preserved only by its proximal endite (1pe). Note the distinct slope of the proximo-lateral margin of the post-mandibular limbs (arrows "b").

B: The antennula of the specimen illustrated in Pl. 6B. Lateral view. The terminal setae plus the subterminal seta are broken off (white arrows). The arthrodial membrane (am) is small compared with the membrane of the succeeding limbs. Note the peduncle-like socket, on which the antennula inserts laterally on the hypostome (black arrow).

C: Image flipped horizontally. The first post-mandibular limb of the specimen illustrated in Pl. 6A. Note the setae arranged at the tip of the proximal endite (pe) and the basipod (bas, arrows "a"). Arrow "b" points to the lateral ridge-like extension of the proximal part of the exopod (ex), while the median part still shows its original annulation (arrow "c"). The exopod is slightly curved to the posterior.

- the valves are about 15% longer than in the previous stage;
- the endopodal portions of the post-mandibular pairs of limbs are more strongly elongated and with a higher number of setae (previous stage: not as elongated, number of setae fewer);
- the second and third post-mandibular pairs of limbs are larger than the first pair (previous stage: first to third post-mandibular pairs of limbs of equal size);
- the latero-proximal margin of the exopods of the second and third post-mandibular pairs of limbs is exposed (previous stage: not exposed, in one line with latero-proximal margin of the basipod);
- the proximal nine annuli of the exopods of the first to third post-mandibular pairs of limbs are laterally fused and bearing nine lateral setae (previous stage: proximal five annuli of the exopods of the first post-mandibular pair of limbs are laterally fused and bearing five lateral setae; proximal eight annuli of the exopods of the second and third post-mandibular pairs of limbs are laterally fused and bearing eight lateral setae).

Shield, interdorsum, doublure, inner lamella

There are no morphogenetic changes concerning the shield and its associated structures such as the interdorsum, doublure and inner lamella. All these structures are slightly larger without changes in the proportions, compared with the previous stage (Tables 10, 15; Fig. 11). The outer margin of the doublure, with a regular row of bottle-like structures developed near the outer margin of the doublure, is restricted to the ventral part of the doublure (Pl. 10B, C). They are about 3 μm wide at their base and about 5 μm long. Pores of about 1 μm in diameter are distributed close to the outer rim of the doublure in between the bottle-like structures (Pl. 10C).

Body

The body proper is longer and more slender compared with the previous stage, comprising at least seven segments as indicated by appendages, dorsally fused to the shield with six segments (Pl. 10A), two segments being free from the shield. Six pairs of limbs are fully developed. There are no significant changes in the hypostome/labrum complex, including the median eye and the sternum, compared with the preceding stage. The hind body of this growth stage posterior to the fourth pair of post-mandibular limbs is not known.

Antennulae

The antennulae are about 75 μm long, about one eighth the shield length (Table 16). There are no changes in the number and arrangement of annuli and setation (Pl. 11A).

PLATE 8

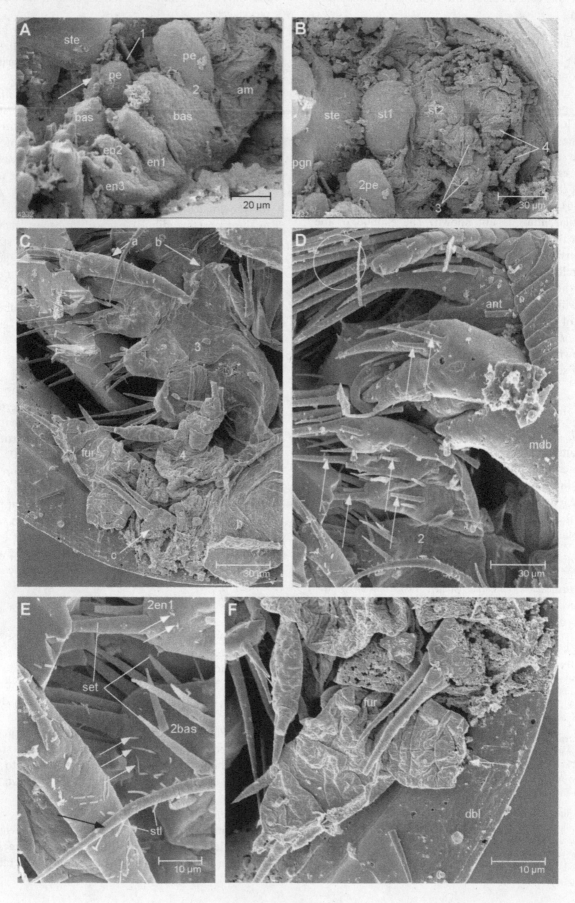

Antennae

The antennae (Pl. 10A) are about 150 μm long (Table 16). Their limb stem is not significantly changed compared with the preceding stage. The first endopodal portion is medially proximo-distally elongated and antero-posteriorly flattened and bears five setae at the median edge. The second portion bears one large seta terminally and two smaller ones medio-distally. Two sensilla are located latero-proximal of the second portion. The exopod is unknown.

Mandibles

The mandibles (Pl. 10A) are not significantly changed compared with the preceding stage. Their limb stem is about 150 μm in length and 150 μm in medio-lateral extension (Table 16). The exopod is unknown.

First pair of post-mandibular limbs

The first pair of post-mandibular limbs is about 150 μm long (Table 16). The enditic protrusions of the whole limbs, i.e. the tip of the proximal endite, the basipod and the three endopodal portions, are more elongated

---

Plate 8. *Hesslandona unisulcata* Müller, 1982; possible growth stage III

A: The same specimen as in Pl. 6B. The first (1) and second (2) post-mandibular limbs. A postero-median view, displaying the slightly medially drawn out enditic protrusions of the proximal endite (pe), the basipod (bas) and the three portions of the endopod (en1–en3). The arrow points to part of a circle of setae around the enditic tip of the proximal endite of the first post-mandibular limb. Note the large arthrodial membrane (am) of the second post-mandibular limb.

B: The same specimen as in Pl. 6B. Hind body. A ventral view, displaying the sternitic plates (st1, st2) posterior to the sternum (ste) and the undifferentiated and poorly preserved limb buds of the third (3) and fourth (4) post-mandibular limbs.

C: The same specimen as in Pl. 7A. The third (3) and fourth (4) post-mandibular limbs from the posterior plus the hind body with furcal rami (fur) from the ventral side of the same specimen as in Pl. 7A (see also Pl. 8F). The lateral edge of the endopod of the third post-mandibular limb is enrolled (arrow "a"). The exopod is basally broken off (arrow "b"). The fourth post-mandibular limb of the left side is strongly damaged (arrow "c").

D: The same specimen as in Pl. 7A. Detail of the antenna (ant), mandible (mdb) and second post-mandibular limb (2) of the same specimen as in Pls. 7A, 8C. Note the dense setation of the limbs with bristles of different sizes (arrows). The long incompletely illustrated setae above (circle) belong to the exopod of the antenna (cf. Pl. 7A).

E: The same specimen as in Pl. 7A. Detail of the basipodal endite (2bas) and the first endopodal portion (2en1) of the second post-mandibular limb of the same specimen as in Pls. 7A, 8C, D. Note the hairs (white arrows) on the basipod and the endopod, distinctly more fragile than the setae (set). The setae display few setulae (stl). The black arrow points to the long seta arising from the distal margin of the proximal endite of the first post-mandibular limb (cf. Pl. 7A, Fig. 23A).

F: The same specimen as in Pl. 7A. The hind body, displaying the furcal rami (fur). The left ramus is broken off (arrow). See also Fig. 20.

---

and more densely setose than in the preceding stage (Table 16). The antero-posteriorly compressed exopod is about 160 μm long and comprises at least 17 annuli, the nine most proximal annuli being partly fused and bearing nine laterally located setae (Pl. 10A, Fig. 24B). The eight distal annuli are adorned with one medially inserting seta each, the terminal one bearing two setae.

Second and third pairs of post-mandibular limbs

The second and third pairs of post-mandibular limbs are subequal. They are about 170 μm long, larger than the first pair (Table 16). The enditic protrusions of the whole limbs are more elongated and more densely setose than in the preceding stage. The exopod is about 180 μm long. It is a plate compressed in antero-posterior profile and somewhat drop-shaped with a latero-proximally exposed, setose edge (Pls. 9D, 10A) – in contrast to the first pair of post-mandibular limbs (Pl. 10A). Its maximum width is distal of the basipod. The exopodal plate is not segmentated proximally, neither laterally nor medially; its distal part is annulated with about eight annuli, which are oval in profile, antero-posteriorly compressed. Towards the distal end the annuli change their profile from long-oval to circular. The exopod is armed with regular marginal setation, ranging on the inner side from halfway to the top; the lateral side is armed with nine setae proximally (cf. Pl. 10B). The setae of the exopod are lightly adorned with irregularly arranged fine secondary setulae.

Fourth pair of post-mandibular limbs

The fourth pair of post-mandibular limbs is a bifid protrusion (Pl. 11B, Fig. 26B). It is about 70 μm long and consists of a basipod, a medio-proximally inserting proximal endite and two rami. The whole limb is antero-posteriorly compressed. The proximal endite is half-circular in anterior aspect, setation is unknown. The basipod and the endopod form more or less a single unit which is subtrapezoidal in shape and slightly medially extended into a setose median edge which slopes from distal to medio-proximal. The median edge of the basipod bears four setae inserting just above each other. The endopod is not clearly demarcated from the basipod. It seems to be a single triangular element with four setae at its sloping median edge. The exopod arises distally from the basipod just lateral to the endopod and is directed slightly over the endopod towards the medial side. It consists of at least four oval annuli. It cannot be stated whether setation is present.

Possible growth stage VI

Material. – Eleven specimens with the shield preserved, among these a few specimens with more or less well-preserved limbs (Pl. 11C), and three additional isolated limbs (Pls. 10B, 11D, F). The length of the valves ranges

PLATE 9

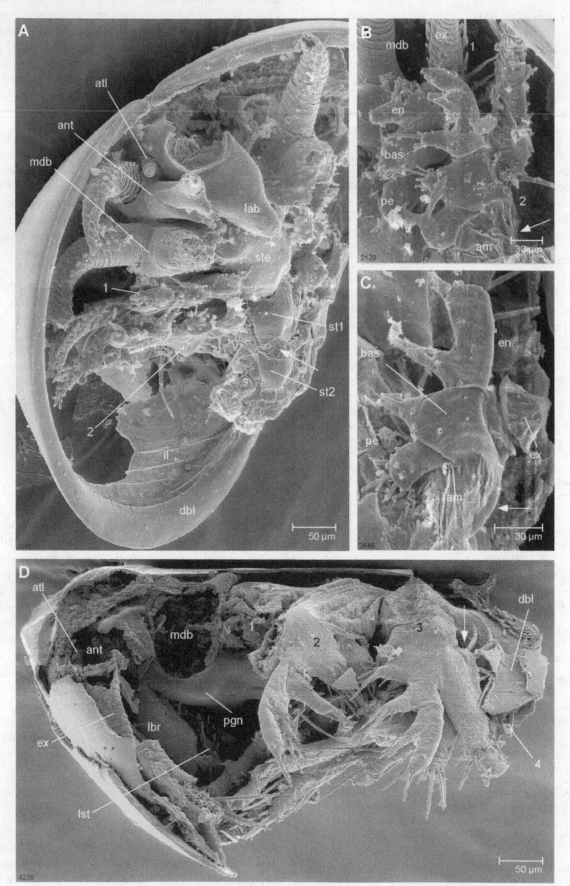

Table 9. List of specimens assigned to the fourth growth stage ordered by the length and the height of the valves (see Fig. 6a). Right column: proportion of the mean length to the mean height by height set to 1.

	Specimen								
	1	2	3	4	5	6	7	8	Proportion
Length	510	525	525	535	540	540	555	560	1.43
Height	355	405	355	355	350	370	410	400	1

Table 10. List of representative specimens assigned to the fifth growth stage ordered by the length and the height of the valves (see Fig. 6a). Right column: proportion of the mean length to the mean height by height set to 1.

	Specimen								
	1	2	3	4	5	6	7	8	Proportion
Length	580	590	610	620	620	630	630	650	1.43
Height	380	400	430	420	440	450	460	450	1

from 670 to 750 µm (Table 11). This stage is distinguished from the previous stage because (Table 16):

- the valves are about 15% longer than in the previous stage;
- the body proper has eight segments (previous stage: seven segments);
- the fourth post-mandibular pair of limbs is completely developed (previous stage: bifid protrusion);
- the endopodal portions of the post-mandibular pairs

Plate 9.   *Hesslandona unisulcata* Müller, 1982

Possible growth stage IV

A: UB W 155. A specimen of about 540 µm in length. Antero-ventral view. Only the first two post-mandibular limbs (1, 2) are preserved on the right side. The hind body is completely missing. The two sternitic plates (st1, st2) posterior to the sternum are well developed, separated by a membranous area (arrow).

B: Image flipped horizontally. The first two post-mandibular limbs (1, 2). Posterior view. Both limbs are more or less equal to each other. The arrow points to the sclerotised spatula proximo-lateral of the third post-mandibular limb.

C: Image flipped horizontally. The same specimen as in Pl. 9A. The third post-mandibular limb. Posterior view. Note the extensive arthrodial membrane (am) and the sclerotised spatula-like lateral extension of the basipod (arrow).

Possible growth stage V

D: Image flipped horizontally. The same specimen as in Pl. 3E, with a damaged and, for the most part, missing shield, probably 470 µm in length. Lateral view. The antenna (ant), mandible (mdb) and first post-mandibular limb (1) of the right side are missing and the insertion areas are accessible. The exopod of the second post-mandibular limb (2) is missing. The arrow points to the free proximo-lateral margin of the exopod of the third post-mandibular limb (3). A fourth post-mandibular limb (4) is probably represented by a set of setae arising from behind the doublure (dbl).

---

of limbs are more strongly elongated and with a higher number of setae (previous stage: not as elongated, number of setae fewer);

- the lateral margin of the exopod of the first post-mandibular pair of limbs is completely setose bearing 10 setae proximally (previous stage: lateral margin proximally setose with nine setae);
- the lateral margin of the exopods of the second to third post-mandibular pairs of limbs is completely setose bearing 13 setae from proximal to distal (previous stage: proximal nine annuli of the exopods of the first to third bearing nine lateral setae).

Shield, interdorsum, doublure, inner lamella
There are no morphogenetic changes concerning the shield and its associated structures such as the interdorsum, doublure and inner lamella. All these structures are slightly larger without changes in the proportions, compared with the previous stage (Tables 15, 16; Fig. 11).

Body
The body proper is longer and more slender compared with the previous stage, comprising at least eight segments, as indicated by appendages, dorsally fused to the shield with six segments, two segments being free from the shield (Pl. 11C). The body curves ventrally even from the sixth segment onwards. The membrane which expands between the shield and the sixth segment is probably very flexible (Pl. 11B, cf. Pl. 12B for possible growth stage VIII). At least seven pairs of limbs are fully developed. There are no significant changes in the hypostome/labrum complex, including the median eye and the sternum, compared with the preceding stage. The hind body of this growth stage posterior to the fifth pair of post-mandibular limbs is not known.

Antennulae
Antennulae are not known from this stage.

Antennae
The antennae (Pl. 11C) are only known by their exopods in this growth stage. The exopods are about 180 µm long and consist of about 24 annuli.

Mandibles
The mandibles (Pl. 11C) are only known by their exopods in this growth stage. The exopod consists of about 24 annuli.

First post-mandibular pair of limbs
The first post-mandibular pair of limbs is slightly larger, about 160 µm long (Table 16), and the endopodal portions have numerous enditic setae (Fig. 24C) compared with the corresponding limbs in the previous stage. The exopod is 180 µm long and is slightly antero-posteriorly compressed forming a flat plate, only its distal part being annulated with about 10 annuli. The exopodal plate is

PLATE 10

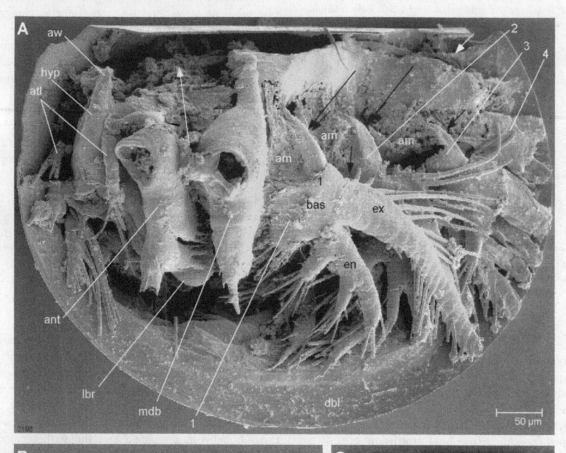

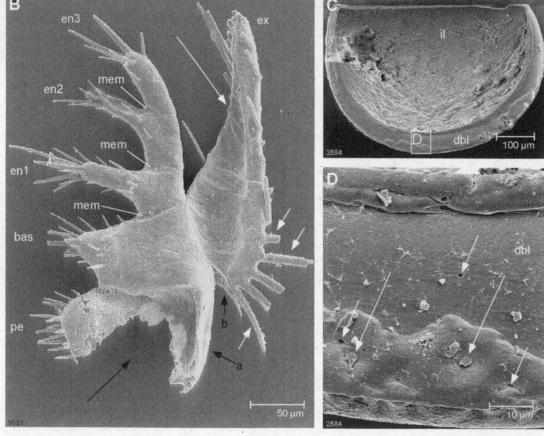

armed with regular marginal setation, ranging on the inner side from halfway to the top; the lateral side is armed with 10 setae laterally, regularly inserting from proximal to distal and not corresponding to the original annuli. The setae of the exopod are lightly adorned with irregularly inserting fine secondary setulae, at least 6 μm long and at a distance of about 4 μm relative to each other.

Second and third post-mandibular pairs of limbs

The second and third post-mandibular pairs of limbs are significantly larger than the first pair, both about 220 μm long. The endopod now has numerous setae, which are not restricted to the enditic tips, but which are also located on the basal part and along the sides of the elongated enditic tips (Pl. 11E, F, Fig. 25C). The exopod is about 200 μm long and is strongly antero-posteriorly compressed forming a flat plate, only its distal part being annulated with about eight annuli. The exopodal plate is armed with regular marginal setation, ranging on the inner side from halfway to the top; the lateral side is armed with 13 setae laterally, regularly inserting from proximal to distal and not corresponding to the original annuli.

Fourth post-mandibular pair of limbs

The fourth post-mandibular pair of limbs is functionally developed, compared with the first three pairs of

---

Plate 10. *Hesslandona unisulcata* Müller, 1982; possible growth stage V

A: UB 670 (Müller 1982a, pl. 6, fig. 1a, b). The specimen is about 600 μm in length. Lateral view. The antero-lateral sides (long white arrow) and the exopods of the antenna (ant) and mandible (mdb) are not preserved. The short white arrow indicates the membrane right posterior of the third post-mandibular limb (3), which connects the body proper with the shield; the hind body after this membrane is free from the shield. Note the extensive arthrodial membranes (am) of the first three post-mandibular limbs (1, 2, 3) and the lateral extensions (long black arrows) of the basipods (bas). The proximo-lateral margin of the exopod of the first post-mandibular limb continues gradually into the lateral margin of the basipod (short black arrow).

B: UB W 156. An isolated post-"maxillulary" limb. Posterior view. The arthrodial membrane is missing (long black arrow). The short black arrow "a" points to the lateral extension of the basipod. Less than half of the exopod is annulated terminally, the long white arrow points to the first annulus distally on the undivided proximal exopodal part. Note the proximo-lateral extension of the exopod (short black arrow "b"), which does not occur in the first post-mandibular limb in this growth stage (cf. Pl. 10A) and the fine setulation on the setae of the exopod (short white arrows).

C: UB W 157. Image flipped horizontally. A specimen of about 580 μm in length, presumably the smallest member of this growth stage. A lateral view inside the empty left valve. The rectangle marks an area magnified in Pl. 10D.

D: Close-up of the outer rim of the ventral doublure (dbl) of the same specimen as in Pl. 10C, displaying bottle-like structures within small dents (long arrows) and pores (short arrows).

---

post-mandibular limbs, being somewhat smaller than the second and third but as large as the first pair (Table 16). The morphology is similar to that of the first post-mandibular pair of limbs.

Fifth post-mandibular pair of limbs

The fifth post-mandibular pair of limbs is known from one specimen only (Pl. 11C). It is not known if the limb is represented by a bud only or if it is more or less well developed. Its preservation is too poor to give a detailed description.

Possible growth stage VII

Material. – Eight specimens with the shield preserved, among these one specimen with more or less well-preserved limbs (Pl. 12A, Table 16), and two additional isolated limbs. The length of the valves is between 760 and 840 μm (Table 12). This stage is distinguished from the previous stage because (Table 16):

- the valves are about 13% larger than in the previous stage;
- the endopodal portions of the post-mandibular pairs of limbs are more strongly elongated and with a greater number of setae (previous stage: not as elongated, number of setae fewer).

Shield, interdorsum, doublure, inner lamella

There are no morphogenetic changes concerning the shield and its associated structures such as the interdorsum, doublure and inner lamella. All these structures are slightly larger without changes in the proportions, compared with the previous stage (Table 12).

Body

The body proper is longer compared with the previous stage, comprising at least eight segments, as indicated by appendages, dorsally fused to the shield with seven segments (Pl. 12A), one segment being free from the shield. An unknown number of post-mandibular pairs of limbs are developed. Seven pairs of limbs are fully developed. There are no significant changes in the hypostome/labrum complex, including the median eye and the sternum, compared with the preceding stage. The hind body of this growth stage posterior to the fifth pair of post-mandibular limbs is not known.

Antennulae

Antennulae are not known from this stage.

Antennae

The antennae (Pl. 12A) are too poorly known for comparisons with preceding growth stages.

Mandibles

The mandibles (Pl. 12A) are too poorly known for comparisons with preceding growth stages.

PLATE 11

Table 11. List of all specimens assigned to the sixth growth stage ordered by the length and the height of the valves (see Fig. 6a). Right column: proportion of the mean length to the mean height by height set to 1.

| | Specimen | | | | | | | | | | | |
	1	2	3	4	5	6	7	8	9	10	11	Proportion
Length	670	680	700	700	730	730	730	740	740	750	750	1.41
Height	510	470	500	530	520	520	530	510	520	500	520	1

First pair of post-mandibular limbs

The first pair of post-mandibular limbs (Pl. 12A, Fig. 24D) are about 180 μm long. The proximal endite is medially elongated, bearing three strong and two smaller setae in the middle, and a row of nine setae inserts on the anterior edge of the basis of the enditic outgrowth (Pl. 12A). A set of an unknown number of setae inserts at the posterior edge of the endite. The median endite of the basipod is strongly elongated medially with five medially located setae and one medio-distal one. Enditic outgrowths of the three portions of the endopod are more elongated medially and with a higher number of setae compared with the preceding stage. The exopod is unknown.

Succeeding post-mandibular pairs of limbs

The morphology of the succeeding post-mandibular pairs of limbs is not known due to missing information.

Plate 11. *Hesslandona unisulcata* Müller, 1982

Possible growth stage V

A: The same specimen as in Pl. 10A. Detail of the antero-lateral part of the head, displaying the antennulae (atl) and the hypostome with the shrunken protrusion of the median eye (me).

B: The same specimen as in Pl. 10A. Detail of the postero-lateral part of the body, displaying the fourth post-mandibular limb (4).

Possible growth stage VI

C: UB W 158. The specimen is about 750 μm long. Lateral view. The limbs of the left side are almost only represented by their insertion areas (mdb, 1–3). The true nature of the fourth and fifth (4, 5) post-mandibular limbs is not reconstructable due to poor preservation. The right side displays the exopods of the antenna (ex ant) and mandible (ex mdb) only.

D: UB W 159. An isolated first post-mandibular limb from the anterior. Note the latero-proximal margin of the exopod not being curved (arrow). The latero-proximal margin of the basipod is covered by a few dirt particles and the exopod (ex) is distorted.

E: UB 665 (Müller 1982a, pl. 4, fig. 1a, b). The second (2) and third (3) post-mandibular limbs from the posterior, displaying the antero-posteriorly flattened exopod (ex).

F: UB W 160. An isolated post-"maxillulary" limb, displaying the proximo-lateral curved margin of the exopod (arrow "a"; different in comparison with the first post-mandibular limbs – cf. Pl. 11D), and a complete lateral setation of the exopod (arrow "b"). The proximal endite is missing (cf. Pl. 11E).

Possible growth stage VIII

Material. – Fourteen specimens with the shield preserved, among these one specimen with more or less well-preserved limbs (Pl. 12B), and two additional isolated limbs. The length of the valves is between 870 and 950 μm (Table 13). This stage is distinguished from the previous stage because (cf. Table 16):

- the valves are about 14% larger than in the previous stage;
- the proximal endite of at least the second and third post-mandibular pairs of limbs is shovel-shaped (previous stage: drop-shaped);
- the endopodal portions of the post-mandibular pairs of limbs are more strongly elongated and with a greater number of setae (previous stage: not as elongated, number of setae fewer).

Shield, interdorsum, doublure, inner lamella

There are no morphogenetic changes concerning the shield and its associated structures such as the interdorsum, doublure and inner lamella. All these structures are slightly larger without changes in the proportions, compared with the previous stage (Table 12).

Body

The body proper is longer and thicker compared with the previous stage, and there are no changes in composition and shield fusion compared with the previous stage. Five post-mandibular pairs of limbs are developed, the shape of the fifth is unknown. There are no significant changes in the hypostome/labrum complex, including the median eye and the sternum, compared with the preceding stage. The hind body is circular in cross-section. The last two body segments are distinctly demarcated from the preceding body portion (Pl. 12D). The hind body is terminally extended into a pair of paddle-shaped flat triangular elements, the furca (Pl. 12D). They are about 130 μm long and adorned with six setae on each lateral side plus a terminal seta.

Antennulae

The antennulae (Pl. 12B) are, at about 90 μm in length, less than 1/10 the shield length. They are rod-shaped, slightly conical and consist of eight irregularly arranged but weakly defined annuli. The basal five annuli are seta-less. The third last annulus bears one seta medio-distally, the seta having about half the length of the antennula. The penultimate annulus bears one seta medio-distally, the seta being about two thirds the length of the antennula. Three bristles insert on the terminal annulus (ch64:3). The medio-distally and latero-distally inserting ones are about as long as the antennula. The third seta inserts distally between the others. It is distinctly thinner but as long as the other setae.

PLATE 12

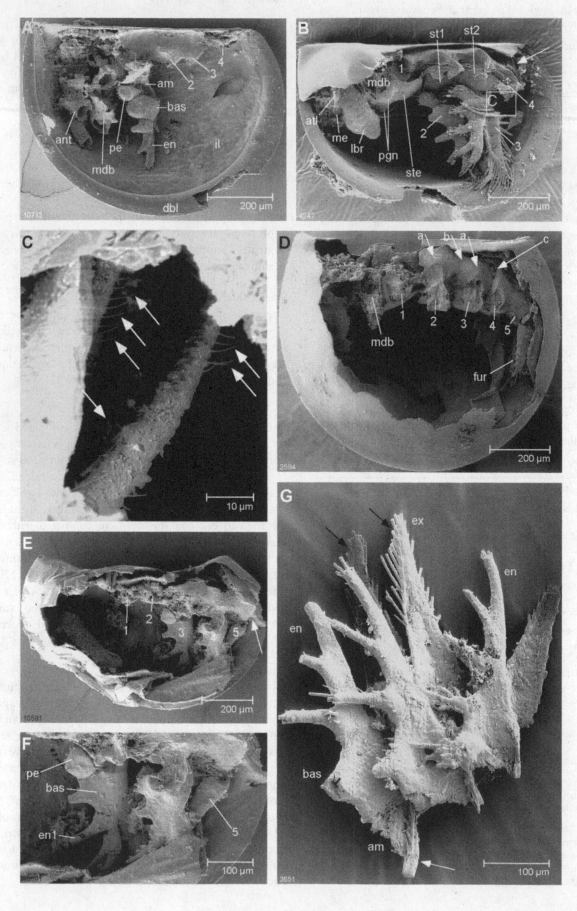

Table 12. List of all specimens assigned to the seventh growth stage ordered by the length and the height of the valves (see Fig. 6a). Right column: proportion of the mean length to the mean height by height set to 1.

	Specimen								Proportion
	1	2	3	4	5	6	7	8	
Length	760	770	790	800	800	820	830	840	1.36
Height	560	600	600	560	580	640	570	590	1

Table 13. List of representative specimens assigned to the eighth growth stage ordered by the length and the height of the valves (see Fig. 6a). Right column: proportion of the mean length to the mean height by height set to 1.

	Specimen								Proportion
	1	2	3	4	5	6	7	8*	
Length	870	880	890	890	910	920	940	950	1.36
Height	680	590	590	680	710	630	670	770	1

* = holotype.

Plate 12. *Hesslandona unisulcata* Müller, 1982

Possible growth stage VII

A: UB W 161. The only specimen with preserved soft parts assignable to possible growth stage VII, about 800 μm in shield length. A lateral view of the inside of the right valve. The soft parts are poorly preserved, apart from the first post-mandibular limb, which shows its composition of portions (labelled) and setation.

Possible growth stage VIII

B: UB W 162. Image flipped horizontally. The only specimen with preserved soft parts assignable to possible growth stage VIII, about 870 μm in shield length. A lateral view of the inside of the left valve. The soft parts are partly well preserved; the antenna and mandible (mdb), as well as the first post-mandibular limb (1), are missing. The arrow points to the membranous area that connects the fourth post-mandibular segment with the shield. The area marked by a rectangle is magnified in Pl. 12C.

C: Image flipped horizontally. Close-up of the area marked in Pl. 12B. A detailed view of the exopodal setae of the third post-mandibular limb, displaying the fine setulae of less than 1μm in diameter.

D: UB 678 (Müller 1982a, pl. 7, fig. 3; pl. 8, fig. 1a, b). A specimen of about 870 μm in length. Lateral view. The limbs (mdb, 1–5) are almost completely missing, but the furca (fur) is preserved at the free hind body. Note the dome-like protrusions of the lateral wall of the body (arrows "a") and the inner-segmental depressions (arrow "b"); the fourth post-mandibular segment is pressed towards the anterior, and the inner-segmental area resembles a distinct segment boundary (arrow "c").

Larger growth stages

E: UB W 163. The largest specimen available displaying soft part preservation, about 1000 μm in shield length. Lateral view. Only the third to fifth post-mandibular limbs are preserved, the hind body (arrow) as well as the head are missing. Some details are shown in Pl. 12F.

F: The same specimen as illustrated in Pl. 12E. A detailed view of the thorax, displaying the strongly elongated endites (en1, en2) of the post-mandibular limbs.

G: UB W 164. An isolated group of three post-mandibular limbs from the right side, displaying the lateral spatula-like outgrowth of the basipod (arrow), the long enditic protrusions of the endopodal (en) podomeres and the terminal annulation of the exopods (ex, black arrows). The length of the limb in the middle of the image is about 330 μm, the length of the exopod is about 330 μm, the latero-median length of the first endopodal podomere is about 130 μm.

Antennae and mandibles
Antennae and mandibles are not known from this stage.

First pair of post-mandibular limbs
The first pair of post-mandibular limbs is unknown.

Succeeding pairs of post-mandibular limbs
The second and third pairs of post-mandibular limbs are subequal (Pl. 12B), about 240 μm long (Table 16). The proximal endite is shovel-shaped. A set of an unknown number of setae inserts at the posterior edge of the endite. The enditic protrusion of the basipod is strongly elongated medially with five medially located setae and one medio-distal one. Enditic outgrowths of the three portions of the endopod are more elongated compared with the preceding stage, the medio-lateral extension of the first endopodal portion being about 110 μm (Table 16). The exopods (Pl. 12B) are flat, long, triangular plates of about 220 μm in length. The setae of the exopods are densely setulated, the setulae being almost 10 μm long (Pl. 12C). The morphology of the fourth pair of post-mandibular limbs, as preserved in one specimen (Pl. 12D), is probably equivalent to that of the preceding post-mandibular pairs of limbs. The fifth pair of post-mandibular limbs is not known (cf. Pl. 12D).

Later growth stages

Material. – Larger stages are almost only represented by valves without soft part preservation and some isolated post-mandibular limbs (Pl. 12E, G; Table 14). The limbs may be part of possible growth stages IX–XI. The only specimen with preserved soft parts (Pl. 13E), about 1000 μm in shield length, does not allow detailed descriptions of significant differences compared with the another stages. These stages are distinguished from the previous stage because:

• the length of the valves is at least 1000 μm (previous stage: about 910 μm in length);
• the endopodal portions of the post-mandibular pairs of limbs are more strongly elongated, the number of setae is presumably not significantly changed (previous stage: not as elongated).

Table 14. List of representative specimens assigned to later growth stages than growth stage VIII ordered by the length and the height of the valves (see Fig. 6a).

	Specimen										Proportion
	1	2	3	4	5	6	7	8	9	10	
Length	970	1030	1100	1120	1190	1230	1390	1490	1580	1650	1.36
Height	750	810	800	830	840	870	980	1090	1160	1220	1

Shield, interdorsum, doublure, inner lamella

There are no morphogenetic changes concerning the shield and its associated structures such as the interdorsum, doublure and inner lamella. All these structures are slightly larger without changes in the proportions, compared with the previous stage (Table 16).

Body

The body proper is longer and thicker compared with the previous stage, and there are no changes in composition and shield fusion compared with the previous stage. Five post-mandibular pairs of limbs are developed, the shape of the fifth is unknown. There are no significant changes in the hypostome/labrum complex, including the median eye and the sternum, compared with the preceding stage. The hind body is circular in cross-section. The last two body segments are distinctly demarcated from the preceding body portion (Pl. 12D). The hind body is terminally extended into a pair of paddle-shaped flat triangular elements, the furca (Pl. 12D). They are about 130 μm long and adorned with six setae on each lateral side plus a terminal seta.

Post-mandibular pairs of limbs

The first three post-mandibular pairs of limbs of later growth stages (Pl. 12E–G) differ from the same limbs of preceding growth stages in their larger size and the more elongated enditic protrusions of the proximal endite, basipod, endopod. They can reach at least about 330 μm in length. The second pair of post-mandibular limbs seems to be the largest. The third might be smaller than the second but slightly larger than the first, and towards the posterior the size decreases gradually. Their detailed morphology is as described for the species. The median outgrowths of the proximal endite, the basipod and especially the three endopodal segments are significantly longer and more slender than in the defined growth stages. They can reach at least 130 μm. They are rod-shaped and circular in cross-section with a small blunt tip (Pl. 12G; Fig. 22). The endites are irregularly adorned with short spine-like setae along their entire lengths (Fig. 22). The median endite of the basipod has seven medially located setae and one medio-distal seta. The exopod is a triangular plate with a proximo-laterally exposed margin. The lateral edges of the exopods are regularly setose along their whole length with at least 17 setae; the annulation is weakly developed only at the tip

of the ramus. The fourth and fifth pairs of post-mandibular limbs are not known in detail from later growth stages. The developmental state of the fifth pair of post-mandibular limbs is not known.

Morphogenesis

Ontogenetic stages. – The ontogeny of *Hesslandona unsiulcata* Müller, 1982 is, in general, anamorphic, i.e. starting with few-segmented larvae. The first larval stage (Pl. 4D) is 230 μm long and has a fully developed bivalved head shield with an interdorsum and a doublure. The body is completely enclosed by the shield and comprises four limb-bearing segments and an undifferentiated hind body, matching the segmental composition of the euarthropod head (cf. Walossek & Müller 1998a; Maas & Waloszek 2001a; Fig. 21, 27A). All four pairs of appendages, the antennulae and the additional three pairs of limbs are functional (ch64:1). The second growth stage is one segment longer than the first one, and one appendage, the second pair of post-mandibular limbs, is initially developed (Fig. 21). Remarkably, the segmental number of the succeeding larvae, growth stages III–V, remains the same with seven segments, and of growth stages VI–VIII, the largest recognised growth stage, with eight segments (Fig. 21, Table 16). Their differentiation could be recognised by the further development of the limbs and overall size. Nothing can be said about whether the segmental number of not more than eight segments is retained in later instars, as individuals with larger shields do not have ventral cuticular details preserved. Even the largest recognised growth stage has not more than eight segments (Table 16). During ontogeny, the shield incorporates progressively more segments, from four in growth stage I to six segments from growth stage III onwards (Table 16). Accordingly, the shield changes from an euarthropodan crustacean head shield, comprising at most the segments of the head, to a cephalothoracic shield at growth stage III. There is a large gap of knowledge from the latest defined stage with a shield of approximately 910 μm in length to the largest empty shells with a length of 1650 μm. Therefore, it can be expected that several more advanced stages are missing, but details are unknown.

Shield. – With increasing shield size in *H. unsulcata*, the proportion of the length to the height of the valves remains more or less the same (Figs. 22, 23); being about 1.4 throughout ontogeny (cf. Tables 7–13). The scatter diagram of the length to the height of the valves (Fig. 22) does not show distinct clusters to discriminate successive growth stages, which could, however, be clearly discriminated by soft part details, particularly the limb differentiation. At the start of ontogeny, the shield grows more strongly than later on during ontogeny. With the first three moults, the shield size increases by about 20%. From the next moult, from possible growth stage V on, the shield grows only at rates of about 15% compared with the size of the respective preceding stage (cf. Tables 15, 16).

The lobe L_1, the doublure and the interdorsum grow gradually without changes in proportions. Pits and pores on the surface of the doublure were first observed in possible growth stage V.

Growth of body and appendages. – In the first growth stage, the body proper is drop-shaped. During the later ontogeny, it progressively elongates and becomes more or less spindle-shaped (Fig. 23). After two moults, segments are added regularly but in groups of three (cf. Table 16). External segmentation is missing; body segments are only indicated by the increase in limb pairs and single sclerotic plates.

During the whole ontogeny observed, the most significant structure of the head, the hypostome, remains almost the same shape, although increasing in size. It grows slightly longer in body axis, corresponding to the growth of the body itself (Fig. 23). The structure present antero-ventrally on the hypostome is regarded as the three cups of a median eye, in accordance with Müller (1982a). It is preserved in different conditions as illustrated by Müller (1982a, pl. 2, figs. 2a, b, 4; pl. 3, fig. 1; pl. 4, figs. 2a, b):

- it is a circular depression of membranous cuticle (Pl. 9A);
- it shows a subdivision of two oval compartments in long body axis, separated by a sclerotised bar plus a long oval depression abaxial and posterior to them (Pl. 3A);
- it is a bulged blister seemingly consisting of three globe-like structures (Pl. 3B, C).

The exact design of the median eye in life is not quite clear. The first two conditions can be regarded as preservational following the death of the animal (Müller 1982a), when the softer cuticle above the cups collapses into the hole of the more sclerotic hypostome. The bulged condition is regarded as closer to the original *in situ* condition, as displayed in a well-preserved specimen (Pl. 3B, C; see also Müller 1982a).

At the start of ontogeny, the labrum is as long as wide and has a blunt tip. Subsequently, the length increases

(Table 16) and the formerly blunt labrum receives a pointed tip from growth stage III onwards. Pores, sensilla, fine hairs and papilliform structures occur on the posterior side of the labrum of later stages (Pl. 3G). They could not be found on the youngest three possible growth stages due to poor preservation.

Throughout the whole ontogeny, the ventral sternitic surface of the segments of the antennae, the mandibles, and the first pair of post-mandibular limbs is a single sclerotic plate, the sternum (Pl. 4D; cf. Pl. 5F; cf. Fig. 23). From growth stage III onwards, the ventral surface of the segment posterior to the sternum forms a sclerotised sternitic plate. From growth stage IV onwards, a further plate is present posterior to the first one, belonging to the segment of the third post-mandibular pair of limbs. Further sternitic plates in later instars were not observed due to poor preservation.

The hind body is known only from a few specimens, but not from every growth stage. In the first two growth stages, the complete body proper is fused dorsally with the shield, and the hind body seems to merge into the postero-dorsal membranous area of the inner lamella. From growth stage III onwards, at least one segment is not dorsally fused to the shield and is exposed, but is also never completely preserved preventing further discussion. The hind body is curved ventrally, possibly to fit into the domocilium. The last segment, which is fused to the shield, is somewhat lifted up. The connecting membrane is very flexible and seems to allow some axial and possibly abaxial movement of the free hind body (cf. Pls. 11B, 12B). An anus was never observed.

The furca is probably absent in the first two growth stages. It is known first from possible growth stage III, where it is flat, paddle-shaped and marginally setose. This morphology is retained throughout the further observable ontogeny. The marginal setation increases from six setae in growth stage III to 13 setae in growth stage VIII, but it is not known from instars in between (cf. Fig. 23).

Limbs appear first as buds and develop sequentially, or they may appear as functional appendages such as the third pair of post-mandibular appendages in possible growth stage III (Pl. 7A, Fig. 21).

The antennulae start as short, irregularly annulated appendages with few setae, and they remain tiny throughout ontogeny. While they increase in size from about 50 µm in possible growth stage III to 90 µm in possible growth stage VIII (cf. Pls. 5F, 10A, 12B), the length of the shield almost trebles in the same ontogenetic phase. During ontogeny, the length of the antennula decreases from about 14% to 10% of the length of the shield (Table 16).

The antennae and the mandibles are similar to each other in general shape and in size increase throughout ontogeny (Table 16). Their limb stem is undivided

PLATE 13

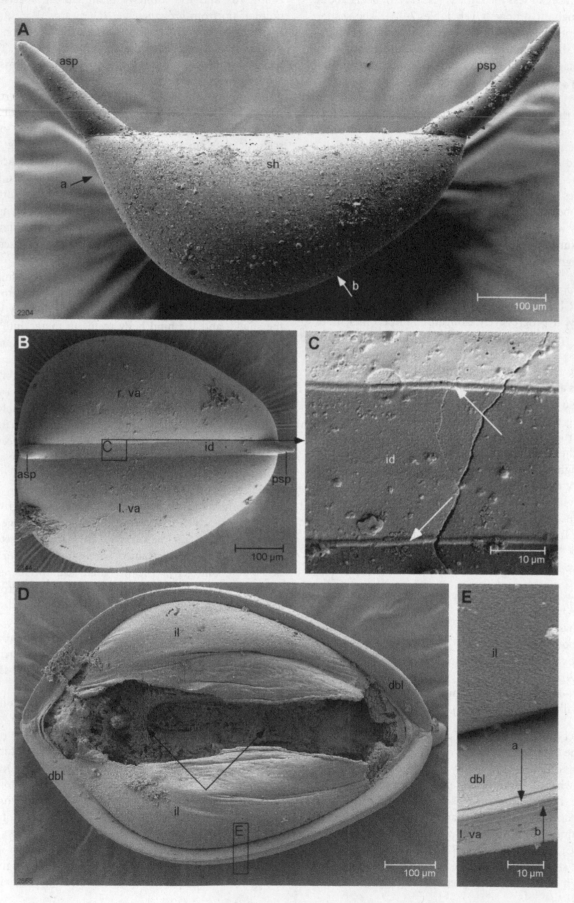

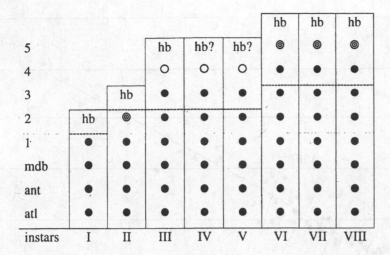

Fig. 21. Life cycle of *Hesslandona unisulcata*. Left column = segments of antennula (atl), antenna (ant), mandible (mdb) and post-mandibular pairs of limbs (1–5); further columns = possible growth stages I–VIII; o = developed and functional appendages; • = incipient, poorly developed, vestigial appendages; ● = appendage present but unclear whether incipient or functional; hb = hind body; hb? = unknown whether the fifth segment is already developed; dotted line = boundary between the head and the appendiferous trunk region in Euarthropoda; dashed line = border of the last segment which is incorporated into the shield.

Table 15. Overview of the size range in all recognised possible growth stages of *Hesslandona unisulcata*. Possible growth stage II is known by one specimen only.

	Growth stage							
	I	II	III	IV	V	VI	VII	VIII
Length	230–315	360	410–470	510–560	580–650	670–750	760–840	870–950
Height	170–240	260	310–350	355–400	380–450	470–530	560–600	590–710

throughout ontogeny, never having a state with a divided limb stem, as known from the earliest larval stages as well as from late larval stages and adults of eucrustaceans. The strong gnathobasic outgrowths of both appendages imply that they were incorporated into the feeding system. The gnathobases were oriented alongside the postero-proximal labral flanks in front of the paragnaths towards the mouth. Both limbs have two-divided endopods throughout ontogeny. The endopods form a combined grinding unit with the gnathobase of the limb stems from the first growth stage onwards. The general shape of the endopods of both limbs does not change during ontogeny, but they become more and more setose. The endopod of the antenna becomes antero-posteriorly flattened from growth stage III onwards, while the exopods of both appendages start as rami, which are composed of about 10 annuli in the possible first growth stage (Pl. 4D). The distal annuli bear one seta each, the terminal annulus two setae. This general morphology does not change throughout ontogeny, but the number of annuli increases to about 24 from possible growth stage III onwards (Pl. 9A; Table 16). From growth stage II onwards, the exopods are very long (Table 16) relative to the length of the body proper, and reach to the antero-ventral inner margin of the shield.

When fully developed, the first to third post-mandibular pairs of limbs comprise a basipod, a proximal endite, a three-segmented endopod and an exopod composed of numerous annuli. They are slightly longer than the antennal and mandibular exopods in all stages. From possible growth stage V onwards, they are different from each other in that the first and third post-mandibular pairs of limbs are smaller than the second (Table 16). The proximal endite increases gradually in size with the whole limb and its armature becomes more elaborate (cf. Pls. 6A, 7C, 10B). Its median set of setae is arranged in three more or less separate groups, one anteriorly, one medially and one posteriorly. The number of setae increases especially in the anterior and posterior groups, while in the median group, fewer setae are added during

---

Plate 13. *Hesslandona necopina* Müller, 1964

A: UB W 165. A specimen representing an advanced growth stage, about 560 µm in length. A lateral view of the left valve. The surface of the valve is covered with many phosphate crystals. Arrow "a" points to the relatively straight antero-ventral margin of the valve, arrow "b" points to the rather straight postero-ventral margin of the valves.

B: UB W 166. A specimen representing an advanced growth stage, about 540 µm long. A dorsal view of the opened shield in the so-called "butterfly" position, displaying the straight interdorsum (id) with smaller spines at both ends than those illustrated in Pl. 13A. A rectangle marks the area magnified in Pl. 13C.

C: Close-up of the area marked in Pl. 13B, displaying the smooth interdorsum apart from preservational pieces of dirt. The arrows point to the pair of membranous dorsal furrows.

D: UB W 167. A specimen representing a late growth stage, about 810 µm in length. A ventral view of an opened shield. The body proper is almost completely torn off (arrows). The inner lamella (il) is swollen due to the decay process. The anterior and posterior spines are broken off. A rectangle marks the area magnified in Pl. 13E.

E: Close-up of the area marked in Pl. 13D, displaying the depressions of the outer margin of the (in this image left) valve (arrow "a") and of the outer rim of the doublure (arrow "b").

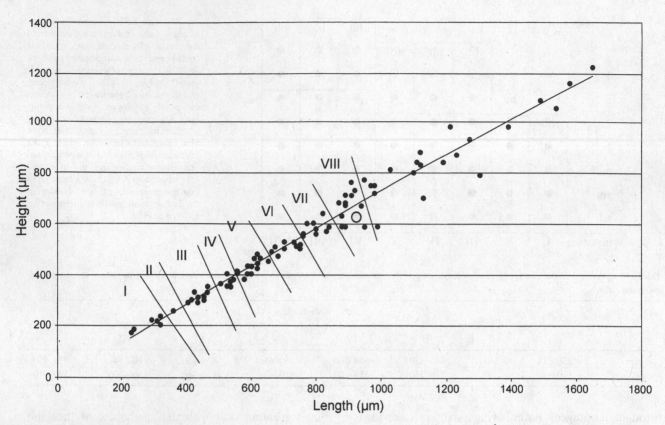

Fig. 22.   Ontogeny of *Hesslandona unisulcata* including those 100 specimens which allow a more or less exact determination of shield length and height. ○ = holotype. The trend line combining all points has the polynomial function $y = 0x^2 + 0.7739x - 24.165$, y = height, x = length. The mean proportion of length to height = 1.37. Instars I–VIII.

ontogeny. Eventually, the strict pattern is lost and the setae are more or less irregularly positioned.

From the beginning of ontogeny onwards, the basipods and especially the endopods, show a protrusion on their median edges (Table 16). This protrusion expands progressively during ontogeny to reach a rather elongated, slightly curved rod-shaped tube with irregularly arranged setae proximally to terminally (cf. Pls. 6B, 11E, F, 12G). The exopod of the first pair of post-mandibular limbs has a somewhat different morphogenesis from those of the succeeding limbs, which will be described as follows.

The first post-mandibular pair of limbs (Fig. 24) starts with an annulated exopod comparable with that of the antenna and the mandible but essentially smaller. At possible growth stage III, the exopod changes into a flap-like structure with apparent fusion of the proximal annuli, five proximo-lateral setae and medio-distal setation. Thereafter, more and more annuli become fused to the proximal part of the exopod and setae are added on the median and lateral margins until, from possible

growth stage VI onwards, the whole lateral margin is setose from proximal to distal (Fig. 24). Even the largest limbs known show at least distal annulation, while the proximal part is strongly flattened and unsegmented (cf. Pl. 12G). The proximo-lateral margin of the exopod remains almost straight (Pl. 10A).

The second and third pairs of post-mandibular limbs (Fig. 25) appear successively (Fig. 21); the second pair of post-mandibular limbs starts as possible limb buds at growth stage II and is fully developed from growth stage III onwards. At this stage the third pair of post-mandibular limbs is also fully developed and the fourth pair appears as limb buds (Fig. 21). The exopods of all "post-maxillulary" pairs of limbs start as slightly flat flaps being composed of numerous annuli, but – in contrast to the exopods of the first post-mandibular limb – the proximal annuli are laterally undivided, the distal part being regularly annulated. Thereafter, similar to the first pair of post-mandibular limbs, more and more annuli become fused to the proximal undivided part, which, thus, extends distally during ontogeny. The proximo-lateral margin of the

---

Fig. 23.   Reconstruction drawings of possible growth stages I–VIII (A–H) of *Hesslandona unisulcata* from inside lateral. Post-antennular limbs are indicated by their insertion areas shaded (cf. Pls. 1–13 for labelling of structures):
◍ functional post-mandibular appendage;
○̧ limb present but either incipient or state of development unclear.

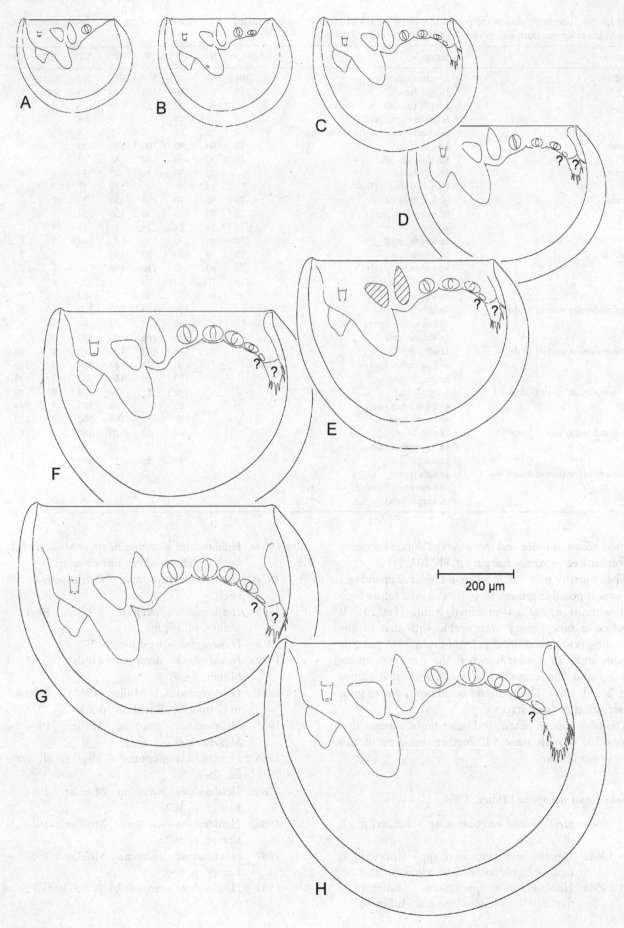

200 μm

Fig. 23

Table 16. Morphometric data of the possible growth stages I–VIII of *Hesslandona unisulcata*. The morphology of the fifth post-mandibular pair of limbs is not known from any specimen. For abbreviations see Table 4.

Structure	Statistics	I	II	III	IV	V	VI	VII	VIII
Shield	Length (μm)	300	360	440	535	615	710	800	910
	Height (μm)	215	260	320	380	440	510	580	680
	Length increase (%)	–	20	22	22	15	15	13	14
Body	Segments (number)	4	45	47	7	7	8	8	8
	Segments fused to shield (number)	4	5	6	6	6	6	6	6
Labrum	Length (μm)	50	60	90	100	110	120	?	150
	Basal width (μm)	50	50	60	80	85	90	?	110
Antennula	Length (μm)	?	50	50	?	75	?	?	90
	Of shield length (%)	?	14	11	?	12	?	?	10
Antenna	lst length (μm)	75	90	110	130	150	?	?	?
	lst extension (μm)	75	80	110	130	150	?	?	?
	ex annuli	≤10	15	24	24	?	24	?	?
	ex length (μm)	50	100	130	150	?	180	?	?
Mandible	lst length (μm)	75	90	110	130	150	?	?	?
	lst extension (μm)	70	80	110	130	150	?	?	?
	ex annuli (number)	≤10	15	24	24	?	24	?	?
	ex length (μm)	50	100	130	150	?	180	?	?
Post-mandibular pair of limbs I	Length (μm)	?	?	140	150	150	160	180	?
	en1 extension (μm)	?	?	60	70	75	80	90	?
	ex length (μm)	?	?	120	160	160	180	?	?
Post-mandibular pair of limbs II	Length (μm)		?	140	150	170	220	?	240
	en1 extension (μm)		?	60	75	80	100	?	110
	ex length (μm)		?	120	160	180	200	?	220
Post-mandibular pair of limbs III	Length (μm)			140	?	170	220	?	240
	en1 extension (μm)			60	?	80	100	?	110
	ex length (μm)			120	?	180	200	?	220
Post-mandibular pair of limbs IV	Length (μm)			60	?	70	160	?	?
	en1 extension (μm)			20	?	30	80	?	?
	ex length (μm)			40	?	?	180	?	?
Post-mandibular pair of limbs V	Length (μm)						?	?	?
	en1 extension (μm)						?	?	?
	ex length (μm)						?	?	?

exopod becomes more and more curved during ontogeny to form a free proximal margin (cf. Pls. 10A, 11F).

The fourth pair of post-mandibular appendages appears at possible growth stage III as a bifid setose limb bud without a distinct proximal endite (Fig. 26). It develops a morphology comparable with that of the preceding post-mandibular pairs of limbs from possible growth stage IV onwards, when the proximal endite appears and the exopod is a multi-annulated ramus (Fig. 26, Pl. 10A). The endopod is unknown due to poor preservation of later stages.

The fifth pair of post-mandibular limbs appears first at possible growth stage VI. Further ontogeny of this limb is not known.

Hesslandona necopina Müller, 1964

v* 1964a   *Hesslandona necopina* n. sp. – Müller, p. 22, pl. 1, fig. 6.

non 1964a   *Hesslandona necopina* n. sp. – Müller, pl. 1, figs. 1, 2 (= *Hesslandona trituberculata*).

non 1964a   *Hesslandona necopina* n. sp. – Müller, pl. 1, figs. 3, 4 (= *Hesslandona kinnekullensis*).

non 1964a   *Hesslandona necopina* n. sp. – Müller, pl. 1, fig. 5 (= *Hesslandona suecica* n. sp.).

.   1974   *Hesslandina necopina* – Martinsson, p. 208 (sic!).

.   1975   *Hesslandona necopina* Müller, 1964 – Müller, pl. 19, fig. 1.

.   1978   *H. necopina* – Rushton, p. 279.

v.   1979a   *Hesslandona necopina* Müller, 1964 – Müller, fig. 7.

.   1981   *H. necopina* K.J. Müller (1964) – Gründel *in* Gründel & Buchholz, p. 63.

v.   1983   *Hesslandona necopina* Müller, 1964 – McKenzie *et al.*, fig. 1.

.   1986   *Hesslandona necopina* – Huo *et al.*, text-fig. 4b–e.

.   1986a   *Hesslandona necopina* Mueller, 1964 – Kempf, p. 400.

.   1986b   *Hesslandona necopina* Mueller, 1964 – Kempf, p. 392.

.   1987   *Hesslandona necopina* Mueller, 1964 – Kempf, p. 436.

.   1987   *Hesslandona necopina* Müller – Tong, p. 433.

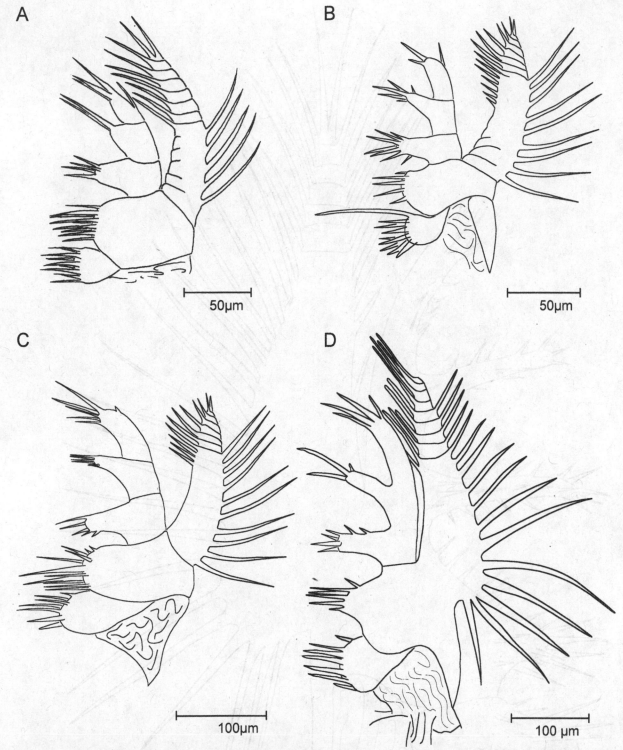

Fig. 24. Morphogenesis of the first post-mandibular limb. A: Growth stage III. B: Growth stage V. C: Growth stage VI. D: Growth stage VII. Cf. Pls. 7C, 10A.

. 1989 *Hesslandona necopina* Müller, 1979 – Zhao & Tong, p. 15 (cop. Müller 1979a, figs. 7a–c).

. 1992a *Hesslandona necopina* Müller, 1964a – Hinz, p. 15 (cop. Müller, pl. 1, fig. 6).

. 1993b *Hesslandona necopina* Müller, 1964 – Hinz-Schallreuter, p. 333.

. 1993c *Hesslandona necopina* Müller, 1964 – Hinz-Schallreuter, p. 396.

. 1998 *Hesslandona necopina* Müller, 1964 – Hinz-Schallreuter, pp. 104, 115.

. 1998 *Hesslandona necopina* Müller, 1964 – Williams & Siveter, p. 30.

. 1999 *Hesslandona necopina* Müller – McKenzie *et al.*, fig. 33.5A.

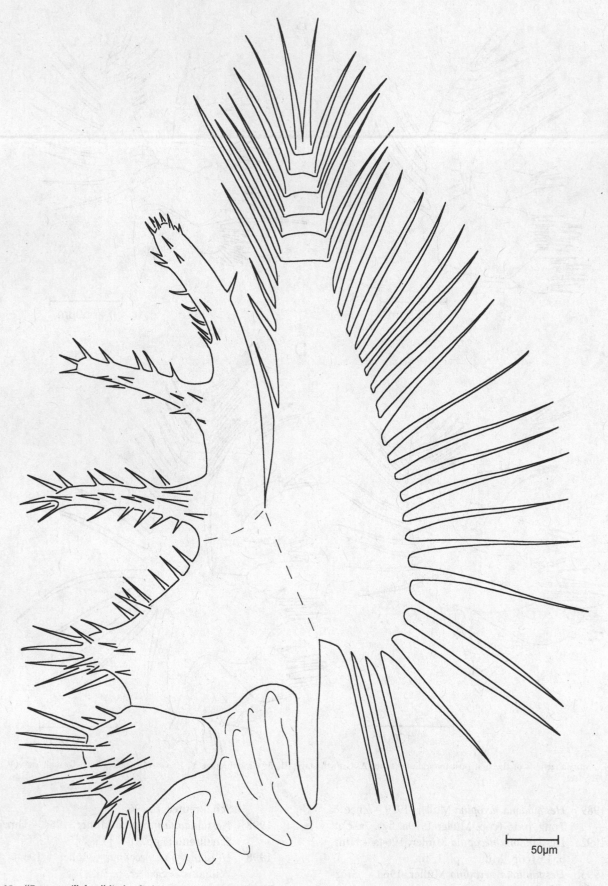

50μm

Fig. 25. "Post-maxillulary" limb of a large growth stage.

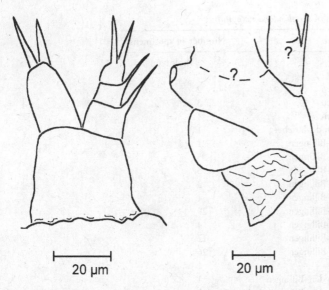

20 µm   20 µm

Fig. 26. Morphogenesis of the fourth post-mandibular limb. A: Growth stage III (cf. Pl. 8C). B: Growth stage V (cf. Pl. 11B).

Name. – The Latin word *necopinus* means "unexpected". Müller's first find of phosphatocopines in the acetic acid residues belongs to this species (cf. Fig. 6).

Holotype. – Slightly opened shield with broken anterior and posterior interdorsal spines, illustrated by Müller (1964a) in his pl. 1, fig. 6a, b. Length 840 µm, height 375 µm.

Remarks. – Müller (1964a) illustrated several specimens of *H. necopina*. In fact, only the holotype [pl. 1, fig. 6 of Müller (1964a)] belongs to this species. All other specimens belong to other species: pl. 1, figs. 1, 2 = *Hesslandona trituberculata* (Lochman & Hu, 1960); pl. 1, figs. 3, 4 = *H. kinnekullensis* Müller, 1964; pl. 1, fig. 5 = *H. suecica* n. sp. See the respective species for additional information.

Type locality. – Near Gudhem, north of the Mösse Mountain, Falbygden, Västergötland, Sweden, 1 km west of the church, in the turning point of the street Falköping–Skara (Fig. 3, number 21).

Type horizon. – Upper Cambrian, *Agnostus pisiformis* Zone (Zone 1).

Material examined. – Two hundred and eighty-six specimens of different stages and from different areas of Zone 1 of the Upper Cambrian of southern Sweden (Table 17).

Dimensions. – Smallest specimen: valves about 450 µm long and about 250 µm high. Largest specimen: valves about 1000 µm long and about 400 µm high.

Additional material. – No additional material of *H. necopina* has been reported so far.

Distribution. – *Agnostus pisiformis* Zone (Zone 1) of the Upper Cambrian, southern Sweden.

Original diagnosis. – (Müller 1964a, translated). "Relatively elongated member of genus with large spines at anterior and posterior end, without sculpture. Three lobes that are fully developed not before later stages [*Remark*: this is due to the confusion with *Hesslandona trituberculata* (Lochman & Hu, 1960)]. Doublure narrow."

Emended diagnosis. – Maximum length of valves corresponds to dorsal rim. Valves pre-plete. Valves close tightly without leaving gaps. Valves smooth, without lobes or spines. Interdorsum drawn out into long spines anteriorly and posteriorly, directing at an angle of about 45 degrees. Doublure widest posteriorly, without any structures.

Description. –

Shield (Fig. 27)
The bivalve shield (ch1:2) has a long and straight dorsal rim (ch2:2) (Pl. 13A, Fig. 27A). The maximum length of the head shield corresponds to the dorsal rim of the head shield (ch3:2, ch5:0) (Pl. 13A, B). The slightly opened shield is a pointed egg shape in dorsal view. The right and left valves are of equal size and symmetrical shape (Pl. 14B). The maximum height of the valves is anterior to the midline (pre-plete, see Fig. 8) (ch4:1). The shield varies from 1.5 in younger stages up to 2.5 times longer than high in older stages. The free anterior part of the shield margin, starting from the relatively straight dorsal rim (ch6:1), curves postero-ventrally (antero-dorsal angle nearly 70 degrees, cf. Fig. 7) towards the ventral maximum (Pl. 13A). In lateral view, the latero-ventral margin curves gently, continuing into an almost straight part (ch7:1) (Pl. 13A) and thereafter curving gently upwards – not as steeply as anteriorly – towards the posterior and curving slightly more steeply right before meeting the postero-dorsal rim (ch8:1, postero-dorsal angle about 80 degrees, cf. Fig. 7). In dorsal view, the closed shield has its maximum width slightly dorsal to the midline. Margins are without outgrowths throughout (ch19:0, ch20:0). The shield valves close tightly without any gaps in all stages (ch9:1). The surface of the valves is smooth, without any lobes or spines (chs10–18:0) (Pl. 1A, B). The outer margin of the valves has a narrow slight furrow close to the doublure (ch21:1) (Pl. 13D, E; Fig. 28).

Interdorsum
The interdorsum is continuous from the anterior to the posterior (ch22:3), being bordered by narrow membranous furrows on both sides (Pl. 13B, C). The interdorsum has the same width apart from both ends, where it

Table 17. Sample productivity of examined specimens of *Hesslandona necopina*.

Sample	Zone	Found at	Number of specimens
955	1	Near Gudhem, Falbygden	1 (holotype)
994	1	Kestad, Kinnekulle	1
6363	1	St. Stolan, Falbygden–Billingen	2
6364	1	St. Stolan, Falbygden–Billingen	3
6367	1	St. Stolan, Falbygden–Billingen	1
6404	1–2	Kestad, between Haggården and Marieberg	1
6408	1	Gum, Kinnekulle, Falbygden–Billingen	21
6409	1	Gum, Kinnekulle, Falbygden–Billingen	25
6410	1	Gum, Kinnekulle, Falbygden–Billingen	5
6411	1	Gum, Kinnekulle, Falbygden–Billingen	3
6412	1	Gum, Kinnekulle, Falbygden–Billingen	1
6414	1	Gum, Kinnekulle, Falbygden–Billingen	29
6415	1	Gum, Kinnekulle, Falbygden–Billingen	4
6416	1	Gum, Kinnekulle, Falbygden–Billingen	12
6417	1	Gum, Kinnekulle, Falbygden–Billingen	26
6483	1	Degerhamn, Öland	1
6730	1	Backeborg, Kinnekulle, Falbygden–Billingen	1
6732	1	Backeborg, Kinnekulle, Falbygden–Billingen	3
6734	1	Backeborg, Kinnekulle, Falbygden–Billingen	2
6735	1	Backeborg, Kinnekulle, Falbygden–Billingen	2
6739	1	Blomberg, Kinnekulle, Falbygden–Billingen	3
6747	1	Stora Stolan, Billingen	3
6749	1	Gum, Kinnekulle, Falbygden–Billingen	2
6750	1	Gum, Kinnekulle, Falbygden–Billingen	6
6751	1	Gum, Kinnekulle, Falbygden–Billingen	2
6752	1	Gum, Kinnekulle, Falbygden–Billingen	4
6754	1	Gum, Kinnekulle, Falbygden–Billingen	5
6755	1	Gum, Kinnekulle, Falbygden–Billingen	9
6757	1	Gum, Kinnekulle, Falbygden–Billingen	9
6758	1	Gum, Kinnekulle, Falbygden–Billingen	2
6760	1	Gum, Kinnekulle, Falbygden–Billingen	14
6761	1	Gum, Kinnekulle, Falbygden–Billingen	15
6764	1	Gum, Kinnekulle, Falbygden–Billingen	9
6765	1	Gum, Kinnekulle, Falbygden–Billingen	3
6768	1	Gum, Kinnekulle, Falbygden–Billingen	4
6771	1	Gum, Kinnekulle, Falbygden–Billingen	1
6772	1	Gum, Kinnekulle, Falbygden–Billingen	5
6773	1	Gum, Kinnekulle, Falbygden–Billingen	3
6774	1	Gum, Kinnekulle, Falbygden–Billingen	6
6776	1	Gum, Kinnekulle, Falbygden–Billingen	12
6777	1	Gum, Kinnekulle, Falbygden–Billingen	2
6780	1	Gum, Kinnekulle, Falbygden–Billingen	5
6782	1	Gum, Kinnekulle, Falbygden–Billingen	6
6783	1	Gum, Kinnekulle, Falbygden–Billingen	6
6784	1	Gum, Kinnekulle, Falbygden–Billingen	6

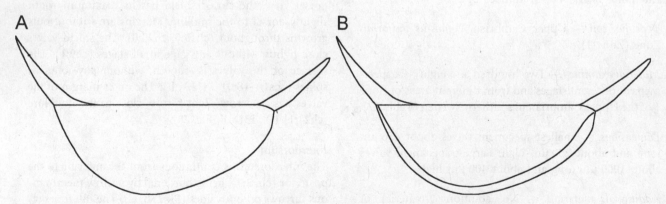

Fig. 27. Schematic drawing of the shield of *Hesslandona necopina* from outside lateral (A) and from inside lateral (B).

narrows. The maximum width is about 1/20 the length of the valves (ch23:1). The interdorsum is convex in axial aspect (ch24:2), without lobes or ornamentation (ch25:1) (Pl. 13B, C). The anterior end is drawn out into a long, sharply pointed spine, basally as thick as the interdorsum (ch26:1, ch27:3). The spine is directed antero-dorsally (ch28:2) at an angle of about 55 degrees, slightly curving dorsally. The posterior end is also drawn out into a long but more bluntly ending spine, basally as thick as the interdorsum (ch29:1, ch30:3). The spine is directed postero-dorsally (ch31:2) at an angle of about 35 degrees, straight. Being part of the interdorsum, the spines are separated from the valves by the membranous furrows that border the interdorsum laterally (Pl. 14A, B). The length of these spines may vary strongly individually (Pls. 13A, B, 14A). The anterior spine is smooth (ch32:1) while the posterior spine shows small spiny scale-like outgrowths (ch33:2) irregularly arranged on the whole surface of the last half of its length (Pl. 14C).

Doublure

A doublure is present (ch34:1) along the inner margin of the valves, being narrowest ventrally (ch35:2), anteriorly slightly wider and posteriorly about one third wider than ventrally (ch36:3) (Pls. 13D, 14D, E; Fig. 27B). Approximating the dorsal rim, the doublures of both sides rapidly narrow slightly to fade out into a membranous area antero- and postero-dorsally (Pl. 15B). The maximum width of the doublure is about 1/16 the length of the valves (ch37:1). On the inner postero-ventral margin of the doublure of more advanced growth stages, five small conical outgrowths are arranged in a row and directly on the surface (ch38:1, ch39:2, ch40:1). They are almost the same distance from each other, except the distance between the third and fourth from ventral is slightly wider (Pl. 15A, C). Pores on the doublure and other ornaments are probably absent (ch41:1, ch42:1). In width, the outer 1/10th of the doublure is slightly depressed relative to the first 9/10th (Pl. 13C, D; Fig. 28).

Inner lamella

The inner lamella is the little sclerotised part of the exoskeleton between the lateral extensions of the shield (Pls. 13D, 14D, E; 15A). It expands along the whole doublure and extends medially to the dorso-lateral side of the body. It shows a wrinkled texture, probably due to its lack of sclerotisation.

Body

The body proper, which is completely enveloped by the bivalved shield, includes at least nine segments (Pl. 15D).

Trunk segmentation of the preserved portion is retained in the insertions of the limbs, distinct segment boundaries or sternites are weakly defined. Minimally, nine segments are dorsally fused to the shield (ch44:7) (Pl. 15D), being a cephalothoracic shield including at least four limb-bearing trunk segments. The area of fusion of the body proper to the shield is very narrow, corresponding to approximately the width of the interdorsum (ch43:1). Starting a few micrometres posterior to the dorsal anterior membranous area of the doublures, the body proper extends along the inner dorsal length of the shield, but the posterior end of the body is unknown (Pl. 15D, E). Its posterior extension and segmental continuation is uncertain, as well as if there is a hind body extending free from the shield into the domicilium. The body proper in ventral view is oval with a blunt anterior and an elongated posterior end. The maximum width of the body proper is at the mandibles. The cross-section of the body proper anterior to the mandibular part is almost trapezoidal with the long axis dorso-ventrally oriented. The cross-section at the mandibular segment is more or less about half-oval with the long axis in dorso-ventral aspect; the height of the body proper in this position is more than one third the height of the valves. Posterior to it the body proper changes its cross-section to about half-circular. The height of the body decreases gradually posteriorly to about one sixth the height of the valves at the posterior end of the body. The limbs are arranged in a more or less regular row at the flanks of the body, being more or less equally set together from the anterior towards the caudal end (Pl. 15D, E).

The anterior part of the body proper is the hypostome/labrum complex. The hypostome forms the anterior sclerotised ventral surface that is somewhat rhomboid and becomes gradually higher to the posterior. Within its antero-lateral edge, the antennulae insert on a distinct lateral slope. The antennae (Pl. 15E) insert postero-laterally to it in a spindle-shaped insertion area located on a lateral slope. The posterior end of the hypostome is drawn out into a lobe-like protrusion, the labrum (ch61:1) (Pl. 15E). Posterior to the hypostome, the ventral surface between the first to third pairs of post-antennular limbs is marked by a sclerotic plate, the sternum (ch62:1) (Pl. 15D, E). It is slightly domed and bears one pair of humps anteriorly, the paragnaths (ch63:1), between the second pair of post-antennular limbs, the mandibles (Pl. 15E). Posterior to the sternum, the ventral body surface becomes progressively softer (Pl. 15D). The post-mandibular pairs of limbs insert on a lateral slope corresponding to the cross-section of the body at the rear of their respective segments (Pl. 15D). The hind body, including the number of segments involved and possible furcal rami, is unknown (ch47:?, ch48:?, ch49:?).

PLATE 14

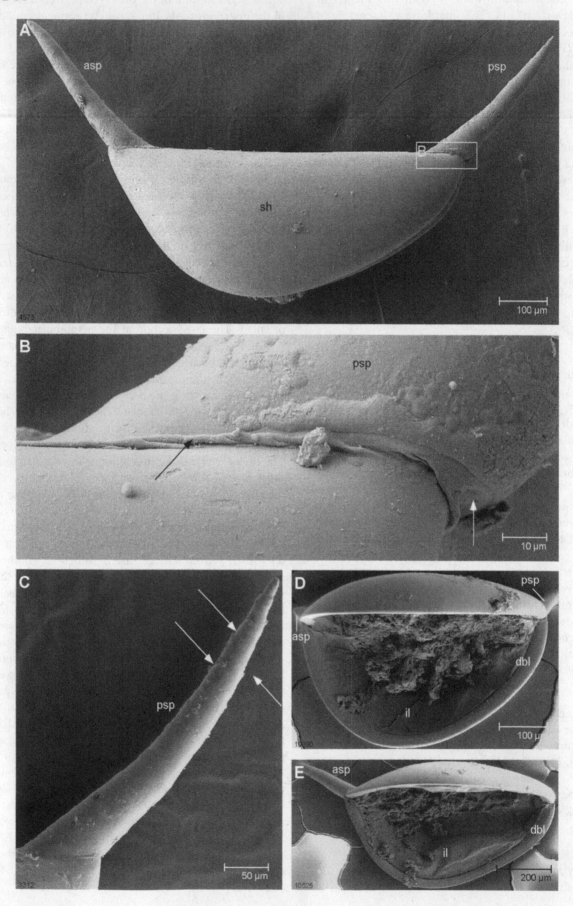

Soft parts

The soft parts are similar to those described in detail for *H. unisulcata*. Because it could not be the scope of this paper to describe the soft parts of *H. necopina* in full detail, only some general remarks are made, important for the discrimination from *H. unisulcata* and the phylogenetic analysis of Phosphatocopina. The antennula is similar to that of *H. unisulcata*, being small and consisting of less than 10 irregular annuli (ch45:1, ch46:1). The tip bears a tuft of four setae, three terminal ones and one additional subterminal seta. A seta on the third last annulus as described for *H. unisulcata* could not be observed. Again, as in *H. unisulcata*, the antenna (Pl. 15E) consists of an undivided limb stem throughout ontogeny, but being a fusion product of the coxa and basipod (see p. 23; ch50:2, ch51:2, ch52:2), a two-divided endopod (ch53:5) and a multi-annulated exopod. In general design, there are no ways to distinguish this limb from that of *H. unisulcata*. The same is true for the mandible (Pl. 15D, E). It has an undivided limb stem throughout ontogeny and an enditic protrusion located medially between the limb stem and the endopod, recognised as remains of the basipod (see p. 27; ch54:2, ch55:2, ch56:2). The endopod is two-divided (ch57:5) and the exopod consists of several annuli, each annulus with one medially projecting seta; the last annulus bears two setae – as in the antenna. The post-mandibular limbs (Pl. 15D, E) are similar to each other (ch58:1), consisting of a basipod, a setae-bearing proximal endite medio-proximally (ch59:1), a three-divided endopod (ch60:4) and an annulated exopod. The endopodal portions are short and slightly projecting medially. They are drop-shaped with a blunt end medially. The exopods of the post-mandibular pairs of limbs (Pl. 15D) are multi-annulated as in the antenna and the mandible but with fewer annuli, they have no lateral setation.

Comparisons

Hesslandona necopina is very similar to *H. trituberculata* and *H. toreborgensis* n. sp. It is distinguished from *H. trituberculata* and *H. toreborgensis* n. sp. by the lack of three dorsal lobes. The spines of the interdorsum of *H. necopina* are distinctly longer and not as curved as in *H. orgensis* n. sp. *Hesslandona trituberculata* has a ventrally curved posterior spine, which does not occur in *H. necopina*. The anterior width of the doublure in relation to the ventral and posterior width is smaller in *H. necopina* than in *H. trituberculata* but comparable with that in *H. toreborgensis* n. sp. *Hesslandona necopina* differs from *H. unisulcata*, *H. suecica* n. sp. and *H. angustata* n. sp. by many features, e.g. the maximum length of the shield is on the dorsal rim and not between the dorsal rim and the midline as in *H. unisulcata*, *H. suecica* n. sp. and *H. angustata* n. sp. *Hesslandona necopina* lacks the prominent lobe that occurs on both valves of *H. unisulcata* and *H. minima* symmetrically. *Hesslandona unisulcata* and *H. angustata* n. sp. have no spines, *H. suecica* n. sp. has only very short spines at the anterior and posterior end of the interdorsum. *Hesslandona ventrospinata* has asymmetrical valves due to asymmetrical outgrowths of the postero-ventral margin of the valves, which do not occur in *H. necopina*. The endopods of all post-mandibular limbs are not strongly medially projecting as in *H. unisulcata*.

Ontogeny

During the ontogeny of *H. necopina*, the length of the valves grows proportionally stronger than the height, such that the shield becomes progressively more and more slender and elongated in long axis (Fig. 29). The anterior and posterior interdorsal outgrowths grow from short protrusions in early larvae to long slender spines in late larvae. The ontogeny shows four to five distinct clusters, which may represent four to five possible growth stages (Fig. 29). The first larval stage recognised consists of a body with at least four limb-bearing segments that are dorsally fused to an all-enclosing bivalved shield (ch64:1).

Plate 14. *Hesslandona necopina* Müller, 1964

A: UB W 168. A specimen representing a late growth stage, about 840 µm in length. A lateral view of the left valve. The surface of the valve is clean. A rectangle marks the area magnified in Pl. 14B.

B: Close-up of the area marked in Pl. 14A, displaying the membranous furrow (arrows) extending along the basal part of the posterior interdorsal spine.

C: UB W 259. A specimen representing a late growth stage, about 870 µm in length. Close-up of the posterior interdorsal spine, displaying small scale-like structures (arrows) distally.

D: UB W 169. Image flipped horizontally. A specimen representing an advanced stage, about 500 µm in length. A lateral view of the inner side of the left valve. The valve is fairly high compared with its length. See Pl. 14E for the difference in the shape of the shield compared with a more advanced stage.

E: UB W 170. Image flipped horizontally. A specimen representing a later stage, about 780 µm in length. A lateral view of the inner side of the left valve. The valve is distinctly more elongated than in a younger specimen (cf. Pl. 14D for the difference in the shape of the shield compared with a less advanced stage).

PLATE 15

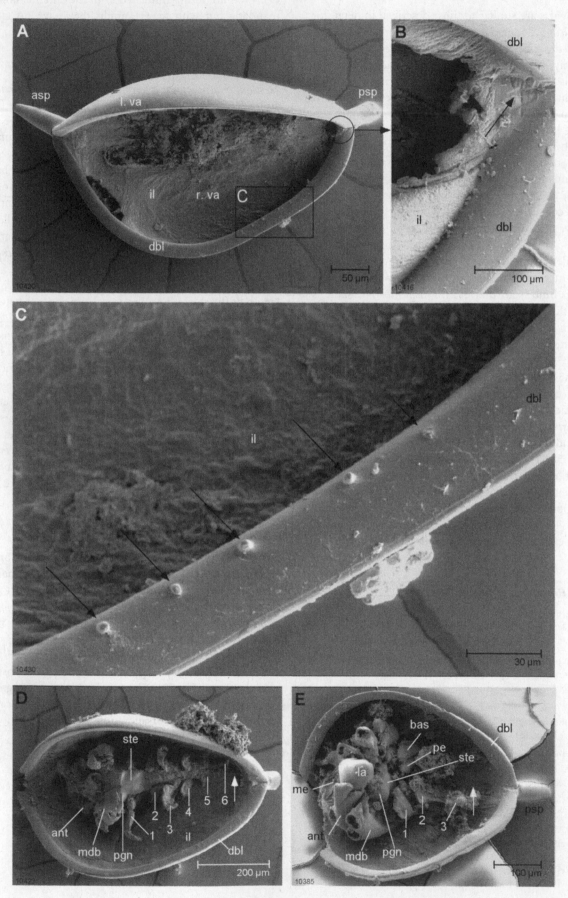

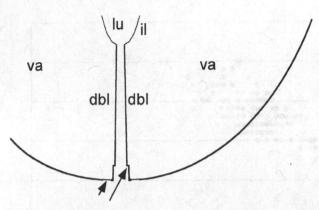

Fig. 28. Schematic drawing of a transversal cut through the ventral part of the shield of *Hesslandona necopina* showing the shallow furrows along the free margin of the valves and on the outer margin of the doublure (arrows). See Table 4 for abbreviations.

Hesslandona kinnekullensis Müller, 1964

v* 1964a *Hesslandona kinnekullensis* n. sp. – Müller, p. 23, pl. 1, figs. 7–9, ?11.
v.   1964a *Hesslandona necopina* – Müller, pl. 1, figs. 3, 4.
.   1974   *H. kinnekullensis* – Martinsson, p. 208.
.   1978   *H. kinnekullensis* Müller – Rushton, p. 279.
non 1979a *Hesslandona kinnekullensis* Müller, 1964 – Müller, fig. 36 (= *Vestrogothia* sp. or *Falites* sp.).
.   1985a *Hesslandona kinnekullensis* Müller, 1964 – Müller & Walossek, fig. 5g.
.   1986a *Hesslandona kinnekullensis* Mueller, 1964 – Kempf, p. 400.
.   1986b *Hesslandona kinnekullensis* Mueller, 1964 – Kempf, p. 308.

Plate 15. *Hesslandona necopina* Müller, 1964

A: UB W 171. A specimen representing a late growth stage, about 780 μm in length. A lateral view of the inner side of the right valve. The postero-dorsal area of a similar specimen, as indicated by a circle, is documented in Pl. 15B. A rectangle marks the area magnified in Pl. 15C.

B: UB W 254. A specimen representing a late growth stage, about 780 μm in length. A lateral view of the postero-dorsal area (see Pl. 15A for orientation), displaying the doublure merging into a membranous area connecting either side (arrow).

C: Close-up of the area marked in Pl. 15A, displaying the five conical outgrowths on the inner postero-ventral margin of the doublure (arrows), the last one on the right is slightly damaged.

D: UB W 172. A specimen representing a late growth stage, about 730 μm in length with preserved body morphology. Ventral view. At least six post-mandibular pairs of limbs are present, all probably fused to the shield, the hind body is missing (arrow). The anterior part of the head is damaged, the hypostome, labrum, antennulae and antennae are destroyed.

E: UB W 173. A specimen representing a young growth stage, about 490 μm in length. Ventral view. The right anterior part of the shield is damaged. At least three post-mandibular pairs of limbs are present, the hind body is missing (arrow). The antennula are not preserved.

.   1987   *Hesslandona kinnekullensis* Mueller, 1964 – Kempf, p. 436.
.   1989b *Hesslandona kinnekullensis* Müller – Zhao, p. 412.
.   1993c *Hesslandona kinnekullensis* Müller, 1964 – Hinz-Schallreuter, p. 396.
.   1998   *Hesslandona kinnekullensis* Müller, 1964 – Hinz-Schallreuter, p. 115, 116.

Derivation of name. – After the Kinnekulle, a hill in Västergötland, near Lidköping at Lake Vänern, southern Sweden (see Fig. 2).

Holotype. – A half-opened shield illustrated by Müller (1964a) in his pl. 1, fig. 8a, b. Length 690 μm, height 390 μm.

Remarks. – Müller (1964a) illustrated several specimens of *Hesslandona necopina* Müller, 1964 but the specimens illustrated on his pl. 1, figs. 3, 4 are in fact *Hesslandona kinnekullensis* Müller, 1964. Müller (1979a, fig. 36) illustrated a phosphatocopine specimen from Zone 5 of the Upper Cambrian with preserved appendages and labelled it *H. kinnekullensis*. This specimen (UB 630), about 200 μm in shield length, is interpreted herein as a larva belonging to a species of *Vestrogothia* Müller, 1964 or *Falites* Müller, 1964.

Type locality. – Near Brattefors at the Kinnekulle, Västergötland, Sweden.

Type horizon. – Upper Cambrian, *Olenus truncatus* Zone (Zone 2) (Fig. 2).

Material. – Seventy-eight specimens of different stages and from different areas of Zone 2 of the Upper Cambrian of southern Sweden (Table 18).

Dimensions. – Smallest specimen: valves about 180 μm long and about 120 μm high. Largest specimen: valves about 950 μm long and about 460 μm high.

Additional material. – No additional material of *H. kinnekullensis* has been reported thus far.

Original diagnosis. – (Müller 1964a, translated). "A member of *Hesslandona* with a thick cone-tooth-like spine anteriorly on the interdorsum that stands vertically on the dorsal rim. Outline more or less symmetrical and highly curved, sides uplifted, without lobes."

Emended diagnosis. – Maximum length slightly ventral to the dorsal rim. Valves pre-plete. Valves close tightly without leaving gaps. Valves smooth, without lobes or spines. Interdorsum, drawn out into spines anteriorly and posteriorly, directed almost straight dorsally. Doublure widest posteriorly, without any ornamentation.

Description. –

Shield (Fig. 30)
The bivalved shield (ch1:2) has a long and straight dorsal rim (ch2:2) (Pl. 16A, B; Fig. 30A). The maximum length

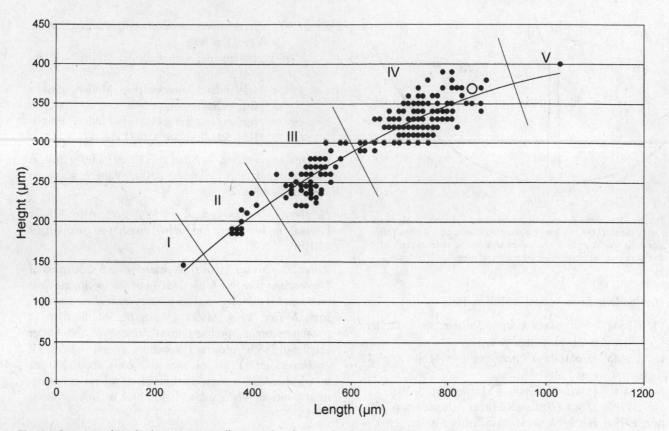

Fig. 29. Ontogeny of *Hesslandona necopina* Müller, 1964: length versus height, displaying four to five clusters possibly representing growth stages. Two hundred and forty-eight specimens with measurable length and height were considered. The trend line shows the stronger increase in length against the increase in the height of the valves during ontogeny. $\bigcirc$ = holotype. The trend line has the polynomial function $y = -0.0003x^2 + 0.6716x - 19.563$, y = height, x = length.

Table 18. Sample productivity of examined specimens of *Hesslandona kinnekullensis.*

Sample	Zone	Found at	Number of specimens
995	2	Brattefors, Kinnekulle, Falbygden–Billingen	4 (including holotype)
1582	2	Kinnekulle	1
5663	2	Degerhamn, Öland	5
6402	2	between Haggården and Marieberg, Kinnekulle, Falbygden–Billingen	1
6404	1–2	Kestad, between Haggården and Marieberg	20
6410	2	Gum, Kinnekulle, Falbygden–Billingen	1
6431	2	Karlsfors, Falbygden–Billingen	2
6432	2	Karlsfors, Falbygden–Billingen	2
6470	2	Degerhamn, Öland	8
6473	2	Degerhamn, Öland	6
6474	2	Degerhamn, Öland	2
6740	2	Blomberg, Kinnekulle, Falbygden–Billingen	1
6745	2	Ledsgården–Gökhem, Falbygden–Billingen	1
6774	2	Gum, Kinnekulle	1
6792	2	Stubbeg, Kinnekulle, Falbygden–Billingen	2
6794	2	Stubbeg, Kinnekulle, Falbygden–Billingen	3
6796	2	Between Haggården and Marieberg, Kinnekulle, Falbygden–Billingen	1
6799	2	Between Haggården and Marieberg, Kinnekulle, Falbygden–Billingen	2
6808	1–2	Toreborg, Kinnekulle, Falbygden–Billingen	4
6810	2	Between Toreborg and Fullösa, Kinnekulle, Falbygden–Billingen	3
6811	2	Between Toreborg and Fullösa, Kinnekulle, Falbygden–Billingen	3
6813	2	Gum, Kinnekulle, Falbygden–Billingen	5

of the head shield is close to the dorsal rim of the head shield (ch3:3). The right and left valves are of equal size and symmetrical shape. The maximum height of the valves is anterior to the midline (pre-plete, see Fig. 8) (ch4:1). The shield varies from 1.7 in younger stages up to 2.0 times longer than high in older stages. The free

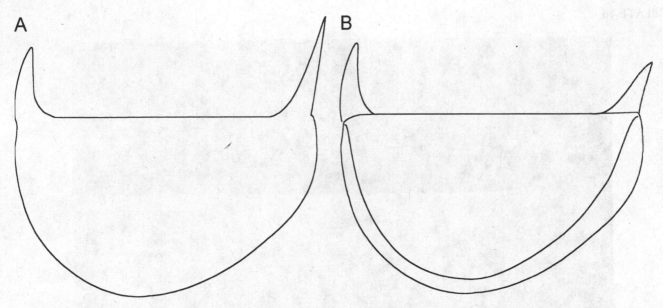

Fig. 30. Schematic drawing of the shield of *Hesslandona kinnekullensis* from outside lateral (A) and from inside lateral (B).

anterior part of the shield margin, starting from the dorsal rim, first curves slightly antero-ventrally (ch6:2) (antero-dorsal angle nearly 75 degrees, cf. Fig. 7) and, after reaching the most anterior part of the shield, curves back postero-ventrally towards the ventral maximum of the valve (ch5:3) (postero-dorsal angle nearly 75 degrees, cf. Fig. 7). The ventral margin is curved (ch7:2) (Pl. 16A, B). More posteriorly, the margin swings gently upwards – as steeply as anteriorly – and towards the most posterior part of the shield, curving slightly anteriorly and meeting the posterior end of the dorsal rim (ch8:2) (postero-dorsal angle about 80 degrees). The maximum width is slightly dorsal to the midline. The margins are without outgrowths throughout (ch19:0, ch20:0). The shield valves close tightly without any gaps in all stages (ch9:1). The surface of the valves is smooth, without any lobes or spines (ch10:0, ch11:0, ch12:0, ch13:0, ch14:0, ch15:0, ch16:0, ch17:0, ch18:0) (Pl. 16A, B). The outer margin of the valves is without any ornamentation (ch21:0).

Interdorsum

The interdorsum is continuous from the anterior to the posterior (ch22:3), being bordered by narrow membranous furrows on both sides (Pl. 16C). The interdorsum is always the same width, apart from both ends, where it becomes narrower. The maximum width is about 1/24 the length of the valves (ch23:1). The interdorsum is convex in abaxial aspect (ch24:2), and the median part is without lobes or any ornamentation (ch25:1) (Pl. 16B). The anterior end is drawn out into a long spine, about one eighth the length of the valves (ch26:1, ch27:3), with a blunt tip, basally as thick as the interdorsum, and directed dorsally at an angle of about 90

degrees (ch28:1) with a slight swing towards the posterior (Pl. 16A–D). The insertion area is somewhat egg-shaped. The anterior edge of the spine is slightly curved backwards and is almost an elongation of the anterior margin of the shield. The posterior margin of the spine slopes steeply into the dorsal rim. The posterior end of the interdorsum is also drawn out into a spine (Pl. 16A–D) slightly longer than the anterior one (ch29:1, ch30:3) with a pointed tip, basally as thick as the interdorsum. The spine is directed almost dorsally (ch31:1) with a slight slope to the posterior at an angle of about 75 degrees. The anterior margin of the spine slopes less steeply into the dorsal rim than the posterior margin of the anterior spine. Being part of the interdorsum, the spines are separated from the valves by the membranous furrows that border the interdorsum laterally. The anterior and posterior spines are smooth (ch32:1, ch33:1).

Doublure

A doublure is present (ch34:1) along the inner margin of the valves, being narrowest ventrally (ch35:2), anteriorly slightly wider, and posteriorly about one third wider than ventrally (ch36:3) (Pl. 16D, E; Fig. 30B). Approximating the dorsal rim, the doublures of both sides rapidly narrow slightly to merge into a membranous area antero- and postero-dorsally. The maximum width of the doublure is about 1/11 the length of the valves (ch37:1). There are no structures or pores present on the postero-ventral margin of the doublure (ch38:1, ch39:1, ch40:0, ch41:1, ch42:1).

Inner lamella

The inner lamella was not observed due to poor preservation (cf. Pl. 16D, E).

PLATE 16

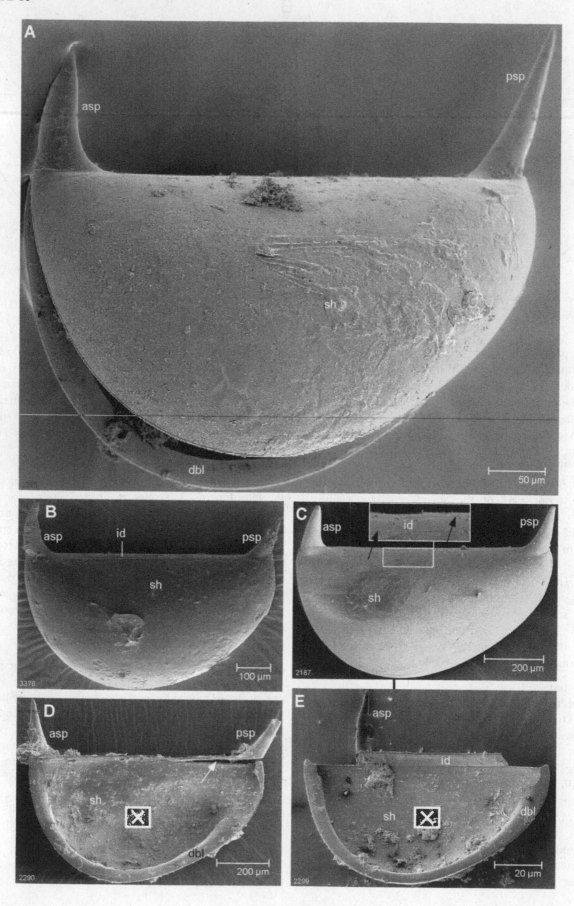

Body

The body is not preserved in any of the specimens of *H. kinnekullensis* available (cf. Pl. 16D, E) (ch43–64:?).

Comparisons

Hesslandona kinnekullensis is similar to *H. necopina* in having a smooth shield and an interdorsum drawn out anteriorly and posteriorly into more or less long spines. The spines of *H. kinnekullensis* are directed dorsally, while the spines of *H. necopina* are directed antero- and/or postero-dorsally. *Hesslandona toreborgensis* n. sp., *H. suecica* n. sp. and *H. angustata* n. sp. also have a smooth shield but the spines of *H. toreborgensis* n. sp. and *H. suecica* n. sp. are distinctly smaller, while *H. angustata* n. sp. does not have spines. All other *Hesslandona* species, i.e. *H. trituberculata*, *H. curvispina* n. sp., *H. ventrospinata*, *H. unisulcata* and *Trapezilites minimus*, are distinguished from *H. kinnekullensis* by the presence of one or three lobes on the valves.

Ontogeny

During the ontogeny of *H. kinnekullensis* the length of the valves grows proportionally greater than the height, such that the shield becomes progressively more and more slender and elongated in its long axis (Fig. 31). The anterior and posterior interdorsal outgrowths grow from short protrusions in early larvae to long slender spines in late larvae (Fig. 32). The ontogeny shows four clusters, which may represent four possible growth stages (Fig. 31).

Plate 16. *Hesslandona kinnekullensis* Müller, 1964

A: UB W 174. A specimen representing a young growth stage, about 590 µm in length. An antero-lateral view of a slightly opened shield, displaying part of the doublure of the right valve. Note the backward swinging anterior spine (asp).

B: UB W 175. Image flipped horizontally. A specimen representing an advanced growth stage, about 730 µm in length. Lateral view.

C: UB W 176. Image flipped horizontally. A specimen representing a late growth stage, about 830 µm in length. A dorso-lateral view displaying the interdorsum (id). Part of the interdorsum, marked with a rectangle, displaying the membranous furrow that borders the interdorsum on both sides (arrows), is magnified in the inset.

D: UB W 177. Image flipped horizontally. A specimen of a late growth stage, about 890 µm in length. A lateral view inside the left valve. The interdorsum is torn off in the posterior part (arrow), the inner lamella is not preserved and the inner surface of the left valve is exposed (⊠).

E: UB W 178. Image flipped horizontally. A specimen representing a late growth stage, about 870 µm in length. A lateral view inside the left valve. The interdorsum (id) is partly missing and the anterior spine (asp) is displaced. The inner lamella is not preserved and the inner surface of the left valve is exposed (⊠).

Hesslandona trituberculata (Lochman & Hu, 1960) Rushton, 1978

*	1960	*Dielymella*? *trituberculata*, n. sp. – Lochmann & Hu, pp. 793, 826, pl. 98, fig. 56.
v.	1964a	*Hesslandona* n. sp. a – Müller, p. 24, pl. 1, fig. 10a, b.
non	1978	*Hesslandona trituberculata* (Lochmann & Hu, 1960) – Rushton, p. 279, pl. 26, fig. 11; text-fig. 2 (= *Hesslandona curvispina* n. sp.).
non	1981	*Hesslandona trituberculata* (Lochmann & Hu) – Gründel, p. 63, pl. 3, fig. 9 (= *Hesslandona curvispina* n. sp.).
non	1981	*Hesslandona trituberculata* (Lochmann & Hu) – Gründel, p. 63, pl. 3, fig. 10 (= *Hesslandona kinnekullensis*).
.	1986a	*Dielymella*? *trituberculata* Lochman & Hu, 1960 – Kempf, p. 316.
non	1986a	*Hesslandona trituberculata* (Lochman & Hu, 1960) Rushton, 1978 – Kempf, p. 400 (= *Hesslandona curvispina* n. sp.).
.	1986b	*Dielymella*? *trituberculata* Lochman & Hu, 1960 – Kempf, p. 610.
non	1986b	*Hesslandona trituberculata* (Lochman & Hu, 1960) Rushton, 1978 – Kempf, p. 610 (= *Hesslandona curvispina* n. sp.).
.	1987	*Dielymella*? *trituberculata* Lochman & Hu, 1960 – Kempf, p. 362.
non	1987	*Hesslandona trituberculata* (Lochman & Hu, 1960) Rushton, 1978 – Kempf, p. 670 (= *Hesslandona curvispina* n. sp.).
.	1993b	*Hesslandona* n. sp. a (Müller 1964) – Hinz-Schallreuter, p. 342.
.	1993c	*Dielymella*? *trituberculata* Lochmann & Hu, 1960 – Hinz-Schallreuter, p. 396 (referred to *Hesslandona*).
.	1994a	*Hesslandona trituberculata* Lochman & Hu – Williams *et al.*, p. 23.
.	1997	*Vestrogothia trituberculata* (Lochmann & Hu) – Siveter & Williams, p. 60, pl. 8, fig. 7.
.	1998	*Dielymella*? *trituberculata* Lochman & Hu, 1960 – Hinz-Schallreuter, pp. 104, 115 (referred to *Hesslandona*).
.	1998	*Hesslandona trituberculata* (Lochman & Hu, 1960) – Williams & Siveter, p. 31.

Name. – Not noted by Lochman & Hu (1960), presumably based on the three dorsally located lobes.

Holotype. – Isolated left valve, illustrated by Lochman & Hu (1960) in their pl. 98, fig. 56 (see also Siveter & Williams 1997, pl. 8, fig. 7); length of shield 910 µm, height of shield 480 µm.

Remarks. – Müller (1964a, pl. 1, fig. 10a, b) illustrated specimens of *H. trituberculata* which he described under

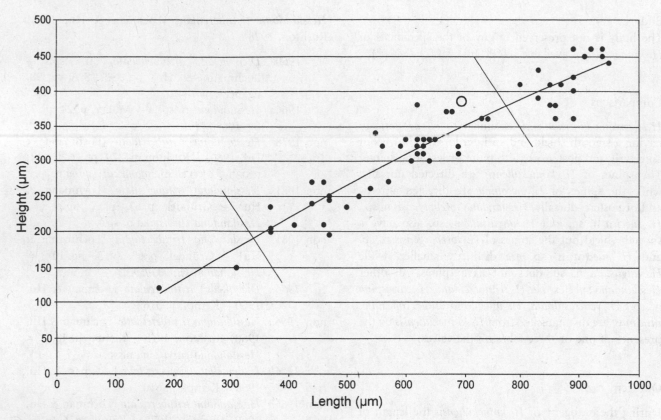

Fig. 31. Ontogeny of *Hesslandona kinnekullensis* Müller, 1964: length versus height. Fifty-nine specimens with measurable length and height were included. ○ = holotype. The curved trend line displays the stronger increase in length against the increase in the height of the valves during ontogeny. The trend line has the polynomial function $y = -0.0001x^2 + 0.5572x + 17.326$; y = height, x = length. Possible growth stages are separated by lines.

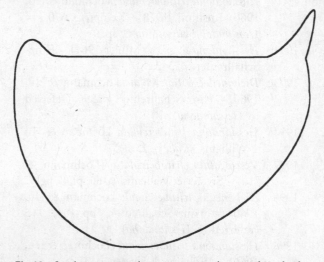

Fig. 32. Specimen representing a young growth stage of *Hesslandona kinnekullensis*, about 440 µm in length. The shield is higher compared with the length than in later stages, the anterior spine is a hump-like protrusion, while the shape of the posterior spine is comparable with that of later stages (cf. Fig. 30).

open taxonomy as *Hesslandona* n. sp. a, probably unaware of the paper by Lochman & Hu (1960), although the illustration given by Lochman & Hu (1960) is very poor (Gründel 1981). Siveter & Williams (1997) re-figured the type specimen at a much higher quality,

so a comparison with other descriptions of specimens supposedly belonging to this species from different areas is possible. A comparison, however, results in an exclusion of the specimens reported by Rushton (1978) and Gründel (1981) assigned to *H. trituberculata*. Their material belongs to both *H. curvispina* n. sp. (see p. 93) and *H. kinnekullensis* Müller, 1964 (see p. 67).

Type locality. – Sheep Mountain Section, Northwest Wind River Mountains, northeastern Sublette County, Wyoming, USA.

Type horizon. – Du Noir limestone, early Upper Cambrian, *Cedaria* Zone.

Material. – Fourteen specimens of different stages and from different areas of Zone 2 of the Upper Cambrian of southern Sweden (Table 19).

Dimensions. – Smallest specimens: valves about 810 µm long and about 410 µm high. Largest specimens: valves about 1460 µm long and about 460 µm high.

Additional material. – The species is known from the type specimen (Lochman & Hu 1960), one single valve from the Upper Cambrian of Wyoming, and Müller (1964a) reported material from the Upper Cambrian of Sweden, which is investigated herein.

Table 19. Sample productivity of examined specimens of *Hesslandona trituberculata.*

Sample	Zone	Found at	Number of specimens
6404	1–2	Between Haggården and Marieberg, Kinnekulle	1
6470	2	Degerhamn, Öland	7
6736	1–2	Backeborg, Kinnekulle	1
6796	2	Between Haggården and Marieberg, Kinnekulle	3
6798	2	Between Haggården and Marieberg, Kinnekulle	1
6801	2	Between Haggården and Marieberg, Kinnekulle	1

Table 20. Complete list of finds of *Hesslandona trituberculata* Lochman & Hu, 1960, reported by different authors, with localities and horizons.

Locality	Horizon	Reference
Sheep Mountain Section, Wyoming, USA	Upper Cambrian, *Cedaria* Zone	Lochman & Hu 1960; Siveter & Williams 1997
Västergötland, southern Sweden	Upper Cambrian, Zone 2	Müller 1964a and herein

Distribution. – *Cedaria* Zone of the Late Cambrian from North America and *Olenus truncatus* Zone (Zone 2) of the Late Cambrian, northern central Europe.

Original diagnosis. – Lochman & Hu (1960) did not give a diagnosis of this species.

Diagnosis. – (given herein). Maximum length of valves corresponds to dorsal rim. Valves pre-plete. Valves leave a gap postero-ventrally. Valves with three subdorsal lobes, two in the anterior, a third in the posterior half. Interdorsum drawn out into short spines anteriorly and posteriorly. Doublure widest posteriorly, with dome-like structures.

Description. –

Shield (Fig. 33)

The bivalved shield (ch1:2) has a long and straight dorsal rim (ch2:2) (Pl. 17A; Fig. 33A). The maximum length of the head shield is slightly ventral to the dorsal rim of the shield (ch3:3) (Pl. 17A, E). The valves are of equal size and the right and left valves are symmetrical. The maximum height of the valves is anterior to the midline (pre-plete, see Fig. 8) (ch4:1). The valves are about 1.9 times longer than high in every stage. The free anterior part of the shield margin, starting from the dorsal rim, curves gradually towards the ventral maximum (ch6:2) (Pl. 17C; antero-dorsal angle nearly 70 degrees). In younger stages the free anterior part curves slightly towards the anterior and after reaching the most anterior part of the shield, curves back postero-ventrally towards the ventral maximum (Pl. 17A). The latero-ventral outline is curved (ch7:2). Thereafter, the margin curves gently upwards – as steeply as anteriorly – towards the most posterior part of the shield, running almost straight anteriorly (ch8:1) to meet the postero-dorsal rim (postero-dorsal angle about 80 degrees). The margin curves back posteriorly slightly more strongly than anteriorly in younger stages (Pl. 17A) and only posteriorly in later stages (ch5:2) (Pl. 17C). The maximum

width of the shield is slightly dorsal to the midline. The margins of the right and left valves are without outgrowths throughout (ch19:0, ch20:0). The shield valves leave a gap posteriorly (ch9:2) (Pl. 17F). The surface of the valves is smooth, but with three dome-like lobes (Pl. 17A, B, D), all of them located close to the dorsal rim. The most anterior one is the largest (L_1) (ch10:1), a second, slightly smaller one (L_2), is located posterior to it just before the antero-posterior midline (ch11:1), a third, inconspicuous one (L_3), is located in the last third of the valves (ch12:1). Other structures are not developed on the valves (chs13–18:0, ch21:0).

Interdorsum

The interdorsum is complete from the anterior to the posterior (ch22:3) (Pl. 17A, B, D), being bordered by narrow membranous furrows on both sides (Pl. 17D). The interdorsum is always the same width, apart from both ends, where it becomes narrower. The maximum width is about 1/14 the length of the valves (ch23:1). The interdorsum is convex in antero-posterior aspect (ch24:2) (Pl. 17B) and – apart from its ends – without any lobes or ornamentation (ch25:1) (Pl. 17D). Its anterior end is drawn out into a short spine, basally as thick as the interdorsum (ch26:1, ch27:2), and directed anteriorly (ch28:3) (Pl. 17C); the spine may be thinner (Pl. 17C) or thicker (Pl. 17D). The posterior end of the interdorsum is drawn out into a short spine, basally as thick as the interdorsum but then essentially thinner than the anterior one (ch29:1, ch30:2), and is directed posteriorly (ch31:3) (Pl. 17F). The spines are – being parts of the interdorsum – separated from the valves by the membranous furrows that border the interdorsum laterally. The anterior and posterior spines are smooth (ch32:1, ch33:1).

Doublure

A doublure is present (ch34:1), of the same narrow width anteriorly and ventrally (ch35:5). It is about twice as wide posteriorly as anteriorly and ventrally (ch36:3)

PLATE 17

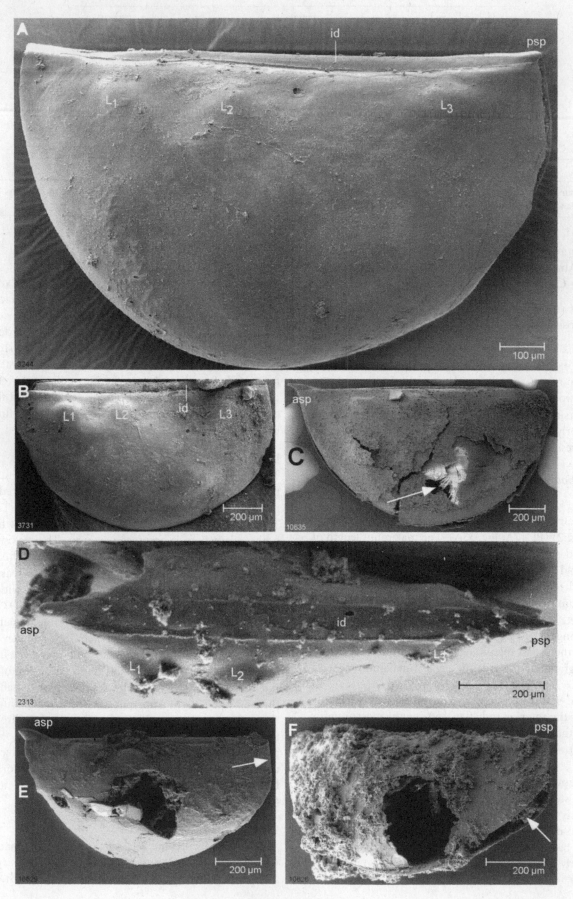

A

B

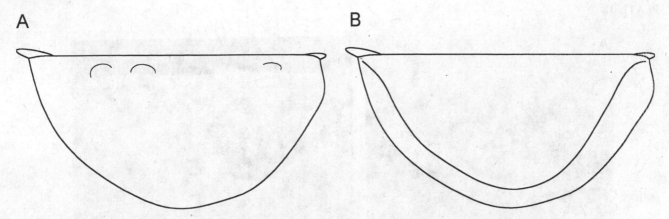

Fig. 33. Schematic drawing of the shield of *Hesslandona trituberculata* from outside lateral (A) and from inside lateral (B).

(Pl. 18A; Fig. 33B). In the largest specimen, the posterior part of the doublure shows a keel-like extension (Pl. 18B). Approximating the dorsal rim, the doublures of both sides rapidly narrow slightly to merge into a membranous area antero- and postero-dorsally. The maximum width of the doublure is about one eighth the length of the valves (ch37:2). On the outer postero-ventral margin of the doublure of more advanced growth stages there are 17 small dome-like outgrowths arranged in a row; they are located directly on the surface (ch38:1, ch39:2, ch40:1). The outgrowths have a blunt tip pointing outside of the domicilium. The distance between each of them varies (Pl. 18C–E). Pores on the doublure and other ornaments could not be seen (ch41:1, ch42:1).

Plate 17.　*Hesslandona trituberculata* (Lochman & Hu, 1960)

A: UB W 179. A specimen representing an advanced growth stage, about 1180 μm in length. A lateral view of the outside of the left valve. The anterior and posterior interdorsal spines are damaged.

B: UB W 180. A specimen representing a late growth stage, about 1300 μm in length. A lateral view of the left valve, displaying the convex interdorsum and the lobes (L1–L3) on the surface of the valves. The anterior and posterior parts of the shield are destroyed.

C: UB W 181. A specimen representing a late growth stage, about 1390 μm in length. A lateral view of the left valve. The valve is damaged medially so part of a limb is recognisable (arrow).

D: UB W 182. A specimen representing a late growth stage, about 1220 μm in length (see also Pl. 18A). A dorsal view, displaying the interdorsum (id), both interdorsal spines (asp, psp) and lobes L₁–L₃ (L1–L3).

E: UB W 183. A specimen representing an advanced growth stage, about 1050 μm in length. A lateral view of the right valve. Note the straight posterior margin of the valves (arrow) and the short and thick anterior interdorsal spine (asp). The area around the insertion of the posterior interdorsal spine is damaged.

F: UB W 184. A specimen representing a young growth stage, about 870 μm in length. A lateral view of the left valve, displaying the gaping of the postero-ventral to posterior region of the valves (arrow).

Inner lamella

The inner lamella expands along the whole doublure and extends medially to the dorso-lateral side of the body, which is never preserved in any specimen available. The inner lamella is weakly sclerotised and frequently shows a wrinkled texture (Pl. 18B).

Body

The body is not preserved in any of the specimens of *H. trituberculata* available and it is not known from the type specimen either (ch43–64:?).

Comparisons

Hesslandona trituberculata is similar to *H. ventrospinata* and *H. curvispina* n. sp. due to the three subdorsal lobes. It is distinguished from *H. curvispina* by the comparably short interdorsal spines and by the position of the maximum length of the shield slightly ventral to the dorsal rim. *Hesslandona ventrospinata* has asymmetrical valves due to asymmetrical outgrowths of the posteroventral margin of the valves, which do not occur in *H. trituberculata*. The spines of the interdorsum of *H. necopina* are also distinctly longer and the valves of this species as well as those of *H. suecica* n. sp., *H. angustata* n. sp., and *H. toreborgensis* n. sp., do not show lobes at all. The valves of *H. unisulcata* have only one prominent lobe.

Ontogeny

During ontogeny, the length of the valves of *H. trituberculata* grows proportionally slightly greater than the height, such that the shield becomes progressively more and more slender in its long axis (Fig. 34). The observed range of shield lengths does not indicate distinct clusters of particular growth stages; this is possibly due to the low number of specimens (Fig. 34).

PLATE 18

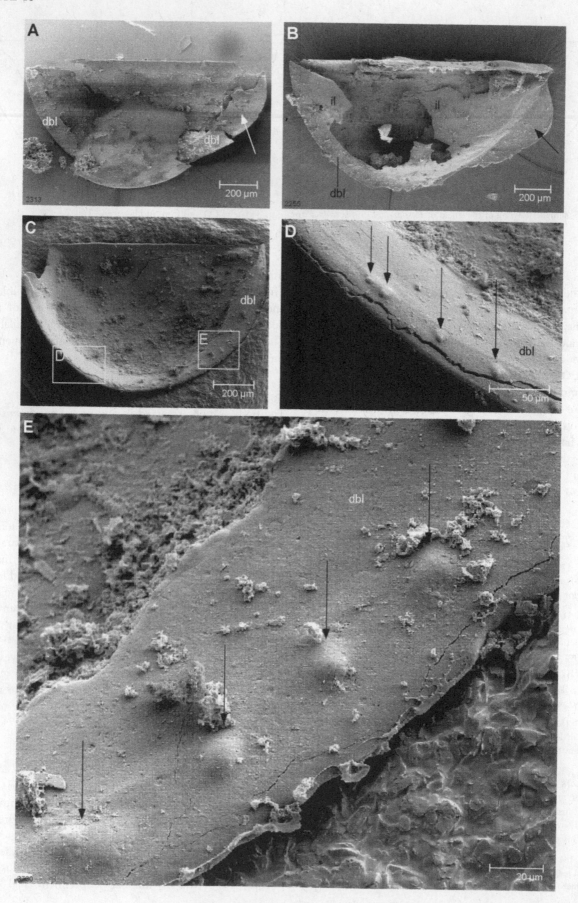

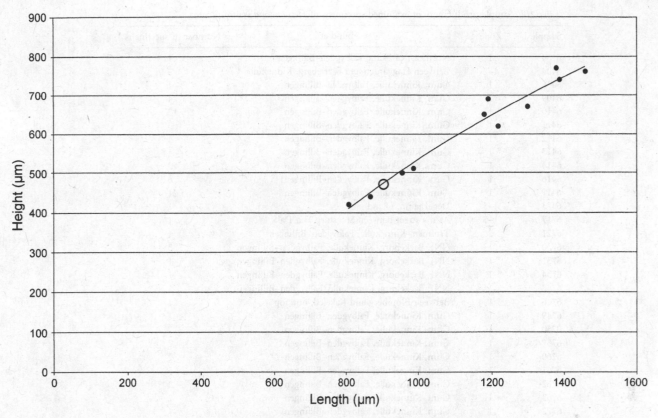

Fig. 34. Ontogeny of *Hesslandona trituberculata* (Lochman & Hu, 1960): length versus height. Eleven specimens with measurable length and height plus the holotype are included. The curved trend line shows the slightly stronger increase in length against the increase in the height of the valves during ontogeny. ○ = holotype. The trend line has the polynomial function $y = -0.0002x^2 + 1.0152x - 284.83$, y = height, x = length.

Hesslandona ventrospinata Gründel *in* Gründel & Buchholz, 1981

*   1981   *Hesslandona? ventrospinata* n. sp. – Gründel, p. 63, pl. 2, fig. 15.
.   *1986a Hesslandona? ventrospinata* Gruendel, 1981 – Kempf, p. 400.

.   *1986b Hesslandona? ventrospinata* Gruendel, 1981 – Kempf, p. 635.
.   *1987   Hesslandona? ventrospinata* Gruendel, 1981 – Kempf, p. 710.
.   1998   *Hesslandona? ventrospinata* Gründel *in* Gründel & Buchholz, 1981 – Hinz-Schallreuter, pp. 115, (126) (referred to *Cyclotron*).

Name. – After the postero-ventral marginal spine (Gründel 1981).

Holotype. – Isolated left valve, illustrated by Gründel (1981) in his pl. 2, fig. 15; length of shield 1300 µm, height of shield 740 µm.

Type locality. – Erratic boulders of Rügen, northeastern Germany.

Type horizon. – *Agnostus pisiformis* Zone (Zone 1), Upper Cambrian.

Material. – One hundred and five specimens of different stages and from different areas of Zone 1 of the Upper Cambrian of southern Sweden (Table 21).

Dimensions. – Smallest specimens: valves about 790 µm long and about 500 µm high. Largest specimen: valves about 1950 µm long and about 1100 µm high.

Plate 18.   *Hesslandona trituberculata* (Lochman & Hu, 1960)

A: The same specimen as illustrated in Pl. 17D, subsequently damaged. A median view of the inside of the right valve, partly covered by the strongly damaged left valve. The inner lamella is not preserved. Note the broad posterior part of the doublure (dbl).

B: UB W 256. The largest specimen available, about 1460 µm in length. A median view of the inside of the right valve. The inner lamella is partly preserved. Note the broad keel-like extension of the posterior shield margin (arrow).

C: UB W 257. Image flipped horizontally. A specimen of an advanced stage, about 1190 µm in length. A median view of the inside of the left valve. The inner lamella is not preserved. Rectangles mark the areas of the doublure magnified in Pl. 18D and E.

D: Image flipped horizontally. Close-up of the area marked in Pl. 18C, displaying the dome-like structures on the antero-ventral part of the doublure (arrows).

E: Image flipped horizontally. Close-up of the area marked in Pl. 18C, displaying the dome-like structures on the postero-ventral part of the doublure (arrows).

Table 21.   Sample productivity of examined specimens of *Hesslandona ventrospinata*.

Sample	Zone	Found at	Number of specimens
6364	1	Northeast of Skara, Falbygden–Billingen	1
6404	1–2	Between Haggården and Marieberg, Kinnekulle	3
6408	1	Gum, Kinnekulle, Falbygden–Billingen	1
6409	1	Gum, Kinnekulle, Falbygden–Billingen	4
6410	1	Gum, Kinnekulle, Falbygden–Billingen	12
6411	1	Gum, Kinnekulle, Falbygden–Billingen	3
6412	1	Gum, Kinnekulle, Falbygden–Billingen	1
6413	1	Gum, Kinnekulle, Falbygden–Billingen	1
6414	1	Gum, Kinnekulle, Falbygden–Billingen	13
6416	1	Gum, Kinnekulle, Falbygden–Billingen	6
6417	1	Gum, Kinnekulle, Falbygden–Billingen	21
6473	1–5	Degerhamn, Öland	1
6717	1–5	At the street between Stenstorp and Dala	1
6721	1	Trolmen, Kinnekulle, Falbygden–Billingen	1
6730	1	NNE Backeborg, Kinnekulle, Falbygden–Billingen	3
6731	1	NNE Backeborg, Kinnekulle, Falbygden–Billingen	1
6734	1	NNE Backeborg, Kinnekulle, Falbygden–Billingen	1
6743	1	NNE Backeborg, Kinnekulle, Falbygden–Billingen	1
6746	1	Between Blomberg and Kakeled, outcrop	1
6749	1	Gum, Kinnekulle, Falbygden–Billingen	1
6750	1	Gum, Kinnekulle, Falbygden–Billingen	7
6757	1	Gum, Kinnekulle, Falbygden–Billingen	2
6760	1	Gum, Kinnekulle, Falbygden–Billingen	2
6761	1	Gum, Kinnekulle, Falbygden–Billingen	5
6762	1	Gum, Kinnekulle, Falbygden–Billingen	3
6763	1	Gum, Kinnekulle, Falbygden–Billingen	2
6772	1	Gum, Kinnekulle, Falbygden–Billingen	1
6783	1	Gum, Kinnekulle, Falbygden–Billingen	2
6784	1	Gum, Kinnekulle, Falbygden–Billingen	3
6788	1	Gum, Kinnekulle, Falbygden–Billingen	1

Additional material. – Apart from the material investigated herein, Gründel (1981) reported two left and right valves plus a doubtful left valve (Table 22).

Distribution. – *Agnostus pisiformis* Zone of the Late Cambrian from northern central Europe.

Original diagnosis. – Gründel (1981): "Valves strongly domed. Both dorsal edges with a small spine directed obliquely dorsal. Postero-ventral margin with obliquely posteriorly directed, basally widened spine."

Emended diagnosis. – Maximum length of valves between dorsal rim and midline. Valves pre-plete. Valves close tightly. Valves with three subdorsal lobes, two in the anterior, a third in the posterior half, plus three lobes in the ventral half of the shield. Interdorsum drawn out into long spines anteriorly, directed antero-dorsally, and posteriorly, directed postero-dorsally. Postero-ventral

margin of both valves with two intercalating triangular outgrowths each. Doublure widest posteriorly.

Description. –

Shield (Fig. 35)

The bivalved shield (ch1:2) has a long and straight dorsal rim (ch2:2) (Pl. 19A, D; Fig. 35A). The maximum length of the head shield is slightly ventral to the dorsal rim of the head shield (ch3:3) (Pl. 19A). The valves are of equal size. The maximum height of the valves is posterior to the antero-posterior midline (post-plete, see Fig. 8) (ch4:3); the valves are about 1.6 times longer than high in every stage. The free anterior part of the shield margin starts straight towards the ventral side from the dorsal rim (ch6:1), thereafter curves gradually towards the ventral maximum (Pl. 19A; antero-dorsal angle 90 degrees). The latero-ventral outline is curved (ch7:2). Thereafter,

Table 22.   Complete list of finds of *Hesslandona ventrospinata* with localities and horizons.

Locality	Horizon	Reference
Erratic boulders, Rügen, Germany	Upper Cambrian, Zone 1	Gründel 1981
Västergötland, southern Sweden	Upper Cambrian, Zone 1	Herein

A   B

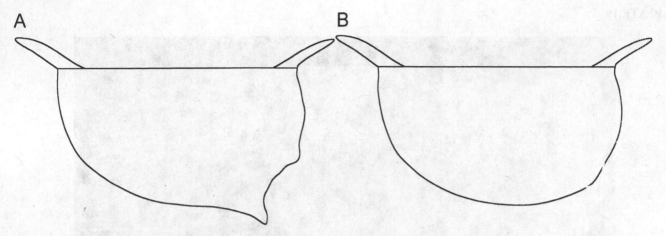

Fig. 35. Schematic drawing of the shield of *Hesslandona ventrospinata* from outside lateral (A) and the right valve from inside lateral (B).

the margin of the left valve is drawn out into a triangular hollow and towards the inside curved outgrowth, a postero-ventral spine with a basal width of one fifth the shield length (ch19:1) (Pl. 19A, D). Posterior to the outgrowth, the margin curves gently upwards and after about another one fifth the shield length, the margin is again drawn out into a second outgrowth – of triangular shape with a broad basal insertion and a blunt tip in the largest specimens at hand. Thereafter, the margin curves towards the most posterior part of the shield, which is located right behind the second outgrowth, running anteriorly in a gradually curved outline (ch5:2) and before meeting the postero-dorsal rim, again slightly changing its direction to the posterior (ch8:2; Pl. 19A; postero-dorsal angle about 80 degrees). The outline of the margin of the right valve is equal to that of the left valve; a triangular, hollow, pointed outgrowth, a postero-ventral spine of smaller size than the spine of the left valve, is located in the gap between both outgrowths of the left valve (Pl. 19A–D). A second outgrowth comparable in size and shape to the second one of the left valve is located on the most posterior part of the shield margin of the right valve (ch20:1) and is slightly curved inwards (Pl. 19A). The maximum width of the shield is slightly dorsal to the midline. The shield valves close tightly (ch9:1). The surface of the valves is smooth, but with three dome-like lobes subdorsally and three further lobes antero-ventrally and centrally on the valves (Pl. 19A). The antero-dorsal lobe (L_1) is small (ch10:1), a slightly larger one (L_2) is located posterior to it just before the antero-posterior midline (ch11:1), a third one (L_3) is located in the last third of the valves (ch12:1). The fourth lobe (L_4) is located anteriorly (ch13:1) and is at least as large as L_1–L_3 together. Even larger is lobe L_5, which, together with the more posteriorly located and L_5-overlapping lobe L_6, represents a doming of the central surface of the valves (ch14:1, ch15:1). The area between L_4 and L_5 is also slightly domed in dorso-ventral elongation (Pl. 19A). Another lobe-like structure is located

postero-ventrally on both valves (Pl. 19D). Other structures are not developed on the valves (ch16:0, ch17:0, ch18:0, ch21:0).

Interdorsum
The interdorsum is continuous from the anterior to the posterior (ch22:3) (Pls. 19E, 20B), being bordered by narrow membranous furrows on both sides. The interdorsum is always the same width, apart from both ends, where it becomes narrower. The maximum width is about 1/14 the length of the valves (ch23:1; Table 36). The interdorsum is convex in cross-section (ch24:2) and – apart from its ends – is without any lobes or ornamentation (ch25:1) (Pl. 19E). Its anterior end is drawn out into a long spine of about one fifth the shield length, basally as thick as the interdorsum (ch26:1, ch27:3; Table 36), and directed antero-dorsally (ch28:2) (Pl. 20A). The posterior end of the interdorsum is drawn out into a spine of about one quarter the shield length, basally as thick as the interdorsum (ch29:1, ch30:3; Table 36), directed postero-dorsally (ch31:2) (Pl. 20A). The spines are – being parts of the interdorsum – separated from the valves by the membranous furrows that border the interdorsum laterally. The anterior and posterior spine is smooth (ch32:1, ch33:1).

Doublure
A doublure is present (ch34:1), being anteriorly slightly wider than ventrally (ch35:2; Table 36) and distinctly wider posteriorly than anteriorly (ch36:3) (Pl. 20B, C; Fig. 35B). Approximating the dorsal rim, the doublures of both sides rapidly narrow to merge into a membranous area antero- and postero-dorsally. The maximum width of the doublure is about 1/11 the length of the valves (ch37:1). On the outer ventral margin of the doublure of more advanced growth stages, nine small dome-like outgrowths are arranged in a row (Pl. 20D); they are located directly on the surface (ch38:1, ch39:2, ch40:1). The distance between each of them becomes gradually smaller towards the posterior (Pl. 20D). Pores

PLATE 19

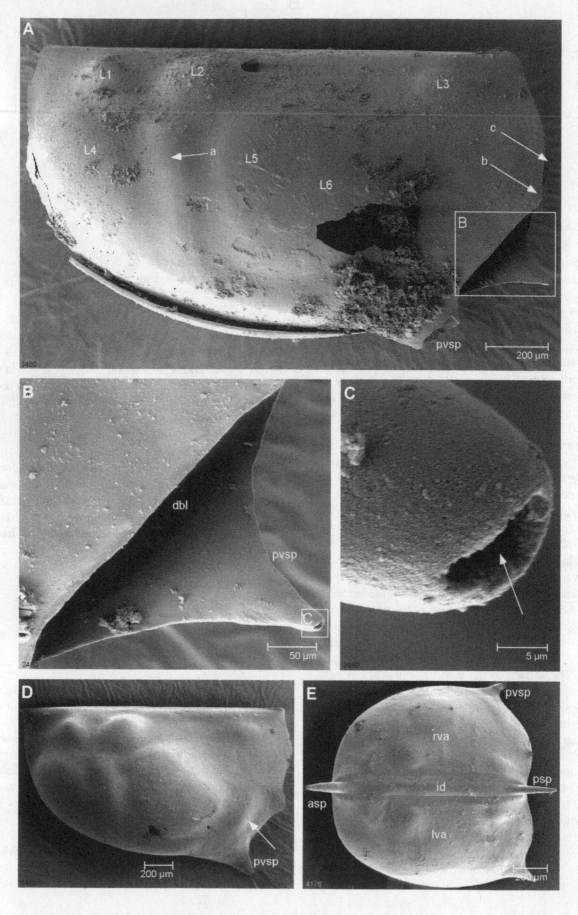

on the doublure and other ornamentation could not be observed (ch41:1, ch42:1).

Inner lamella

The inner lamella expands along the whole doublure and extends medially to the dorso-lateral side of the body, which is not preserved in any specimen to hand. The inner lamella is weakly sclerotised and frequently shows a wrinkled texture (Pl. 20B, C). No scar of a possible closing muscle is preserved.

Body

The body proper, which is completely enveloped by the bivalved shield, includes at least nine segments (Pl. 21A). Trunk segmentation of the preserved portion is retained in the insertions of the limbs, and distinct segment boundaries or sternites are weakly defined. At least eight segments are dorsally fused to the shield (ch44:5) (Pl. 21A). Hence, the shield represents a cephalothoracic shield including at least three limb-bearing trunk segments. The area of fusion of the body proper to the shield is very narrow, approximately corresponding to the width of the interdorsum (ch43:1) (Pls. 20B, 21A). Starting some micrometres posterior to the dorsal anterior membranous area of the doublures, the body proper extends along the inner dorsal length of the shield (Pl. 20B), but the posterior end of the body is unknown. Its posterior extension and segmental continuation remain uncertain, as well as whether there is a hind body extending free from the shield into the domicilium. The body proper, in ventral view, is oval with a blunt anterior and an elongated posterior end (Pl. 20B). The maximum width of the body proper is located at the

---

Plate 19. *Hesslandona ventrospinata* Gründel *in* Gründel & Buchholz, 1981

A: UB W 185. A specimen of a late growth stage, about 1660 μm in shield length. A lateral view of the left valve; the postero-ventral spine of the right valve is partly exposed and magnified in Pl. 19B, as indicated by the rectangle. Arrow "a" points to the elongated dome between lobes L$_4$ and L$_5$ (L4, L5). The posterior margin of both valves is drawn out into blunt triangular outgrowths (arrow "b" for the left valve, arrow "c" for the right valve).

B: Close-up as indicated in Pl. 19A, displaying the postero-ventral spine (pvsp) of the right valve from the inside. The tip of the spine is magnified in Pl. 19C, as indicated by the rectangle.

C: Close-up as indicated in Pl. 19B, displaying the tip of the postero-ventral spine of the right valve. The spine is internally hollow (arrow).

D: UB W 186. A specimen of a late growth stage, about 1950 μm in shield length. A lateral view of the left valve, the right valve is not preserved. For labelling the lobes, see Pl. 19A. Note the elongated node on the postero-ventral part of the shield (arrow) immediately above the postero-ventral spine (pvsp).

E: UB W 187. A specimen of an advanced growth stage, about 1180 μm in shield length. A dorsal view with opened valves, the so-called "butterfly" position, displaying the interdorsum (id) and its anterior (asp) and posterior (psp) extensions into spines (cf. Pl. 20A).

mandibles (Pl. 20B). The cross-section of the body proper anterior to the mandibular part is almost trapezoidal with the long axis dorso-ventrally oriented. The cross-section at the mandibular segment is more or less a half-oval with the long axis in dorso-ventral aspect; the height of the body proper in this position measures less than one third the height of the valves. Posterior to it, the body proper changes its cross-section to a half-circular shape (Pl. 21A). The height of the body decreases gradually towards the posterior to about one sixth the height of the valves at the posterior end of the body. The limbs are arranged in a more or less regular row at the flanks of the body, being more or less equally set together from the anterior towards the caudal end (Pl. 21B).

The anterior part of the body proper is the hypostome/labrum complex. The hypostome forms the anterior sclerotised ventral surface, has a somewhat rhomboid outline in ventral view and becomes gradually higher to the posterior in lateral view. Antero-laterally on the hypostome, the antennulae insert on a shoulder-like slope in a circular joint area. The antennae (Pl. 21A) insert postero-laterally to the hypostome in a more spindle-shaped joint area, the posterior edge touching the labral part of the hypostome/labrum complex. The median eye is located antero-distally on the hypostome (Pl. 21A). The posterior end of the hypostome is drawn out distally into a lobe-like protrusion, the labrum (ch61:1) (Pl. 15E) with the mouth probably posteriorly at its base. The anterior side of the labrum has a pair of slight depressions (Pl. 21A). Posterior to the hypostome/labrum complex, the ventral surface between the first to third pairs of post-antennular limbs is marked by a sclerotic plate, the sternum (ch62:1) (Pl. 21A). The sternum is slightly domed and bears one pair of humps anteriorly, the paragnaths (ch63:1), and between the second pair of post-antennular limbs, the mandibles (Pl. 21A). A second pair of elongated humps is located posterior to the paragnaths between the second pair of post-mandibular limbs (Pl. 21A). A keel-like elongated lobe is located in between the two humps (Pl. 21A). Posterior to the sternum, the ventral body surface becomes progressively softer. The post-mandibular limb pairs insert on a lateral slope corresponding to the cross-section of the body. The hind body, including the number of segments involved and possible furcal rami, is unknown (ch47:?, ch48:?, ch49:?).

Soft parts

The soft parts are somewhat different to those described in detail for *H. unisulcata*. Because it is beyond the scope of this paper to describe the soft parts of *H. ventrospinata* in full detail, only some general remarks are made, important for discriminating it from *H. unisulcata* and for the phylogenetic analysis of Phosphatocopina. The

PLATE 20

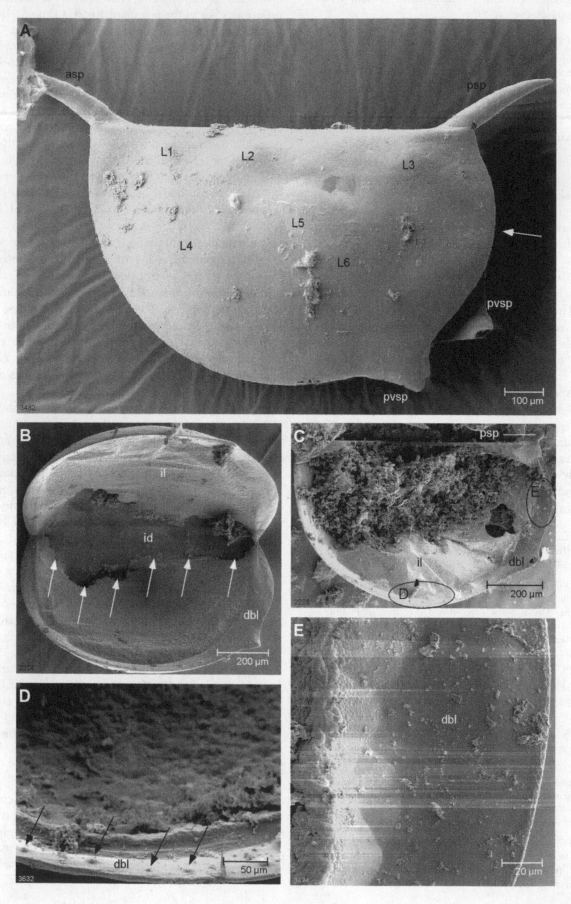

detailed description is postponed for a second paper. The antennula is similar to that of *H. unisulcata*, being small and consisting of less than 10 irregular annuli (ch45:1, ch46:1). The tip bears a tuft of four setae, three terminal ones and one additional subterminal seta. A seta on the third last annulus as described for *H. unisulcata* could not be observed. Again, as in *H. unisulcata*, the antenna (Pl. 15E) consists of an undivided limb stem, at least in later stages, but being a fusion product of the coxa and basipod (see p. 23; ch50:2, ch51:?, ch52:2), a two-part endopod (ch53:5) and a multi-annulated exopod. The same is true for the mandible (Pl. 15D, E). It has an undivided limb stem, at least in later stages, and an enditic protrusion located medially between the limb stem and the endopod (Pl. 21B), recognised as the remains of the basipod (see p. 27; ch54:2, ch55:?, ch56:2). The endopod has two parts (ch57:5) and the exopod consists of several annuli, each annulus with one medially projecting seta; the last annulus bears two setae – as in the antenna. The post-mandibular limbs are separated into two groups of different shape (Pl. 21A). The first three post-mandibular pairs of limbs are serially similar to each other (ch58:1), consisting of a basipod, a setae-bearing proximal endite medio-proximally (ch59:1), a three-part endopod (ch60:4) and an annulated exopod (Pl. 21B, C). The endopodal portions are short and slightly projecting medially. The first two are oval-shaped in posterior view with a blunt, setose end medially. The third portion is small and hump-like with one distal seta

---

Plate 20. *Hesslandona ventrospinata* Gründel *in* Gründel & Buchholz, 1981

A: UB W 188. A specimen of an advanced growth stage, about 1070 μm in shield length. A lateral view of the left valve, displaying the interdorsal spines (asp, psp). Note that lobes L4–L6 (L4–L6) are inconspicuous compared with larger individuals, probably representing later growth stages (cf. Pl. 19A, D). The margin of both valves is drawn out into only one postero-ventral spine (pvsp); a second outgrowth, as in larger specimens, is not present (arrow, cf. Pl. 19A).

B: UB W 189. A specimen of an advanced growth stage, about 1000 μm in shield length. A ventral view of an opened shield in the so-called "butterfly" position. The body is torn off, its lateral extension is retained (arrows) in the margin of the inner lamella (il). The interdorsum (id) is exposed from the inside.

C: UB W 190. A specimen of an advanced growth stage, about 920 μm in shield length. A lateral view inside the right valve. The body is preserved very coarsely and does not display any details. The ellipses mark areas that are displayed in a higher magnification in Pl. 20D and E but from different specimens of about 940 μm in shield length each.

D: UB W 191. Close-up of the ventral part of the doublure as indicated in Pl. 20C, displaying small dome-like outgrowths (arrows). Note the distance between them becomes shorter towards the posterior (= right side in the image).

E: UB W 192. Close-up of part of the posterior doublure, as indicated in Pl. 20C, displaying a smooth doublure without any outgrowths.

(Pl. 21C). The exopods are triangular plates with marginal setation (Pl. 21C); their tip is annulated – very much like the exopods in *H. unisulcata* (cf. Pls. 11F, 12G). The fourth to sixth post-mandibular pairs of limbs (Pl. 21A, B) consist of a single limb stem and two rami (Pl. 21D). The limb stem bears two groups of two spine-like setae medio-proximally and medio-distally. The endopod is located medio-distally on the limb stem. It is distinctly separated from the limb stem by a joint membrane and consists of a single short tube-like portion with a blunt tip that bears a tuft of three setae. The exopod forms more or less the distal extension of the limb stem, without being distinctly separated from the limb stem. It is a flat plate of triangular shape with a pointed terminal tip and with marginal setation (Pl. 21D).

Comparisons

Hesslandona ventrospinata is distinguished from all other phosphatocopine species investigated by the postero-ventral and posterior outgrowths of the shield margin of both valves. However, such kinds of outgrowth are also known from different species of *Bidimorpha*, but their outgrowths are ventral to postero-ventral and not postero-ventral to posterior as in *H. ventrospinata*. *Vestrogothia spinata* has a spine-like outgrowth only on the left valve, but it lacks the interdorsum. *Hesslandona ventrospinata* is similar to *Waldoria rotundata* and the species of *Cyclotron* Gründel *in* Gründel & Buchholz (1981) in having the same distribution of six shield lobes on their valves. *Hesslandona trituberculata* and *H. curvispina* n. sp. only have three subdorsal lobes, i.e. lobes L1–L3 of *H. ventrospinata*. *Hesslandona unisulcata* and *Trapezilites minimus* have only one lobe, L1, as *Falites fala*; and the shields of *H. necopina*, *H. suecica* n. sp., *H. angustata* n. sp. and *H. toreborgensis* n. sp. do not have lobes at all.

Ontogeny

During ontogeny, the proportion of the length to the height of the valves of *H. ventrospinata* remains the same (Fig. 36). The observed range of shield lengths and heights does not indicate distinct clusters of particular growth stages (Fig. 36).

The specimens of small size representing early instars are missing lobes L3, L4, L5, and L6 and the posterior outgrowth of the margin of the right valve (Pl. 21E). L4 and L5 are inconspicuously present in a succeeding ontogenetic stage (Pl. 20A). The interdorsal outgrowths enlarge during ontogeny (cf. Pls. 20A, 21E). The number of limbs present in the first larval stage is unknown (ch64:?).

PLATE 21

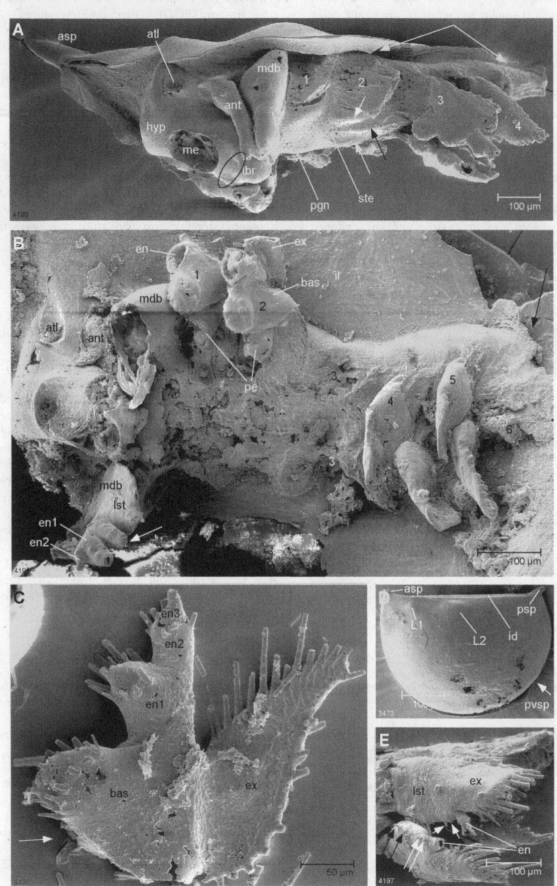

Hesslandona suecica n. sp.

- . 1964a *Hesslandona necopina* – Müller, pl. 1, fig. 5.
- . 1982d *Hesslandona asulcata* – Müller, fig. 3 (nom. nud.).
- . 1996 *Hesslandona* sp. nov. – Hou *et al.*, fig. 9a, d (UB 1629, 1627).

Name. – The Latin word *suecica* means "Swedish"; this species is only known from localities in Sweden.

Holotype. – Opened head shield, 620 μm long and 390 μm high, illustrated in Pl. 12A–C (UB W 258).

Remarks. – Müller (1982d) illustrated the new species and used the name *H. asulcata* in the description of the image. He did not give the information that this species was new, and he did not give a description, a diagnosis or designate a type specimen. Consequently, according to Article 13.1. of the *International Code of* Zoological Nomenclature (fourth edition 1999), the name "*Hesslandona asulcata*" is not available, and we therefore introduce the new name *H. suecica* n. sp. for this species.

Type locality. – Near Gum, Kinnekulle, Falbygden–Billingen, Västergötland, Sweden (Fig. 2, number 9).

Type horizon. – Upper Cambrian, *Agnostus pisiformis* Zone (Zone 1).

Material examined. – Two hundred and eighty specimens of different stages and from different areas of Zone 1 of the Upper Cambrian of southern Sweden (Table 23).

Dimensions. – Smallest specimen: valves about 240 μm long and about 170 μm high. Largest specimen: valves about 790 μm long and about 510 μm high.

Additional material. – No additional material of *H. suecica* n. sp. has been reported so far.

Distribution. – *Agnostus pisiformis* Zone (Zone 1) of the Upper Cambrian, southern Sweden.

Diagnosis. – Maximum length of bivalved shield between dorsal rim and midline. Valves antero-posteriorly symmetrical. Valves amplete, without lobes or spines. Interdorsum wide, with short spines anteriorly and posteriorly. Doublure widest posteriorly, without any structures.

Description. –

Shield (Fig. 37)

The bivalve shield (ch1:2) has a long and straight dorsal rim (ch2:2) (Pl. 22A, Fig. 37A). The maximum length of the shield is between the dorsal rim and the midline (ch3:3). The right and left valves are of equal size and symmetrical shape (Pl. 22A). The maximum height of the valves is at the antero-posterior midline (amplete, see Fig. 8) (ch4:2). The proportion of the length to the height of the valves varies from about 1.4 in younger stages up to 1.5 times longer than high in older stages. The free anterior part of the shield margin starts from the dorsal rim, curves antero-ventrally (Pl. 22A; antero-dorsal angle nearly 70 degrees) and equally postero-ventrally towards the ventral maximum of the valve (ch6:2). The ventral part of the margin is gently curved (ch7:2). More posteriorly, the margin swings gently upwards in the same way as anteriorly (ch5:3, angle 70 degrees, Pl. 22A) to recurve slightly anteriorly to meet the posterior end of the dorsal rim (ch8:2). In dorsal view, the closed shield has its maximum width slightly dorsal to the midline. The margins are without outgrowths throughout (ch19:0, ch20:0). The shield valves close tightly without any gaps in all stages (ch9:1). The surface of the valves is smooth, without any lobes or spines (chs10–21:0) (Pl. 22A).

Plate 21. *Hesslandona ventrospinata* Gründel *in* Gründel & Buchholz, 1981

A: UB W 193. Image flipped horizontally. A specimen of unclear shield size but distinctly longer than 1000 μm; the shield is mostly not preserved. A latero-ventral view, displaying the sternum with a pair of humps (short white arrows) between the second post-mandibular limbs of the left side (2). The black arrow points to the keel-like elongated node between the two humps. The median eye (me) is represented by a circular depression. Note the paired depressions (ellipse) on the anterior side of the labrum (lbr), the different antero-posterior width of the antenna (ant) and the mandible (mdb), and the difference in shape of the third (3) and fourth (4) pairs of post-mandibular limbs. The long white arrows point to the recurvation of the lateral body wall towards the dorso-median, implying a narrow area of fusion of the body proper with the shield.

B: UB W 194. Image flipped horizontally. A specimen of an advanced stage, about 1070 μm in shield length. Ventral view. The body proper is dorsally fused to the shield up to at least the eighth segment (fifth post-mandibular limb; 5), thereafter the body proper is damaged (black arrow). The third post-mandibular pair of limbs is only represented by its insertion areas (3). The mandible (mdb) has an enditic protrusion (white arrow) between the limb stem (lst) and the first endopodal portion (en1).

C: UB W 195. An isolated post-mandibular limb 1, 2 or 3 of the same specimen, whose hind body is illustrated in Pl. 21E. The proximal endite is not preserved (arrow). Note the similarity of the exopod with the exopods of post-mandibular limbs in *Hesslandona unisulcata* (cf. Pl. 11F) and the similarity of this limb with one of the anterior post-mandibular limbs of *Waldoria rotundata* (see Pl. 36C).

D: UB W 196. Image flipped horizontally. The smallest specimen available, representing an early growth stage, about 790 μm in shield length. A lateral view of the right valve, displaying the initial postero-ventral spine (pvsp) on the right valve, while the margin of the left valve is smooth (arrow).

E: UB W 195. Part of the hind body including one pair of post-mandibular limbs 4, 5 or 6 (another limb is illustrated in Pl. 21C), displaying the tiny endopod (en) and two pairs of setae medio-proximally (black arrows) and medio-distally (white arrows) of the limb stem (lst).

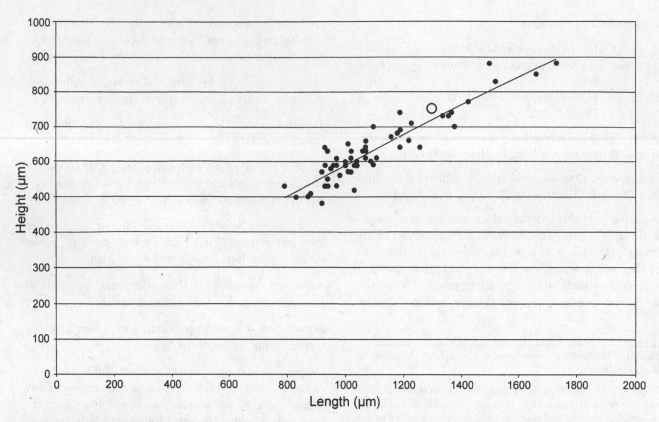

Fig. 36. Ontogeny of *Hesslandona ventrospinata*: length versus height. Sixty specimens with measurable length and height plus the holotype are included. The curved trend line shows the slightly stronger increase in length against the increase in the height of the valves during ontogeny. ○ = holotype. The trend line has the polynomial function $y = 0x^2 + 0.496x + 121.83$, y = height, x = length.

Interdorsum

The interdorsum is continuous (Pl. 22A) from the anterior to the posterior (ch22:3), being bordered by narrow membranous furrows on both sides (Pl. 22B). The interdorsum is always the same width, apart from both ends, where it narrows (Pl. 22A–C). The maximum width is about 1/30 the length of the valves (ch23:3). The interdorsum is flat (ch24:1), without lobes or ornamentation in its median part (ch25:1) (Pl. 22A). The anterior end is drawn out into a short spine-like protrusion (Pls. 22B, C, 23A–C) of about 1/15 the length of the valves (ch26:1, ch27:2), a blunt tip, basally as thick as the interdorsum (ch28:0). The insertion area is somewhat egg-shaped. The anterior edge of the spine is almost straight and is almost an elongation of the anterior margin of the shield. The posterior margin of the spine slopes gradually into the dorsal rim. The posterior end of the interdorsum is drawn out into a spine (Pls. 22B, C, 23A–C), distinctly longer than the anterior one, about one ninth the shield length (ch29:1, ch30:2) with a pointed tip, basally as thick as the interdorsum. The spine is directed postero-dorsally (ch31:2). The anterior margin of the spine slopes less steeply into the dorsal rim than the posterior margin of the anterior spine. Being part of the interdorsum, the spines are separated from the valves by the membranous furrows

that border the interdorsum laterally. The anterior and posterior spines are smooth (ch32:1, ch33:1) (Pls. 22B, C, 23A–C).

Doublure

A doublure is present (ch34:1) along the inner margin of the valves, being narrowest ventrally (ch35:2), anteriorly slightly wider and posteriorly about one third wider than ventrally (ch36:3) (Pl. 23D, E; Fig. 37B). Approximating the dorsal rim, the doublures of both sides rapidly narrow slightly to merge into a membranous area antero- and postero-dorsally (Pl. 23E). The maximum width of the doublure is about 1/10 the length of the valves (ch37:1). Ornaments or pores on the doublure are probably absent (ch38:1, ch39:1, ch40:0, ch41:1, ch42:1).

Inner lamella

The inner lamella (Pl. 23E) is similar to that described for *H. unisulcata* and *H. necopina.*

Body

The body proper is equal to that of *H. unisulcata* and *H. necopina.* It is completely enveloped by the bivalved shield and comprises at least nine segments. The segmentation of the body is only retained by the insertions of the limbs. At least the first eight segments are dorsally

Table 23. Sample productivity of examined specimens of *Hesslandona suecica* n. sp.

Sample	Zone	Found at	Number of specimens
6364	1	Between Skara and Stolan, Falbygden–Billingen	3
6365	1	Between Skara and Stolan, Falbygden–Billingen	4
6404	1–2	Between Haggården and Marieberg, Kinnekulle, Falbygden–Billingen	13
6408	1	Gum, Kinnekulle, Falbygden–Billingen	7
6409	1	Gum, Kinnekulle, Falbygden–Billingen	32
6410	1	Gum, Kinnekulle, Falbygden–Billingen	15 (holotype included)
6411	1	Gum, Kinnekulle, Falbygden–Billingen	9
6412	1	Gum, Kinnekulle, Falbygden–Billingen	1
6414	1	Gum, Kinnekulle, Falbygden–Billingen	42
6415	1	Gum, Kinnekulle, Falbygden–Billingen	8
6416	1	Gum, Kinnekulle, Falbygden–Billingen	4
6417	1	Gum, Kinnekulle, Falbygden–Billingen	43
6418	1	Northeast Trolmen, Kinnekulle, Falbygden–Billingen	1
6419	1	Northeast Trolmen, Kinnekulle, Falbygden–Billingen	1
6722	1	Trolmen, Kinnekulle, Falbygden–Billingen	2
6728	1	NNE Backeborg, Kinnekulle, Falbygden–Billingen	1
6729	1	NNE Backeborg, Kinnekulle, Falbygden–Billingen	2
6730	1	NNE Backeborg, Kinnekulle, Falbygden–Billingen	8
6731	1	NNE Backeborg, Kinnekulle, Falbygden–Billingen	3
6733	1	NNE Backeborg, Kinnekulle, Falbygden–Billingen	1
6734	1	NNE Backeborg, Kinnekulle, Falbygden–Billingen	3
6747	–	Northeast Stora Stolan, Billingen	1
6748	1	Gum, Kinnekulle, Falbygden–Billingen	4
6749	1	Gum, Kinnekulle, Falbygden–Billingen	2
6750	1	Gum, Kinnekulle, Falbygden–Billingen	10
6752	1	Gum, Kinnekulle, Falbygden–Billingen	1
6755	1	Gum, Kinnekulle, Falbygden–Billingen	1
6754	1	Gum, Kinnekulle, Falbygden–Billingen	1
6755	1	Gum, Kinnekulle, Falbygden–Billingen	1
6757	1	Gum, Kinnekulle, Falbygden–Billingen	1
6759	1	Gum, Kinnekulle, Falbygden–Billingen	2
6760	1	Gum, Kinnekulle, Falbygden–Billingen	5
6761	1	Gum, Kinnekulle, Falbygden–Billingen	2
6763	1	Gum, Kinnekulle, Falbygden–Billingen	2
6764	1	Gum, Kinnekulle, Falbygden–Billingen	7
6765	1	Gum, Kinnekulle, Falbygden–Billingen	2
6771	1	Gum, Kinnekulle, Falbygden–Billingen	2
6772	1	Gum, Kinnekulle, Falbygden–Billingen	3
6773	1	Gum, Kinnekulle, Falbygden–Billingen	3
6774	1	Gum, Kinnekulle, Falbygden–Billingen	5
6776	1	Gum, Kinnekulle, Falbygden–Billingen	7
6780	1	Gum, Kinnekulle, Falbygden–Billingen	6
6782	1	Gum, Kinnekulle, Falbygden–Billingen	6
6784	1	Gum, Kinnekulle, Falbygden–Billingen	3

fused to the shield (ch44:5) (Pl. 24A), being a cephalo-thoracic shield including at least three limb-bearing trunk segments. The area of fusion of the body proper to the shield is very narrow, corresponding to approximately the width of the interdorsum. Starting a few micrometres posterior to the dorsal anterior membranous area of the doublures (ch43:1), the body proper extends along the inner dorsal length of the shield, but the posterior end of the body is unknown, so its posterior extension and segmental continuation remain uncertain, as well as if there is a hind body extending free from the shield into the domicilium. The body proper in ventral view is oval with a blunt anterior and an elongated posterior end. The maximum width of the body

proper is at the mandibles. The cross-section of the body proper anterior to the mandibular part is almost trapezoidal with the long axis dorso-ventrally oriented. The cross-section at the mandibular segment is more or less half-oval with the long axis in dorso-ventral aspect; the height of the body proper in this position is more than one third the height of the valves. Posterior to it the body proper changes its cross-section to half-circular. The height of the body decreases gradually towards the posterior to about one sixth the height of the valves at its posterior end. The limbs are arranged in a more or less regular row at the flanks of the body, becoming more closely set together on the ventral side towards the caudal end (Pl. 24A).

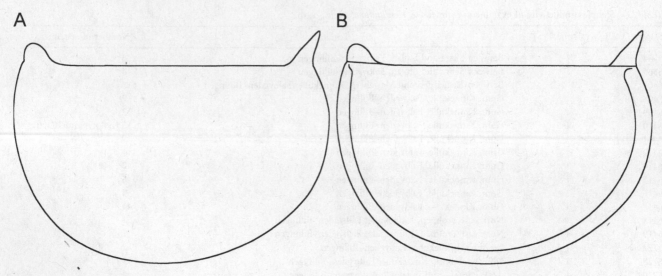

A   B

Fig. 37.   Schematic drawing of the shield of *Hesslandona suecica* n. sp. from outside lateral (A) and from inside lateral (B).

Soft parts

The soft parts are similar to those described in detail for *H. unisulcata* and more or less equal to those of *H. necopina*. Because it could not be the scope of this paper to describe the soft parts of *H. suecica* n. sp. in full detail, only some general remarks are made, important especially for the phylogenetic analysis of Phosphatocopina. The anterior part of the body proper is the hypostome/labrum complex (ch61:1) (cf. Pls. 24A, 25A, B). A bulging structure, interpreted as the median eye, is located antero-ventrally on the hypostome/labrum complex (Pl. 24A). The sternites of the three post-antennular segments are fused to form a single sternum (ch62:1) (Pls. 24A, 25B). The mandibular part has a pair of humps, the paragnaths (ch63:1). Another, slightly less distinct pair of humps is located posterior to the paragnaths and belongs to the segments of the first pair of post-mandibular limbs (Pl. 24A). The antennula is similar to that of *H. unisulcata*, being small, less than one eighth the length of the shield (ch46:1), and consisting of less than 10 irregular annuli (ch45:1). The tip bears a tuft of four setae, three terminal ones and one additional subterminal seta. A seta on the third last annulus as described for *H. unisulcata* could not be observed. Again, as in *H. unisulcata*, the antenna consists of an undivided limb stem, composed of a coxa and a basipod (see p. 23) throughout ontogeny (ch50:2, ch51:2, ch52:2), a two-part endopod (ch53:5) and a multi-annulated exopod (Fig. 65A, B). In general design, there are no ways to distinguish this limb from that of *H. unisulcata*. The same is true for the mandible. It has an undivided limb stem (Pl. 24A) throughout ontogeny (ch54:2, ch55:2, ch56:2); an enditic protrusion is located medially between the limb stem and the endopod (Pl. 24A), recognised as the remains of the basipod (see p. 27). The endopod is two-divided (ch57:5) and the

exopod consists of several annuli, each annulus with one medially projecting seta; the last annulus bears two setae – as in the antenna. The post-mandibular limbs are similar to each other (ch58:1), consisting of a basipod, a setae-bearing proximal endite medio-proximally (ch59:1), a three-divided endopod (ch60:4) and an annulated exopod (Pl. 24A–C). The endopods of all post-mandibular limbs do not consist of strongly medially projecting portions as in *H. unisulcata*. The endopodal portions are only slightly projecting medially. They are drop-shaped with the pointed tip medially, and not curved elongated tubes as in *H. unisulcata*. The exopods of the post-mandibular pairs of limbs are multi-annulated as in the antenna and the mandible, but with fewer annuli. They are not flattened, their proximal annuli are not fused and they have no lateral setation as in *H. unisulcata* (Pl. 24A). The hind body is not known (ch47:?, ch48:?, ch49:?) (cf. Pl. 25B).

Comparisons

Hesslandona suecica is very similar to *H. angustata* n. sp. in the outline of the valves, but it differs from it in the presence of anterior or posterior spines of the interdorsum. The interdorsum is distinctly proportionally wider than in any other species of *Hesslandona* (cf. Table 39). The valves of *H. suecica* are lobe-less as in *H. angustata* n. sp., *H. necopina*, *H. kinnekullensis*, and *H. toreborgensis* n. sp., but *H. suecica* differs from these species by the length and shape of the interdorsal spines and by the outline of the valves. All other phosphatocopine species investigated differ from *H. suecica* n. sp. by, e.g. the presence of one or more lobes on the valves.

Ontogeny

During the ontogeny of *H. suecica* n. sp., the length of the valves grows proportionally slightly greater than the height (Fig. 38), such that the shield becomes progressively more and more slender in its long axis. The ontogeny shows four to five more or less distinct clusters that may represent four to five possible growth stages (Fig. 38). The smallest individual known from this species is a larva with four fully developed pairs of appendages, representing a head-larva (ch64:1).

Hesslandona angustata n. sp.

Derivation of name. – *Angusta*, Latin for "narrow", refers to the relatively narrow interdorsum compared with the interdorsum of, e.g. *H. suecica* n. sp. or *H. necopina*.

Holotype. – Opened head shield, 490 μm long and 330 μm high, illustrated in Pl. 26A–D (UB W 203).

Type locality. – Near Gum, Kinnekulle, Falbygden–Billingen, Västergötland, Sweden (Fig. 2, number 9).

Type horizon. – Upper Cambrian, *Agnostus pisiformis* Zone (Zone 1).

Material examined. – Thirty-one specimens of different stages from Gum, Kinnekulle, Falbygden–Billingen, Västergötland, southern Sweden, *Agnostus pisiformis* Zone (Zone 1) of the Upper Cambrian of southern Sweden (Table 24).

Dimensions. – Smallest specimen: valves about 330 μm long and about 230 μm high. Largest specimen: valves about 700 μm long and about 450 μm high.

Additional material. – No additional material of *H. angustata* n. sp. has been reported so far.

Distribution. – As for the type specimen.

Diagnosis. – Maximum length of bivalved shield between dorsal rim and midline. Valves antero-posteriorly symmetrical. Valves close tightly without leaving gaps. Valves amplete, without lobes or spines. Interdorsum narrow, without thickenings or spines. Doublure widest posteriorly, without any structures.

Description. –

Shield (Fig. 39)

The bivalve shield (ch1:2) has a long and straight dorsal rim (ch2:2) (Pl. 26A, Fig. 39A). The maximum length of the shield is between the dorsal rim and the midline (ch3:3). The right and left valves are of equal size and symmetrical shape (Pl. 26A). The maximum height of the valves is on the antero-posterior midline (amplete, see Fig. 8) (ch4:2). The proportion of the length to the height of the valves is about 1.45 (Fig. 40). The free anterior part of the shield margin starts from the dorsal rim, curves antero-ventrally (Pl. 26A; antero-dorsal angle nearly 70 degrees) and equally postero-ventrally towards the ventral maximum of the valve (ch6:2). The ventral

part of the margin is gently curved (ch7:2). More posteriorly, the margin swings gently upwards in the same way as anteriorly (ch5:3, angle 70 degrees, Pl. 26A) to recurve slightly anteriorly, meeting the posterior end of the dorsal rim (ch8:2). In dorsal view, the closed shield has its maximum width slightly dorsal to the midline.

The margins are without outgrowths throughout (ch19:0, ch20:0). The shield valves close tightly without any gaps in all stages (ch9:1). The surface of the valves is smooth, without any lobes or spines (chs10–21:0) (Pl. 26A).

Interdorsum

The interdorsum is continuous (Pl. 26A) from the anterior to the posterior (ch22:3), being bordered by narrow membranous furrows on both sides (Pl. 26C). The interdorsum is the same width, apart from both ends, where it narrows (Pl. 26B, D). The maximum width is about 1/30 the length of the valves (ch23:2). The interdorsum is flat (ch24:1), without lobes or ornamentation in the median part (ch25:1) (Pl. 26C). The anterior and posterior ends are without any thickenings or spines (Pl. 26B, D) (chs26–33:0).

Doublure

A doublure is present (ch34:1) along the inner margin of the valves, being narrowest ventrally (ch35:2), anteriorly slightly wider and posteriorly about one third wider than ventrally (ch36:3) (Pl. 27A; Fig. 39B). Approximating the dorsal rim, the doublures of both sides rapidly narrow slightly to fade out into a membranous area antero- and postero-dorsally (Pl. 27A). The maximum width of the doublure is about 1/10 the length of the valves (ch37:1). Ornaments or pores on the doublure are probably absent (ch38:1, ch39:1, ch40:0, ch41:1, ch42:1).

Inner lamella

The inner lamella (Pl. 27A) is similar to that described for *H. unisulcata* and *H. necopina*.

Body

The body proper is equal to that of *H. unisulcata* and *H. necopina*. It is completely enveloped by the bivalved shield and comprises at least eight segments. The segmentation of the body is only retained by the insertions of the limbs. At least the first six segments are dorsally fused to the shield (ch44:3), being a cephalothoracic shield including at least one limb-bearing trunk segment. The area of fusion of the body proper to the shield is very narrow, corresponding to approximately the width of the interdorsum. Starting a few micrometres posterior to the dorsal anterior membranous area of the doublures (ch43:1), the body proper extends along the inner dorsal length of the shield, but the posterior end of the body is unknown, so its posterior extension and segmental continuation remain uncertain, as well as if

PLATE 22

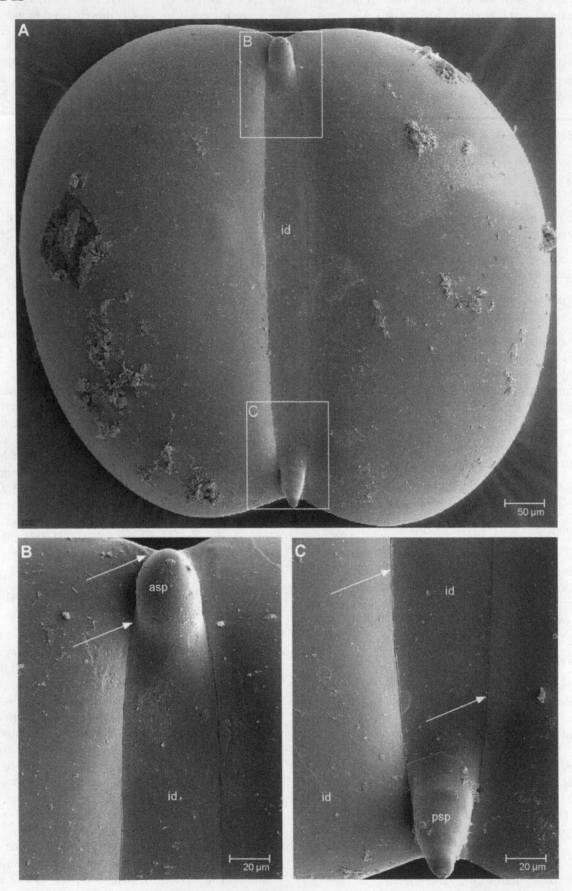

there is a hind body extending free from the shield into the domocilium. The body proper in ventral view is oval with a blunt anterior and an elongated posterior end. The maximum width of the body proper is at the second post-antennular appendages, the so-called mandibles. The cross-section of the body proper anterior to the mandibular part is almost trapezoidal with the long axis dorso-ventrally oriented. The cross-section at the mandibular segment is more or less half-oval with the long axis in dorso-ventral aspect; the height of the body proper in this position is more than one third the height of the valves. Posterior to it the body proper changes its cross-section to half-circular. The height of the body decreases gradually towards the posterior to about one sixth the height of the valves at its posterior end. The limbs are arranged in a more or less regular row at the flanks of the body, becoming more closely set together on the ventral side towards the caudal end.

Soft parts

The soft parts are similar to those described in detail for *H. unisulcata* and more or less equal to those of *H. necopina* and *H. suecica* n. sp. Because it could not be the scope of this paper to describe the soft parts of *H. angustata* n. sp. in full detail, only some general remarks are made, important especially for the phylogeny analysis of Phosphatocopina. The anterior part of the body proper is the hypostome/labrum complex (ch61:1) (cf. Pl. 27A). The sternites of the three post-antennular segments are fused to form a single sternum (ch62:1) (Pl. 27A). The mandibular part has a pair of humps, the paragnaths (ch63:1) (Pl. 27A). The antennula is similar to that of *H. unisulcata*, being small, less than one eighth the length of the shield (ch46:1), and consists of less than 10 irregular annuli (ch45:1). The tip bears a tuft of four setae, three terminal ones and one additional subterminal seta. A seta on the third last annulus as described for *H. unisulcata* could not be observed. Again, as in *H. unisulcata*, the antenna consists of an undivided limb stem (Pl. 27B), composed of a coxa and

a basipod (see p. 23) throughout ontogeny (ch50:2, ch51:2, ch52:2), a two-divided endopod (ch53:5) and a multi-annulated exopod (Pl. 27B), in which seta-less annuli occur irregularly (Pl. 27B). In general design, there are no ways to distinguish this limb from that of *H. unisulcata*. The same is true for the mandible. It has an undivided limb stem throughout ontogeny (ch54:2, ch55:2, ch56:2); an enditic protrusion is located medially between the limb stem and the endopod (Pl. 27C), recognised as the remains of the basipod (see p. 27). The limb stem and enditic protrusion show anterior and posterior guiding setae (cf. Pl. 27C). The endopod is two-divided (ch57:5) and the exopod consists of several annuli, each annulus with one medially projecting seta; the last annulus bears two setae – as in the antenna. The post-mandibular limbs are similar to each other (ch58:1), consisting of a basipod, a setae-bearing proximal endite medio-proximally (ch59:1), a three-divided endopod (ch60:4) and an annulated exopod. The endopods of all post-mandibular limbs do not consist of strongly medially projecting portions as in *H. unisulcata*. The endopodal portions are only slightly projecting medially. They are drop-shaped with the pointed tip medially, and not elongated tubes as in *H. unisulcata*. The exopods of the post-mandibular pairs of limbs are multi-annulated as in the antenna and the mandible but with fewer annuli. They are not flattened, their proximal annuli are not fused and they have no lateral setation as in *H. unisulcata*. The hind body is not known (ch47:?, ch48:?, ch49:?).

Comparisons

Hesslandona angustata is very similar to *H. suecica* n. sp. in the outline of the valves, but it differs from it and all other species of *Hesslandona* in the lack of anterior or posterior thickenings, respectively spines, on the interdorsum. The interdorsum is distinctly proportionally narrower than in any other species of *Hesslandona*.

Ontogeny

With increasing shield size in *H. angustata* n. sp., the proportion of the length to the height of the valves remains more or less the same (Fig. 40); being about 1.45 throughout ontogeny. A group of small individuals is distinctly separated from a group of larger ones, both forming clusters in the ontogeny (Fig. 40, circles), perhaps representing at least two growth stages. At least one other possible growth stage (Fig. 40, arrows), including the holotype (Fig. 40), is again distinctly separated by the clusters. The smallest individual known from this species is a larva with four fully developed pairs of appendages, representing a head-larva (ch64:1).

Plate 22. *Hesslandona suecica* n. sp.

A: Holotype (UB W 258). A specimen of an advanced growth stage, about 620 μm in shield length. A dorsal view with opened valves, displaying the interdorsum (id) and its anterior (asp) and posterior (psp) extensions into spines. The rectangles mark areas of the doublure magnified in Pl. 22B and C.

B: Close-up of the area marked in Pl. 22A, displaying the anterior end of the interdorsum. The dorsal furrows extend (arrows) at both sides of the anterior spine (asp).

C: Close-up of the area marked in Pl. 22A, displaying the posterior end of the interdorsum. The arrows point to the paired dorsal furrows that extend along both sides of the interdorsum.

PLATE 23

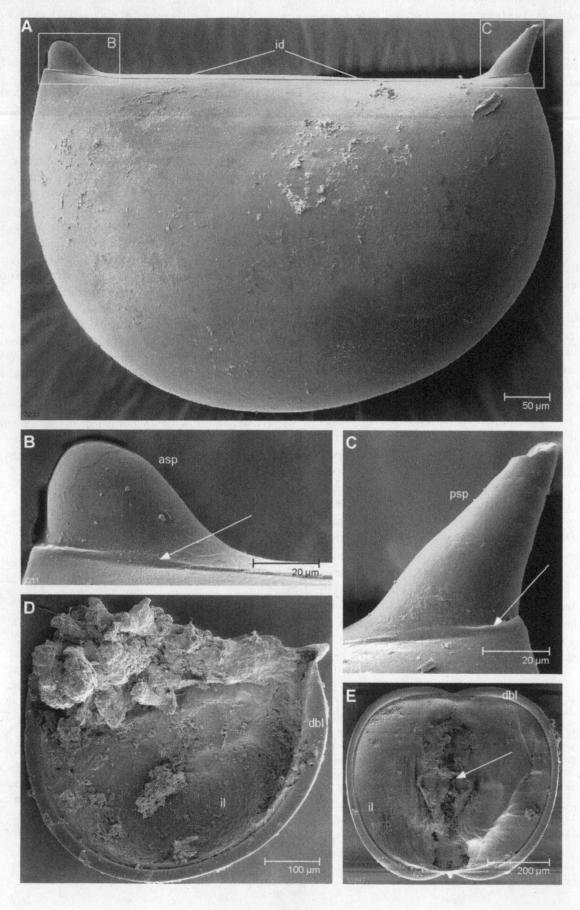

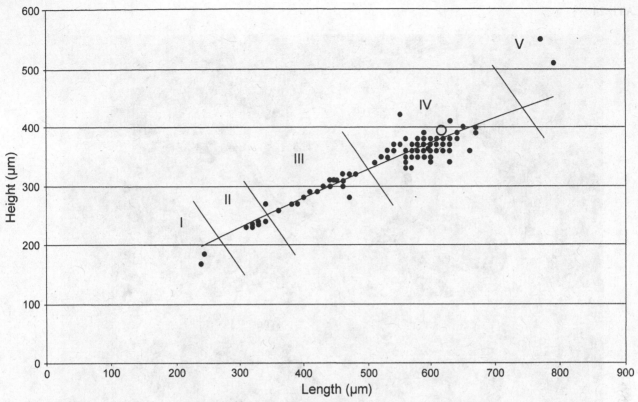

Fig. 38. Ontogeny of *Hesslandona suecica* n. sp.: length versus height, displaying four to five clusters possibly representing growth stages, separated by lines and numbered with Roman numbers. ○ = holotype. The curved trend line displays the stronger increase in length against the increase in the height of the valves during ontogeny. The trend line has the polynomial function $y = 0.0001x^2 + 0.6082x + 59.624$, $y =$ height, $x =$ length.

Hesslandona curvispina n. sp.

p. 1972 Undeterminable species of *Vestrogothia* – Taylor & Rushton, p. 18.

. 1978 *Hesslandona trituberculata* (Lochmann & Hu, 1960) – Rushton, p. 279, pl. 26, fig. 11; text-fig. 2.

Plate 23. *Hesslandona suecica* n. sp.

A: UB W 197. A specimen of an advanced growth stage, about 590 µm in shield length. A lateral view of the left valve. Rectangles mark areas of the antero- and postero-dorsal ends magnified in Pl. 23B and C.

B: Close-up of the area marked in Pl. 23A, displaying the antero-dorsal end of the shield with its anterior spine (asp) from the lateral side. The arrow points to the furrow that extends along the spine (cf. Pl. 22B).

C: Close-up of the area marked in Pl. 23A, displaying the postero-dorsal end of the shield with its anterior spine (asp) from the lateral side. The arrow points to the furrow that extends along the spine (cf. Pl. 22B).

D: UB W 198. A specimen of an advanced growth stage, about 550 µm in shield length. A lateral view inside the right valve, displaying the inner lamella (il) and the doublure (dbl). The body is preserved very coarsely and does not display any details.

E: UB W 199. A specimen of an advanced growth stage, about 590 µm in shield length. A ventral view with opened valves, displaying the inner lamella (il) and the doublure. The body proper is not preserved such that a scar in the inside of the shield indicates the original extension (arrow).

. 1981 *Hesslandona trituberculata* (Lochmann & Hu) – Gründel, p. 63, pl. 3, fig. 9 (non fig. 10).

. 1986a *Hesslandona trituberculata* (Lochman & Hu, 1960) Rushton, 1978 – Kempf, p. 400.

. 1986b *Hesslandona trituberculata* (Lochman & Hu, 1960) Rushton, 1978 – Kempf, p. 610.

. 1987 *Hesslandona trituberculata* (Lochman & Hu, 1960) Rushton, 1978 – Kempf, p. 670.

p. 1992a *Hesslandona trituberculata* Gründel – Hinz, pp. 13, 15 [only fig. 9 of Gründel (1981)].

Derivation of name. – Due to the strongly curved posterior interdorsal spine.

Holotype. – A complete shield illustrated in Pl. 28A, B; length of shield 1530 µm, height of shield 650 µm (UB W 205).

Remarks. – Müller (1964a) and Taylor & Rushton (1972) reported few specimens of the *Agnostus pisiformis* Zone from the Outwoods Formation, Merevale no. 3 borehole, Upper Cambrian of the Nuneaton District, Warwickshire, Great Britain, which they assigned to *Vestrogothia.* Rushton (1978) re-investigated the same area and found four right valves and one left valve (BDA 1476, 1479, 1782, 1785, 1817) belonging to the same species as the material from the same time zone of

PLATE 24

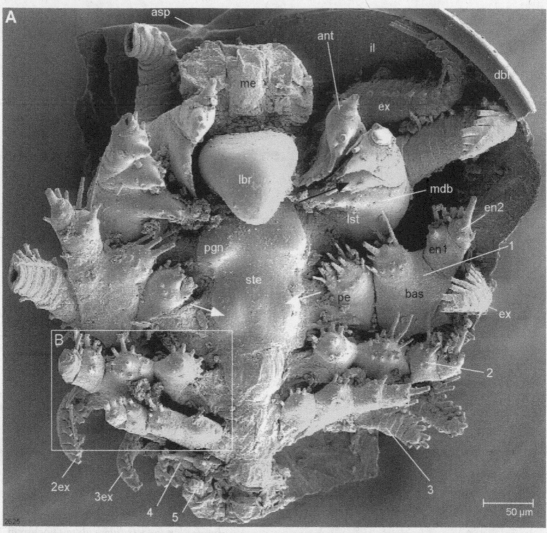

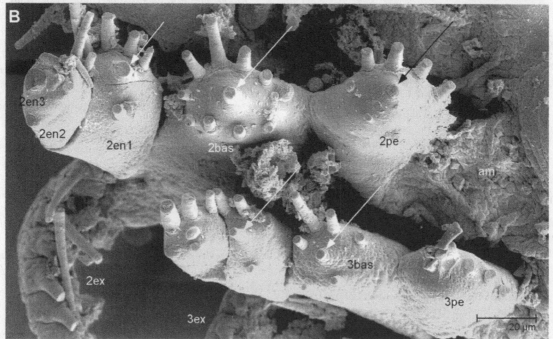

Table 24. Sample productivity of examined specimens of *Hesslandona angustata* n. sp.

Sample	Zone	Found at	Number of specimens
6408	1	Gum, Kinnekulle, Falbygden–Billingen	2
6409	1	Gum, Kinnekulle, Falbygden–Billingen	7
6410	1	Gum, Kinnekulle, Falbygden–Billingen	3 (holotype included)
6414	1	Gum, Kinnekulle, Falbygden–Billingen	1
6415	1	Gum, Kinnekulle, Falbygden–Billingen	3
6416	1	Gum, Kinnekulle, Falbygden–Billingen	7
6417	1	Gum, Kinnekulle, Falbygden–Billingen	2
6755	1	Gum, Kinnekulle, Falbygden–Billingen	2
6761	1	Gum, Kinnekulle, Falbygden–Billingen	1
6773	1	Gum, Kinnekulle, Falbygden–Billingen	1
6782	1	Gum, Kinnekulle, Falbygden–Billingen	1
6783	1	Gum, Kinnekulle, Falbygden–Billingen	1

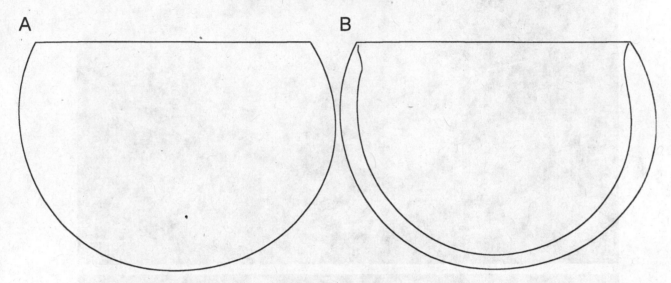

A   B

Fig. 39. Schematic drawing of the shield of *Hesslandona angustata* n. sp. from outside lateral (A) and from inside lateral (B).

Taylor & Rushton (1972). He assigned it to *Dielymella*? *trituberculata* Lochman & Hu, 1960. Siveter & Williams (1997), who re-illustrated the type specimen of Lochman & Hu (1960), and Williams & Siveter (1998) considered the specimens found by Rushton (1978) to be indeterminate juveniles of *Cyclotron* Rushton, 1969 or "other vestrogothiids". The specimen illustrated by Rushton (1978), however, resembles the material found in the Upper Cambrian of Sweden and dealt with herein, in details like the outline of the shield and the position and

shape of the three lobes. Therefore, it seems likely that the material of Taylor & Rushton (1972) and of Rushton (1978) belongs to the same species as the material at hand. Gründel (1981) found eight left and four right valves in erratic boulders of northern East Germany, which he assigned to Zone 4 of the Middle Cambrian and Zone 1 of the Upper Cambrian. He illustrated two specimens from the Upper Cambrian. The specimen of his pl. 3, fig. 10 shows a completely different outline from the specimen illustrated by Rushton (1978) and the material investigated herein. The maximum length of the shield is located ventral to the dorsal rim but not on the dorsal rim as in his specimen illustrated in pl. 3, fig. 9 of Gründel (1981). Therefore, it is assumed that Gründel actually illustrated two different species, pl. 3, fig. 10 probably being a specimen of *Hesslandona kinnekullensis* Müller, 1964.

Type locality. – Near Backeborg, Kinnekulle, Västergötland, Sweden.

Type horizon. – Upper Cambrian, *Agnostus pisiformis* Zone (Zone 1).

Plate 24. *Hesslandona suecica* n. sp.

A: UB W 200. A specimen of an advanced growth stage, about 600 µm in shield length. A ventral view of the opened shield with the body preserved. The large parts of the shield and hind body are missing. Note the enditic protrusion (black arrow) between the mandibular (mdb) limb stem (lst) and its endopod (en). The white arrows point to the paired humps on the sternite (ste) posterior to the paragnaths (pgn). The rectangle marks the area magnified in Pl. 24B.

B: Close-up of the area marked in Pl. 24A. A median view of the second and third post-mandibular limbs. Note the central spine (arrows) on the enditic protrusions of the proximal endite (pe), the basipod (bas) and the first endopodal portion (en1).

PLATE 25

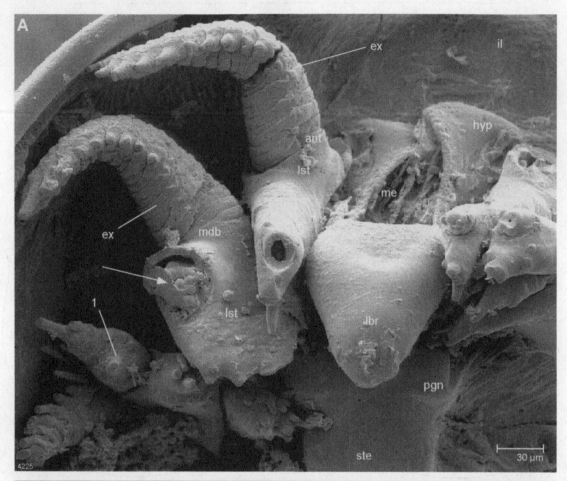

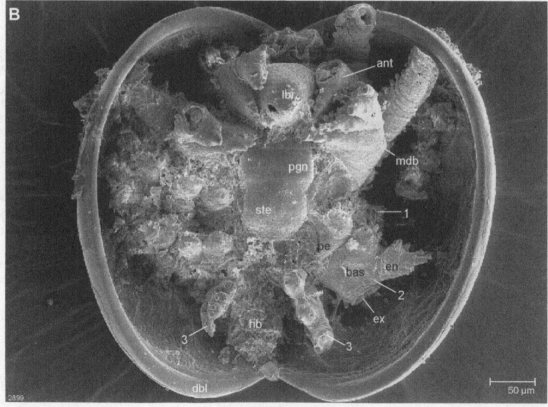

Material examined. – Eighty specimens of different stages and from different areas of Zone 1 of the Upper Cambrian of southern Sweden (Table 25).

Dimensions. – Smallest specimen: valves about 640 μm long and about 330 μm high. Largest specimen: valves about 1460 μm long and about 760 μm high.

Additional material. – Taylor & Rushton (1972) and Rushton (1978) reported material from the Outwoods Formation, Upper Cambrian of Great Britain. Gründel (1981) found eight left and four right valves in erratic boulders of northern East Germany, which he assigned to Zone 4 of the Middle Cambrian and Zone 1 of the Upper Cambrian. The assignment of the Middle Cambrian specimens to this species is uncertain since Gründel (1981) did not illustrate and describe them in detail.

Diagnosis. – Bivalved shield with straight dorsal rim, maximum length slightly ventral to dorsal rim. Valves pre-plete. Valves with three dorsally located lobes, two in the first half, one in the second half. Interdorsum complete, with short spines anteriorly and posteriorly, directing antero-dorsally, respectively posteriorly. Doublure width posteriorly twice as wide as anteriorly or ventrally, with dome-like structures near the outer margin. Shield closes tightly in all known semaphoronts.

Description. –

Shield (Fig. 41)

The bivalve shield (ch1:2) has a long and straight dorsal rim (ch2:2) (Pl. 28A, B; Fig. 41A). The maximum length of the head shield corresponds to the dorsal rim of the shield (ch3:2) (Pl. 28A, B). The valves are of equal size and the right and left valves are symmetrical. The maximum height of the valves is anterior to the midline (preplete) (ch4:1). The proportion of the length to the height of the valves is about 2.15. The free anterior part of the shield margin, starting from the dorsal rim, is almost straight (ch6:1) (antero-dorsal angle nearly 80 degrees) ventrally, then it curves gradually towards the ventral maximum (Pl. 28B). The latero-ventral outline is curved (ch7:2). Thereafter, the margin curves gently upwards –

not as steeply as anteriorly – and becomes almost straight postero-ventrally. At about one third the height of the shield, before reaching the dorsal rim, the margin curves again and thereafter runs almost straight (ch5:0; ch8:1) to meet the postero-dorsal rim (postero-dorsal angle about 80 degrees). The maximum width of the shield is slightly dorsal to the midline. The margins of the right and left valves are without outgrowths throughout (ch19:0, ch20:0). The shield valves close tightly (ch9:1). The surface of the valves is smooth, but with three dome-like lobes (Pl. 28B, C), all located close to the dorsal rim. The most anterior one is the largest (L_1) (ch10:1), a second, slightly smaller, one (L_2), is located posterior to it just before the antero-posterior midline (ch11:1), a third, inconspicuous, one (L_3) is located in the last third of the valves (ch12:1). Other structures are not developed on the valves (chs13–18:0, ch21:0).

Interdorsum

The interdorsum is complete from the anterior to the posterior (ch22:3) (Pl. 28C), being bordered by narrow membranous furrows on both sides. The interdorsum is always the same width, apart from its anterior and posterior ends, where it narrows. The maximum width is about 1/18 the length of the valves (ch23:1). The interdorsum is convex in antero-posterior aspect (ch24:2) (Pl. 28C) and – apart from its ends – is without any lobes or ornamentation (ch25:1) (Pl. 28C). Its anterior end is drawn out into a long spine (one third to one half the shield length), basally as thick as the interdorsum (ch26:1, ch27:3), and directing straight antero-dorsally (ch28:2) (Pl. 28A). The posterior end of the interdorsum is drawn out into a long spine (one third to one half the shield length), basally as thick as the interdorsum (ch29:1, ch30:3), that curves postero-ventrally along its whole length (ch31:4) (Pl. 28A). The spines are – being parts of the interdorsum – separated from the valves by the membranous furrows that border the interdorsum laterally. The anterior spine is smooth (ch32:1). The posterior spine bears small scale-like outgrowths distally (ch33:2) (Pl. 29A, B).

Doublure

A doublure is present (ch34:1), being narrowest ventrally (ch35:2), anteriorly slightly wider and posteriorly again distinctly wider than anteriorly (ch36:3) (Pl. 29C; Fig. 41B; Table 37). Approximating the dorsal rim, the doublures of both sides rapidly narrow slightly to merge into a membranous area antero- and postero-dorsally. The maximum width of the doublure is about 1/11 the length of the valves (ch37:1). Numerous small dome-like outgrowths arranged in a row are located on the outer ventral margin of the doublure of more advanced growth stages (Pl. 30A, B); they are located directly on the surface (ch38:1, ch39:2, ch40:1). The outgrowths have a blunt tip pointing outside of the domicilium. The

---

Plate 25. *Hesslandona suecica* n. sp.

A: UB W 201. A specimen of an advanced growth stage, about 590 μm in shield length. A ventral view of the anterior body, displaying the antenna, mandible and first post-mandibular limb. The hypostome (hyp) is damaged and the median eye (me) is depressed. The distal portions of the right mandible (mdb) are torn off.

B: UB W 202. A specimen of a young growth stage, about 440 μm in shield length. A ventral view of the opened shield, displaying the complete body. Note the limbs stand in an oval around the sternum (ste). The hind body (hb) is coarsely preserved, probably due to its soft cuticle.

PLATE 26

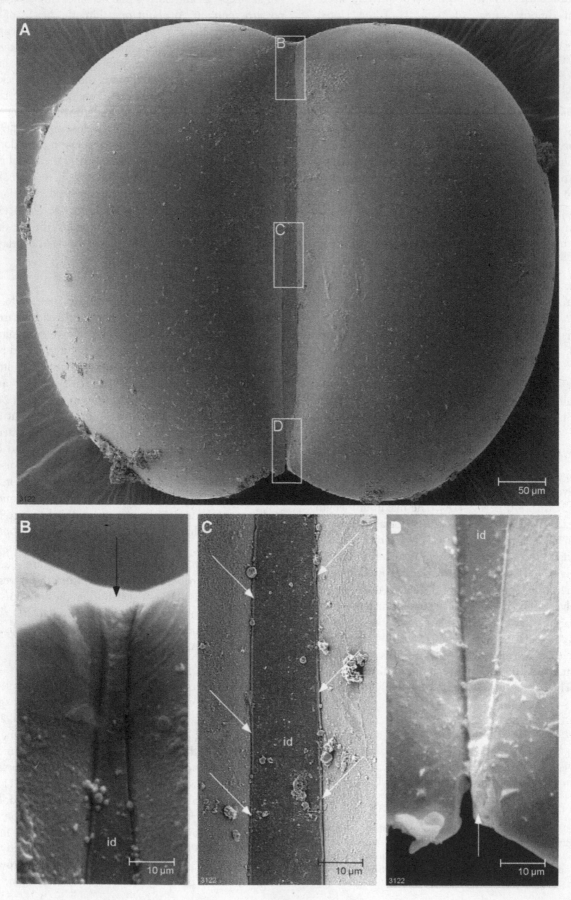

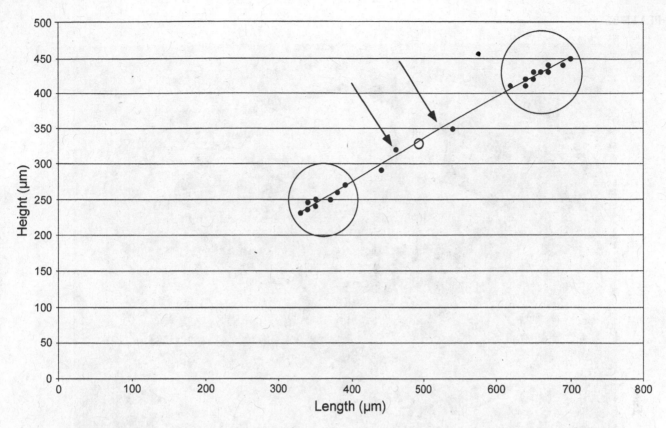

Fig. 40. Ontogeny of *Hesslandona angustata* n. sp.: length versus height. Twenty-nine specimens with measurable length and height are included. ○ = holotype. The trend line has the polynomial function $y = 0x^2 + 0.6694x + 17.991$, y = height, x = length. Circles may include members of possible growth stages, the arrows point to another set of individuals, presumably belonging to another possible growth stage. Mean proportion of length to height = 1.45.

distance between each of them varies (Pl. 30A, B). The anterior and posterior parts of the doublure are smooth (Pl. 29D, E). Pores and other ornaments could not be observed (ch41:1, ch42:1).

Inner lamella
The inner lamella (Pl. 29C) is similar to that described for *H. unisulcata* and *H. necopina*.

Body
The body proper, which is completely enveloped by the

Plate 26.   *Hesslandona angustata* n. sp.

A: Holotype (UB W 203). A specimen of a young growth stage, about 490 μm in shield length. A dorsal view of the shield with opened valves, displaying the smooth, narrow interdorsum (id). Rectangles mark areas of the doublure magnified in Pl. 26B, C and D.

B: Close-up of the area marked in Pl. 26A, displaying the antero-dorsal end of the shield. The interdorsum (id) tapers towards its end without any thickenings or spine-like extensions (arrow).

C: Close-up of the area marked in Pl. 26A, displaying the median part of the dorsal rim. The arrows point to the paired dorsal furrows that extend along both sides of the interdorsum.

D: Close-up of the area marked in Pl. 26A, displaying the postero-dorsal end of the shield. As anteriorly, the interdorsum (id) tapers towards its end without any thickenings or spine-like extensions (arrow).

bivalved shield, includes at least nine segments (Pl. 30C). Trunk segmentation of the preserved portion is retained in the insertions of the limbs and distinct segment boundaries or sternites are weakly defined. At least nine segments are dorsally fused to the shield (ch44:6) (Pl. 30C). Hence, the shield represents a cephalothoracic shield including at least four limb-bearing trunk segments. The area of fusion of the body proper to the shield is very narrow, corresponding to approximately the width of the interdorsum (ch43:1) (Pl. 30C). Starting some micrometres posterior to the dorsal anterior membranous area of the doublures, the body proper extends along the inner dorsal length of the shield (Pl. 30C), but the posterior end of the body is unknown. Its posterior extension and segmental continuation remain uncertain, as well as whether there is a hind body extending free from the shield into the domicilium. The body proper, in ventral view, is oval with a blunt anterior and an elongated posterior end. The maximum width of the body proper is located at the mandibles. The cross-section of the body proper anterior to the mandibular part is almost trapezoidal with the long axis dorso-ventrally oriented. The cross-section at the mandibular segment is more or less half-oval with the long axis in dorso-ventral aspect; the height of the body proper in this position measures less

PLATE 27

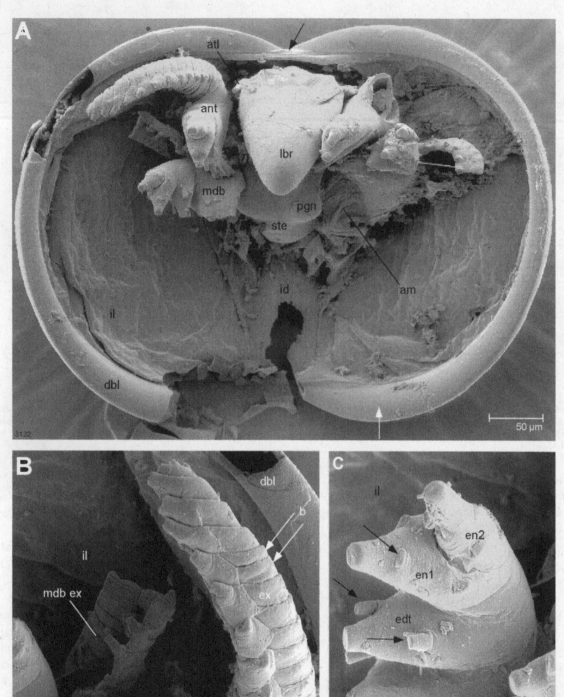

Table 25. Sample productivity of examined specimens of *Hesslandona curvispina* n. sp.

Sample	Zone	Found at	Number of specimens
6250	1	Ödegården, 500 m east of Varvboholm	1
6364	1	St. Stolan, Falbygden–Billingen	1
6404	1–2	Between Haggården and Marieberg (Kinnekulle)	2
6408	1	Gum, Kinnekulle, Falbygden–Billingen	2
6409	1	Gum, Kinnekulle, Falbygden–Billingen	4
6410	1	Gum, Kinnekulle, Falbygden–Billingen	3
6411	1	Gum, Kinnekulle, Falbygden–Billingen	1
6414	1	Gum, Kinnekulle, Falbygden–Billingen	2
6415	1	Gum, Kinnekulle, Falbygden–Billingen	1
6416	1	Gum, Kinnekulle, Falbygden–Billingen	2
6417	1	Gum, Kinnekulle, Falbygden–Billingen	10
6730	1	Gum, Kinnekulle, Falbygden–Billingen	2
6731	1	Backeborg, Kinnekulle, Falbygden–Billingen	1 (holotype)
6732	1	Backeborg, Kinnekulle, Falbygden–Billingen	1
6734	1	Backeborg, Kinnekulle, Falbygden–Billingen	2
6735	2	Backeborg, Kinnekulle, Falbygden–Billingen	2
6736	1	Backeborg, Kinnekulle, Falbygden–Billingen	2
6749	1	Gum, Kinnekulle, Falbygden–Billingen	1
6750	1	Gum, Kinnekulle, Falbygden–Billingen	1
6751	1	Gum, Kinnekulle, Falbygden–Billingen	3
6755	1	Gum, Kinnekulle, Falbygden–Billingen	1
6757	1	Gum, Kinnekulle, Falbygden–Billingen	1
6760	1	Gum, Kinnekulle, Falbygden–Billingen	4
6761	1	Gum, Kinnekulle, Falbygden–Billingen	5
6763	1	Gum, Kinnekulle, Falbygden–Billingen	2
6765	1	Gum, Kinnekulle, Falbygden–Billingen	2
6771	1	Gum, Kinnekulle, Falbygden–Billingen	2
6772	1	Gum, Kinnekulle, Falbygden–Billingen	1
6773	1	Gum, Kinnekulle, Falbygden–Billingen	1
6774	1	Gum, Kinnekulle, Falbygden–Billingen	2
6776	1	Gum, Kinnekulle, Falbygden–Billingen	1
6779	1	Gum, Kinnekulle, Falbygden–Billingen	1
6781	1	Gum, Kinnekulle, Falbygden–Billingen	1
6782	1	Gum, Kinnekulle, Falbygden–Billingen	1
6783	1	Gum, Kinnekulle, Falbygden–Billingen	4
6784	1	Gum, Kinnekulle, Falbygden–Billingen	4
6785	1	Gum, Kinnekulle, Falbygden–Billingen	2

than one third the height of the valves. Posterior to it, the body proper changes its cross-section to a half-circular shape. The height of the body decreases gradually towards the posterior to about one sixth the height of the valves

---

Plate 27. *Hesslandona angustata* n. sp.

A: UB W 204. A specimen of an early growth stage, about 350 μm in shield length. A ventral view of the opened shield, the body proper being slightly displaced. The maximum width of the doublure is located posteriorly (white arrow). The black arrow points to the anterior membranous area in between both doublures. The antenna and mandible of this specimen are illustrated in Pl. 27B, respectively C.

B: The same specimen as in Pl. 27A. The antenna from the anterior. Note the seta on the median edge of the limb stem (arrow "a"). The exopod shows setae-bearing and seta-less annuli (arrow "b").

C: The same specimen as in Pl. 27A. The mandible from the distal side. Note the undivided limb stem (lst), the enditic protrusion (edt) squeezed between the limb stem (lst) and the endopod (end). The median edges of the limb stem, the endite and at least the first endopodal portion are guided by setae on either side (arrows).

at the posterior end of the body. The limbs are arranged in a more or less regular row at the flanks of the body, being more or less equally set together from the anterior towards the caudal end.

The anterior part of the body proper is the hypostome/labrum complex. The hypostome forms the anterior sclerotised ventral surface, has a somewhat rhomboid outline in ventral view and becomes gradually higher to the posterior in lateral view. Antero-laterally on the hypostome, the antennulae insert on a shoulder-like slope in a circular joint area (Pl. 30C). The antennae (Pl. 30C) insert postero-laterally to the hypostome in a more spindle-shaped joint area, the posterior edge touching the labral part of the hypostome/labrum complex. The median eye is located antero-distally on the hypostome. The posterior end of the hypostome is drawn out distally into a lobe-like protrusion, the labrum (ch61:1); the mouth opening could not be observed. Posterior to the hypostome/labrum complex, the ventral surface between the first to third pairs of post-antennular

Table 26. Complete list of finds of *Hesslandona curvispina* n. sp., reported by different authors, with localities and horizons.

Locality	Horizon	Reference
Outwoods Formation, Merevale, UK	Upper Cambrian, Zone 1	Taylor & Rushton 1972
Outwoods Formation, Merevale, UK	Upper Cambrian, Zone 1	Rushton 1978
Erratic boulders, northeastern Germany	Middle Cambrian, Upper Cambrian, Zone 1	Gründel 1981

limbs is marked by a sclerotic plate, the sternum (ch62:1) (Pl. 30C). The sternum is slightly domed and bears one pair of humps anteriorly, the paragnaths (ch63:1), between the mandibles. A second pair of elongated humps is located posterior to the paragnaths between the second pair of post-mandibular limbs (Pl. 30C). A keel-like elongated lobe is located in between the two humps (Pl. 30C). Posterior to the sternum, the ventral body surface becomes progressively softer. The post-mandibular limb pairs insert on a lateral slope corresponding to the cross-section of the body. The hind body, including the number of segments involved and possible furcal rami, is unknown (ch47:?, ch48:?, ch49:?).

Soft parts

The soft parts are somewhat different to those described in detail for *H. unisulcata*. Because it is beyond the scope of this paper to describe the soft parts of *H. curvispina* n. sp. in full detail, only some general remarks are made, important for the discrimination from *H. unisulcata* and the phylogenetic analysis of Phosphatocopina. The detailed description is postponed to a second paper. The antennula is not known (ch45:?, ch46:?). As in *H. unisulcata*, the antenna consists of an undivided limb stem throughout ontogeny, but is a product of fusion of the coxa and basipod (see p. 23; ch50:2, ch51:2, ch52:2), a two-divided endopod (ch53:5) and a multi-annulated exopod. The same is true for the mandible (Pl. 30C). It has an undivided limb stem throughout ontogeny, and an enditic protrusion is located medially between the limb stem and the endopod, which is recognised as the remains of the basipod (see p. 27; ch54:2, ch55:2, ch56:2). The endopod has two parts (ch57:5) and the exopod consists of several annuli, each annulus with one medially projecting seta; the last annulus bears two setae – as in the antenna. The post-mandibular limbs are – as in *H. ventrospinata* – separated into two groups of different shape (Pl. 30C, D). The first three post-mandibular pairs of limbs are serially similar to each other (ch58:1), consisting of a basipod, a setae-bearing proximal endite medio-proximally (ch59:1), a three-part endopod (ch60:4) and an annulated exopod (Pl. 30D). The endopodal portions are short and slightly projecting medially. The exopods are triangular plates with marginal setation (Pl. 30D); their tip is annulated – very much like the exopods in *H. unisulcata* (cf. Pls. 11F, 12G), but distinctly smaller in proportion to the body size. At least the second post-mandibular

limb bears one prominent seta proximo-laterally (Pl. 30D). The fourth to sixth post-mandibular pairs of limbs (Pl. 30D) consist of a single limb stem and two rami. The endopod is located medio-distally on the limb stem. It is distinctly separated from the limb stem by a joint membrane and consists of a single short tube-like portion with a blunt tip. The exopod forms more or less the distal extension of the limb stem, without being distinctly separated from the limb stem. It is a flat plate of triangular shape with a pointed terminal tip and with marginal setation (Pl. 30D).

Comparisons

Hesslandona curvispina n. sp. is similar to *H. necopina* in the general shape of the shield. The maximum length corresponds in both species to the dorsal rim; in all other investigated species, the maximum length is located ventral to the dorsal rim. *Hesslandona curvispina* n. sp. is distinguished from *H. necopina* by the possession of three lobes. Three lobes also occur in *H. trituberculata*, from which *H. curvispina* n. sp. is distinguished not only by the outline of the shield, but also by the anterior and posterior interdorsal spines. These spines are distinctly longer in proportion to the shield in *H. curvispina* n. sp. than in any other known phosphatocopine species. *Hesslandona curvispina* n. sp. is similar to *H. ventrospinata* in the separation of the post-mandibular pairs of limbs into two different morphological groups. It differs from this species by the possession of only three lobes and the lacking of marginal outgrowths of the valves. There is as yet too little information to be able to evaluate the similarity in limb morphology in a phylogenetic approach. This will be included in the second paper on the Phosphatocopina.

Plate 28.　*Hesslandona curvispina* n. sp.

A: Holotype (UB W 205). A specimen of a late growth stage, about 1530 μm in shield length. A lateral view of the left valve. The valve without anterior and posterior spines is illustrated in Pl. 28B, as indicated by the rectangle.

B: Close-up of the shield illustrated in Pl. 28A, displaying a smooth shield with three subdorsal lobes (L1–L3).

C: UB W 206. A specimen of a late growth stage, about 1250 μm in shield length. An antero-dorsal view of the slightly opened shield, displaying the slightly convex interdorsum.

PLATE 28

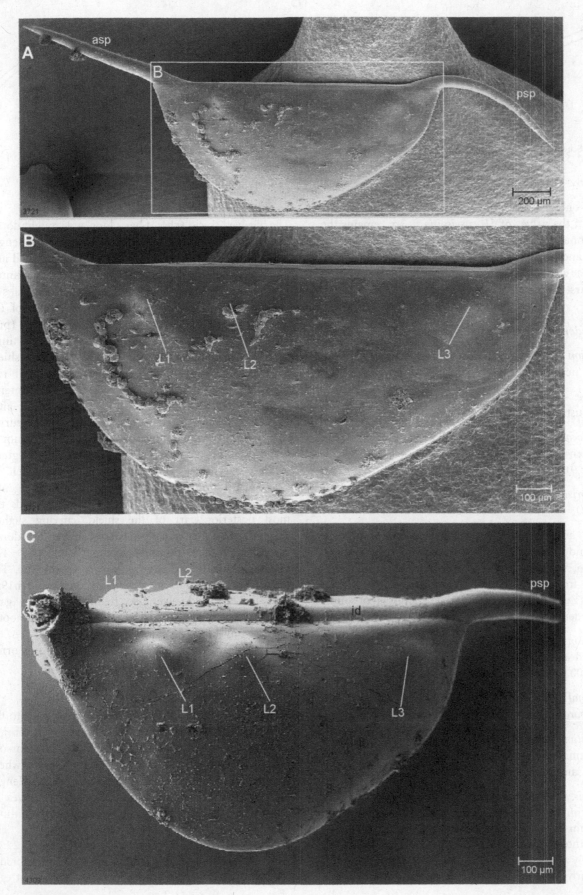

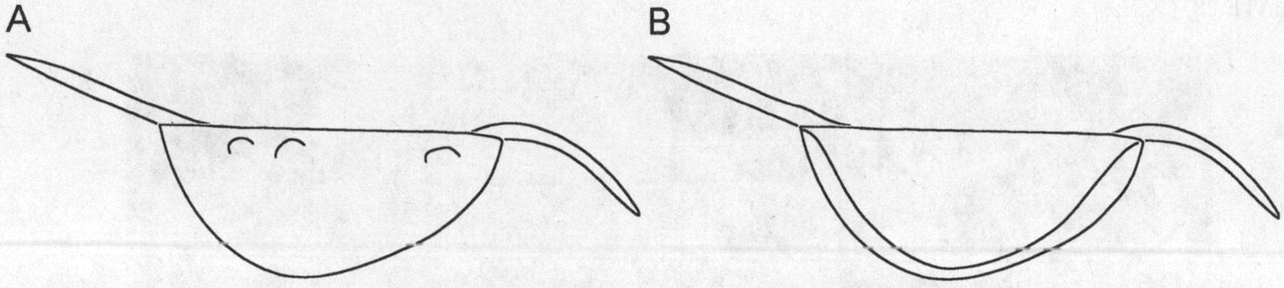

A

B

Fig. 41. Schematic drawing of the shield of *Hesslandona necopina* from outside lateral (A) and from inside lateral (B).

Ontogeny

During ontogeny, the proportion of the length to the height of the valves of *H. curvispina* n. sp. remains the same, about 2.15 (Fig. 42). The observed range of shield lengths and heights does not indicate distinct clusters of particular growth stages (Fig. 42). The segmental amount of the first larva is uncertain (ch64:?).

Hesslandona toreborgensis n. sp.

Derivation of name. – Toreborg is the village in southern Sweden near where the holotype was discovered.

Holotype. – Complete shield, 680 µm long and 330 µm high, illustrated in Pl. 31A (UB W 214).

Type locality. – North of Toreborg, Kinnekulle, Falbygden–Billingen, Västergötland, Sweden (Fig. 2, number 9).

Type horizon. – Upper Cambrian, *Olenus gibbosus* Zone (Zone 2).

Material examined. – Eighty-eight specimens of different stages and different areas at the Kinnekulle, Falbygden–Billingen, Västergötland, southern Sweden, plus 10 specimens from Degerhamn, Öland, from the *Olenus gibbosus* Zone (Zone 2) of the Upper Cambrian of southern Sweden (Table 27).

Dimensions. – Smallest specimen: valves about 320 µm long and about 190 µm high. Largest specimen: valves about 900 µm long and about 460 µm high.

Additional material. – No additional material of *H. toreborgensis* n. sp. has been reported thus far.

Distribution. – Kinnekulle, Falbygden–Billingen and Degerhamn, Isle of Öland, southern Sweden, of the Upper Cambrian Zone 2.

Diagnosis. – Maximum length of bivalved shield between dorsal rim and midline, dorsal rim almost as long as maximum length. Valves pre-plete, without lobes or spines. Interdorsum narrow, with dome-like outgrowth anteriorly and a short spine posteriorly. Doublure widest posteriorly, without any structures.

Description. –

Shield (Fig. 43)

The bivalved shield (ch1:2) has a long and straight dorsal rim (ch2:2) (Pl. 31A, B; Fig. 43A). The maximum length of the shield is close to the dorsal rim of the shield and the dorsal length is almost as long as the maximum length (ch3:3). The right and left valves are of equal size and symmetrical shape. The maximum height of the valves is anterior to the antero-posterior midline (pre-plete, see Fig. 8) (ch4:1). The shield is about 2.1 times longer than high. The free anterior part of the shield margin, starting from the rather straight dorsal rim (ch6:1), curves first slightly antero-ventrally (antero-dorsal angle nearly 75 degrees, cf. Fig. 7) and, after reaching the anterior-most part of the shield, curves back postero-ventrally towards the ventral maximum of the valve (postero-dorsal angle nearly 75 degrees, cf. Fig. 7). The ventral margin is curved (ch7:2) (Pl. 31A, B). More posteriorly, the margin swings gently upwards – as steeply as anteriorly – and towards the most posterior part of the shield, curving slightly anteriorly and meeting the posterior end of the dorsal rim (ch8:2, postero-dorsal angle about 80 degrees). The maximum width is slightly dorsal to the midline. The margins are without outgrowths throughout (ch19:0, ch20:0). The shield valves close tightly without any gaps in all stages (ch9:1). The surface of the valves is smooth, without any lobes or spines (chs10–18:0) (Pl. 31A, B). The outer margin of the valves is without any ornamentation (ch21:0).

Interdorsum

The interdorsum is continuous from the anterior to the posterior (ch22:3), being bordered by narrow membranous furrows on both sides (Pl. 31C). The interdorsum is always the same width, apart from both ends, where it narrows. The maximum width is about 1/24 the length of the valves (ch23:1). The interdorsum is convex in axial aspect (ch24:2), and the median part is without lobes or any ornamentation (ch25:1) (Pl. 31B). The anterior end has a dome-like thickening (ch26:1, ch27:1), with a blunt tip, basally as thick as the interdorsum (ch28:0) (Pl. 31A). The insertion area is somewhat egg-

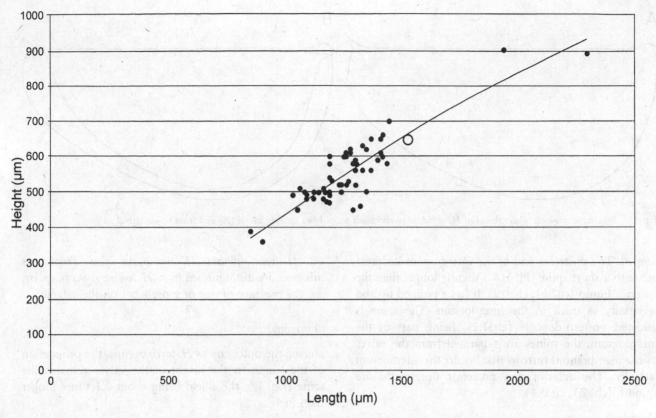

Fig. 42. Ontogeny of *Hesslandona curvispina* n. sp.: length versus height. Fifty-eight specimens with measurable length and height are included. O = holotype. The trend line has the polynomial function $y = 0x^2 + 0.5751x - 76.575$, y = height, x = length. Mean proportion of length to height = 2.15.

Table 27. Sample productivity of examined specimens of *Hesslandona toreborgensis* n. sp. (cf. Figs. 1–4).

Sample	Zone	Found at	Number of specimens
6361	2b	Near Stolan, Falbygden–Billingen	7
6362	–	Near Stolan, Falbygden–Billingen	3
6377	1–2	Near Sandtorp, Kinnekulle, Falbygden–Billingen	2
6402	2a–b	Between Haggården and Marieberg, Kinnekulle, Falbygden–Billingen	1
6404	1–2	Kestad, between Haggården and Marieberg, Kinnekulle	19
6431	2	South of Karlsfors, Falbygden–Billingen	3
6432	2	South of Karlsfors, Falbygden–Billingen	2
6474	–	Degerhamn, Öland	10
6740	1–2b	Blomberg, Kinnekulle, Falbygden–Billingen	1
6745	2a–b	At the road to Ledsgården north-northeast of Gökhem	3
6747	–	Northeast of Stora Stolan, Billingen	2
6789	1–2b	West of Stubbeg, southeast slope of the Kinnekulle	1
6792	2	West of Stubbeg, southeast slope of the Kinnekulle	4
6794	2	West of Stubbeg, southeast slope of the Kinnekulle	4
6796	2	Between Håggarden and Marieberg, Kinnekulle	3
6798	2	Between Håggarden and Marieberg, Kinnekulle	2
6800	2	Between Håggarden and Marieberg, Kinnekulle	1
6802	2	Between Håggarden and Marieberg, Kinnekulle	1
6805	1–2	East Österplana, Kinnekulle, old outcrop	2
6806	–	East Österplana, Kinnekulle, old outcrop	2
6807	1–2b	East Österplana, Kinnekulle, old outcrop	2
6808	1–2	West Toreborg, Kinnekulle, old outcrop	4
6809	–	West Toreborg, Kinnekulle, old outcrop	3
6811	2	North Toreborg, Kinnekulle, west Fullösa	2 (holotype included)
6813	2	North Toreborg, Kinnekulle, west Fullösa	4

A

B

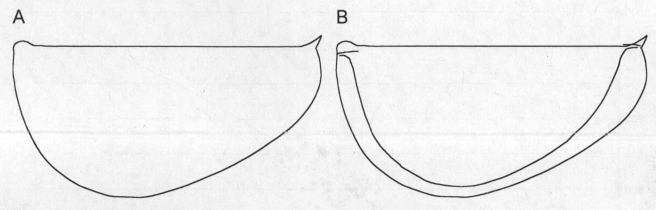

Fig. 43. Schematic drawing of the shield of *Hesslandona toreborgensis* n. sp. from outside lateral (A) and from inside lateral (B).

shaped. The posterior end of the interdorsum is drawn out into a short spine (Pl. 31A), slightly longer than the anterior hump (ch29:1, ch30:2). It has a pointed tip and is basally as thick as the interdorsum. The spine is directed postero-dorsally (ch31:2). Being part of the interdorsum, the spines are separated from the valves by the membranous furrows that border the interdorsum laterally. The anterior and posterior outgrowths are smooth (ch32:1, ch33:1).

Doublure

A doublure is present (ch34:1) along the inner margin of the valves, being narrowest ventrally (ch35:2), anteriorly slightly wider, and posteriorly about one third wider than ventrally (ch36:3) (Pl. 31D). Approximating the dorsal rim, the doublures of both sides rapidly narrow slightly to merge into a membranous area antero- and postero-dorsally. The maximum width of the doublure is about 1/11 the length of the valves (ch37:1) (cf. Table 39). There are no structures or pores present on the postero-ventral margin of the doublure (ch38:1, ch39:1, ch40:0, ch41:1, ch42:1).

Inner lamella

The inner lamella was not observed due to poor preservation (cf. Pl. 31D).

Body

The body is not preserved in any of the specimens of *H. toreborgensis* available (ch43–64:?).

Comparisons

Hesslandona toreborgensis is similar to *H. kinnekullensis*, *H. suecica* n. sp., *H. angustata* n. sp., and *H. necopina* in having a smooth shield. It differs from *H. necopina*, *H. suecica* n. sp. and *H. kinnekullensis* in the small posterior spine and in the dome-like protrusion anteriorly on the interdorsum. The interdorsum of *H. angustata* n. sp. does not have any outgrowths, and its interdorsum is significantly narrower than that of *H. toreborgensis* n. sp. All other *Hesslandona* species, i.e. *H. trituberculata*, *H. curvispina* n.

sp., *H. ventrospinata*, *H. unisulcata*, plus *Trapezilites minimus*, are distinguished from *H. toreborgensis* n. sp. by, e.g. the presence of one or more lobes on the valves.

Ontogeny

During the ontogeny of *H. toreborgensis*, the proportion of the length to the height of the valves remains the same (Fig. 44), the shield being about 2.1 times longer than high.

Trapezilites Hinz-Schallreuter, 1993

1993c *Trapezilites* n. g. – Hinz-Schallreuter, pp. 399, 402.
1998   *Trapezilites* Hinz-Schallreuter – Williams & Siveter, p. 29.

Derivation of name. – After the subtrapezoidal outline of the shield (Hinz-Schallreuter 1993c).

Type species. – *Aristozoe ? minima* Kummerow, 1931.

Original diagnosis. – (Hinz-Schallreuter 1993c). "Outline trapezoidal with rounded ventral corners, subamplete. Distinct node [lobe in the terminology applied herein] in centro-dorsal field just in front of mid-height."

Emended diagnosis. – Phosphatocopine with subtrapezoidal valves of equal size, close tightly. Valves with one prominent antero-dorsal lobe. Maximum length of valves around the dorso-ventral midline. Valves amplete. Interdorsum with hump-like thickenings at both ends. Doublure wide, maximum width ventrally, about one sixth the length of valves.

Species referred to genus. – The genus *Trapezilites* encompasses only one species.

Trapezilites minimus (Kummerow, 1931) Hinz-Schallreuter, 1993

.   1924   *Aristozoë primordialis* Linnss. sp. – Kummerow, pp. 406, 445, 446.

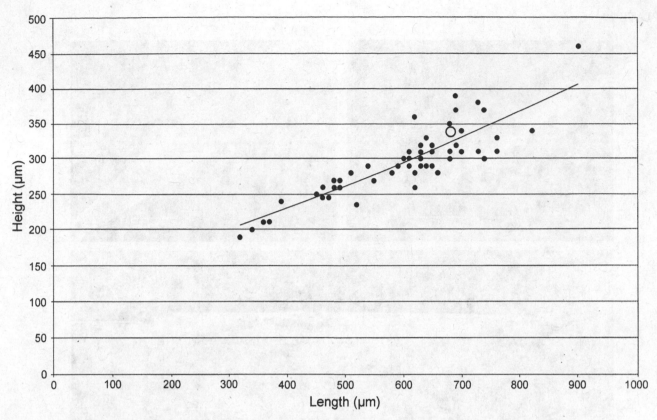

Fig. 44. Ontogeny of *Hesslandona toreborgensis* n. sp.: length versus height. Fifty-four specimens with measurable length and height are included. ○ = holotype. The trend line has the polynomial function $y = 0x^2 + 0.2358x + 121.68$; y = height, x = length. Mean proportion of length to height about 2.1.

. 1928 *Aristozoe* ? cf. *primordialis* Linnss. sp. – Kummerow, pp. 42, 59, pl. 2, fig. 19.

. 1931 *Aristozoe* ? *minima* n. sp. – Kummerow, p. 254–255, text-fig. 18.

. 1934 "*Aristozoe*" *minima* (Kummerow), 1931 – van Straelen & Schmitz, pp. 176, 209, 228, 236, 238.

v. 1964a *Falites* (?) *minima* (Kummerow) – Müller, p. 29, pl. 4, figs. 8–12, 16.

. 1965 *Falites minima* (?) (Kummerow) – Adamczak, p. 28, pl. 1, figs. 3a, b; text-fig. 2.

. 1972 *Falites? minimus* (Kummerow) – Taylor & Rushton, p. 13, pl. 4 (borehole record).

. 1974 *Falites minimus* – Martinsson, p. 208.

p. 1978 *Falites? minimus* (Kummerow, 1931) – Rushton, p. 277 [*partim*: specimens BGS BDA 1167/1168, BGS BDA 1276/1277, BDA 1452/1453 (pl. 26, figs. 9, 10; text-fig. 2), *non* specimen BGS BDA 1771/1774 (= *Waldoria rotundata*)].

v. 1979a *Hesslandona* n. sp. – Müller, fig. 8.

v. 1979a *Falites? minima* (Kummerow) – Müller, p. 11.

. 1981 *Falites? minima* (Kummerow) – Gründel, p. 63, pl. III, figs. 7, 8.

. 1986a *Aristozoe? minima* Kummerow, 1931 – Kempf, p. 65.

. 1986a *Falites? minimus* (Kummerow, 1931) Mueller, 1964 – Kempf, p. 355.

. 1986b *Aristozoe? minima* Kummerow, 1931 – Kempf, p. 369.

. 1986b *Falites? minimus* (Kummerow) Mueller, 1964 – Kempf, p. 370.

. 1987 *Aristozoe? minima* Kummerow, 1931 – Kempf, p. 167.

. 1987 *Falites? minimus* (Kummerow, 1931) Mueller, 1964 – Kempf, p. 436.

. 1993c *Aristozoe? minima* Kummerow, 1931 – Hinz-Schallreuter, pp. 388, 402 (referred to *Trapezilites* n. g.).

. 1994 *Trapezilites minimus* – Hinz & Jones, p. 368.

. 1996a *Trapezilites minimus* (Kummerow) – Hinz-Schallreuter, pp. 85–88.

. 1998 *Aristozoe? minima* Kummerow, 1931 – Hinz-Schallreuter, p. 103 (historical review).

. 1998 *Trapezilites minimus* (Kummerow, 1931) – Williams & Siveter, pp. 29, 30, pl. 5, figs. 5, 6.

Name. – Not specified by Kummerow (1931).

Holotype. – One shield (?), originally deposited at the "Bundesanstalt für Geowissenschaften und Rohstoffe", Berlin, Germany, right valve illustrated by Kummerow

PLATE 29

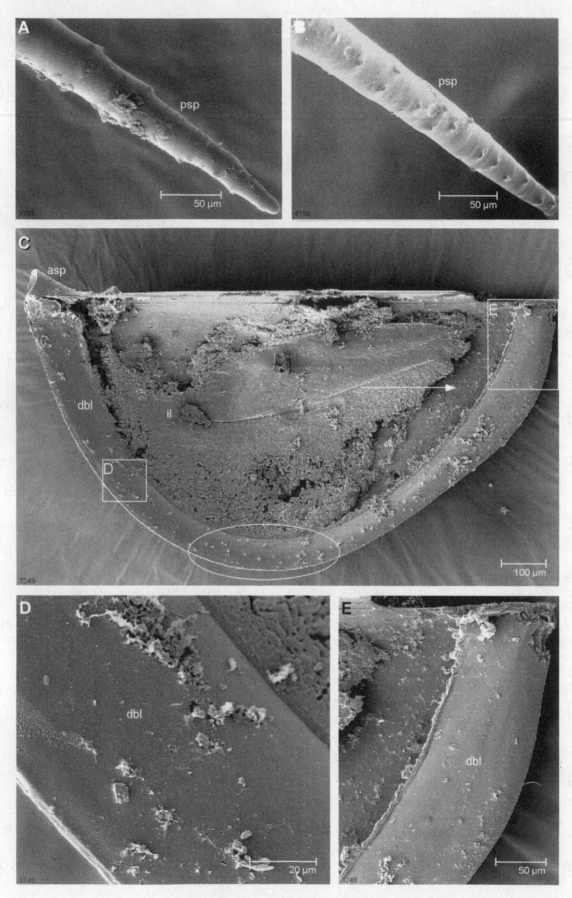

(1931) in his fig. 18; shield about 1200 µm long and about 1000 µm high. According to Hinz-Schallreuter (1996a), the holotype is lost (cf. Williams & Siveter 1998).

Remarks. – Kummerow (1924) reported this species from material from erratic boulders of northern Europe, but believed it to represent young of *Leperditia primordialis* Linnarsson, 1869. However, due to morphological differences with other species of *Leperditia* Rouault, 1851, Kummerow (1924) assigned his specimens to the genus *Aristozoe* Barrande, 1868, which was regarded as a phyllocarid. But the material found by Kummerow (1924) does not belong to the same species as the material of Linnarsson (1869a, b; see also van Straelen & Schmitz 1934). Kummerow (1931) was aware of this fact and established a new species for the material described by him in 1924.

Type locality. – Degerhamn, Öland, Sweden.

Type horizon. – Not specified by Kummerow (1931); probably Upper Cambrian, *Agnostus pisiformis* Zone (Zone 1).

Material examined. – Twenty-six specimens of different growth stages and from different areas, *Agnostus pisiformis* Zone (Zone 1) and Zone 2a of the Upper Cambrian of Sweden (Table 28).

Dimensions. – Smallest specimen: valves about 440 µm long and about 410 µm high. Largest specimen: valves about 1010 µm long and about 870 µm high.

Additional material. – Kummerow (1924) (Table 29) found material from erratic boulders, mostly isolated specimens without additional fauna. In one case from Jeserig near Brandenburg, he found *Trapezilites minimus* together with *Agnostus laevigatus* Dalman, 1828 (Euarthropoda) and *Acrotreta socialis* von Seebach (Brachiopoda). In 1928, Kummerow described more material belonging to this species from erratic boulders of Voigtsdorf, Mecklenburg-Vorpommern, northern

Germany. Later on, Kummerow (1931) (Table 29) described new material, including the type material of a newly established species from Degerhamn, Öland, Sweden [see Müller (1964a) and Hinz-Schallreuter (1998) for a historical review]. Rushton (1978) (Table 29) reported material from the Outwoods Shale Formation, *Agnostus pisiformis* Zone, Merioneth Series, Merevale no. 3 borehole, Warwickshire, UK (see also Williams & Siveter 1998) (Table 29). Gründel (1981) (Table 29) mentioned that he had discovered numerous valves in 20 different erratic boulders from northern East Germany, dated to Zones 1, 2 and 5. Because no other material of *H. minima* has been reported from Zone 5, the data of Gründel (1981) are doubted.

Distribution. – Zone 1 (*Agnostus pisiformis* Zone) and Zone 2 of the Late Cambrian, England, southern Sweden, northern Germany. The report of this species in Zone 5 (Gründel 1981) is uncertain as stated above.

Original diagnosis. – (Müller 1964a, translated). "Phosphatocopine with simple hinge line, valves of equal size, with only one, very large and anteriorly located lobe. Outline about equally high as long."

Emended diagnosis. – As for *Trapezilites*.

Description. – The material consists of different growth stages of *Trapezilites minimus* (Kummerow, 1931). Investigated material is almost completely restricted to isolated valves of shields.

Shield (Fig. 45)
The bivalved shield (ch1:2) has a long and straight dorsal rim (ch2:2) (Pl. 32A, B; Fig. 45A). The maximum length of the head shield is on the midline (ch3:1). The valves are of equal size and the right and left valves are symmetrical. The maximum height of the valves is on the midline (amplete) (ch4:2), about 1.1 times longer than high. The free anterior part of the shield margin starts from the dorsal rim, curves antero-ventrally (Pl. 32A, B; antero-dorsal angle nearly 70 degrees) and equally postero-ventrally towards the ventral maximum of the valve (ch6:2). The ventral part of the margin is gently curved (ch7:2).

More posteriorly, the margin swings gently upwards more rapidly than anteriorly (ch5:2, angle 60 degrees) to recurve even slightly anteriorly, meeting the posterior end of the dorsal rim (ch8:2). The maximum width of the shield is on the lobe. The margins are straight throughout, without outgrowths (ch19:0, ch20:0), and the shield valves close tightly along the whole margin (ch9:1). The surface of the valves is smooth, but with one large lobe, L_1 (ch10:1) (Pl. 32A, B) on the anterior area closer to the interdorsum than to the midline of the valves. The lobe is elongated in the antero-dorsal/postero-ventral direction (Pl. 32A). The valves are characterised by three further nodes, which are significantly

Plate 29. *Hesslandona curvispina* n. sp.

A: UB W 207. Close-up of the posterior interdorsal spine (psp) of a specimen of 1120 µm shield length. The spine is armed with scale-like structures distally (cf. Pl. 29B).

B: UB W 208. Close-up of the posterior interdorsal spine (psp) of a specimen of 1000 µm shield length. The spine is armed with numerous scale-like structures distally (cf. Pl. 29A).

C: UB W 209. Image flipped horizontally. A specimen representing an advanced stage, about 1190 µm in length. A lateral view of the inner side of the left valve. Rectangles mark areas magnified in Pl. 29D and E. The ellipse indicates the position of the part of the doublure that is shown in Pl. 30A, B from different individuals.

D: Image flipped horizontally. Close-up of the area indicated in Pl. 29C, displaying a smooth antero-ventral doublure.

E: Image flipped horizontally. Close-up of the area indicated in Pl. 29C, displaying a smooth posterior doublure.

PLATE 30

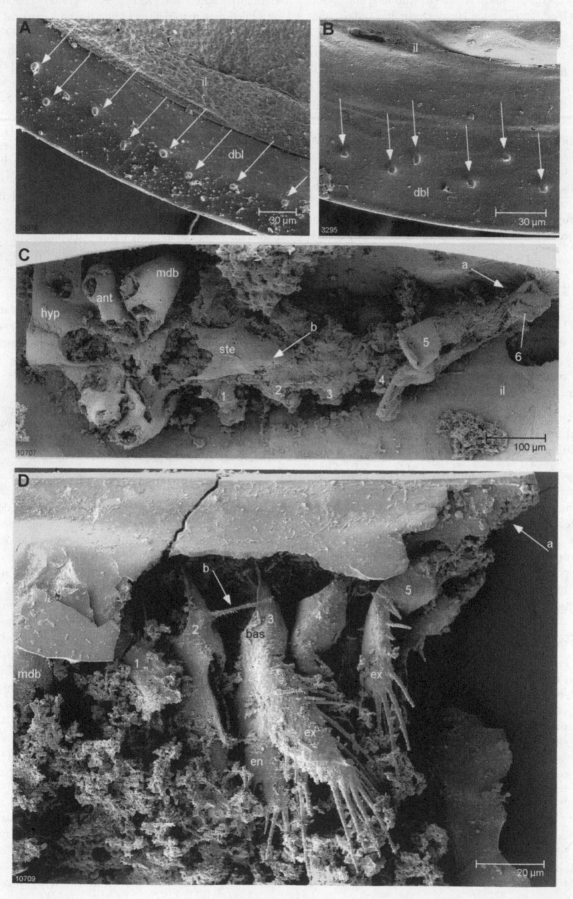

less distinct than lobe L_1. One circular node is located posteriorly (Pl. 32C, arrow a), extending into an elongated one centrally (Pl. 32C, arrow b), and the third node extends into a bow close to the antero-ventral margin of the valves (Pl. 32C, arrow c). These nodes are not coded as their position is not comparable with the lobes of other species (chs11–18:0, ch21:0).

Interdorsum

The interdorsum is continuous from the anterior to the posterior end of the shield (ch22:3), being laterally bordered by narrow furrows on both sides. From the anterior, the interdorsum widens first and then remains of nearly the same width until it tapers to its posterior end (Pl. 32C). The maximum width is about 1/14 the length of the valves (ch23:1). The interdorsum is almost flat (ch24:1) with no ornamentation in the middle part (ch25:1) but with small hump-like thickenings at the anterior (ch26:1, ch27:1, ch28:0) and posterior ends (Pl. 32A) (ch29:1, ch30:1, ch31:0, ch32:0, ch33:0).

Doublure

A doublure is present (ch34:1), being narrowest anteriorly and posteriorly (ch35:4), widening increasingly towards the medio-ventral rim (ch36:2) (Pl. 32B; Fig. 45B). The maximum width of the doublure is about one sixth the length of the valves (ch37:2). Approximating the dorsal rim, the doublure narrows rapidly to fade out into a membranous area. The doublure is entirely smooth (ch38:1, ch39:1, ch40:0, ch41:1, ch42:1).

Plate 30.   *Hesslandona curvispina* n. sp.

A: UB W 210. Close-up of the ventral part of the doublure (see Pl. 29C for orientation) of a specimen of 1280 μm in shield length, displaying irregularly arranged conical humps (arrows).

B: UB W 211. Image flipped horizontally. Close-up of the ventral part of the doublure (see Pl. 29C for orientation) of a specimen of 1290 μm in shield length, displaying irregularly arranged conical humps (arrows).

C: UB W 212. A specimen representing a late growth stage, about 1300 μm in length. An antero-lateral view of the body. Preservation is very poor. The fifth post-mandibular limb is very similar to the posterior limbs of *Hesslandona ventrospinata* and *Waldoria rotundata* (cf. Pl. 21B, E). Arrow "a" points to the terminal end of the preserved body, indicating that the body is fused to the shield up to at least the sixth post-mandibular segment (6). Note the keel-like elongated lobe between two humps at the rear of the sternum (arrow "b").

D: UB W 213. A specimen representing a late growth stage, about 1400 μm in length. A lateral view of the hind body, displaying the post-mandibular limbs from the lateral side. The specimen is partly covered with dirt. Note the difference in morphology of the third (3) and fourth (4) post-mandibular limbs. The body is missing the posterior to the fifth post-mandibular segment (arrow "a"). Note the prominent seta proximo-laterally on the second post-mandibular limb (arrow "b").

Inner lamella

The inner lamella is the little sclerotised part of the exoskeleton between the lateral extensions of the shield. It expands along the whole doublure and extends medially to the dorso-lateral side of the body (Pl. 32D). The antero-dorsal and postero-dorsal areas of the inner lamella extend into the membranous parts of the doublure. It is concave, fitting to the inner concave surface of the valves. The inner lamella frequently shows a wrinkled texture. There are no structures on the inner lamella.

Body and soft parts

The two specimens of this species at hand show poor soft part preservation (Pls. 32D, 33A). It is not the scope of this paper to describe all details. Only some remarks that are important for the phylogenetic analysis are made. The general body morphology is similar to that of *H. unisulcata* (see p. 16). The body proper is completely enveloped by the bivalved shield and includes at least seven segments (Pl. 33A). At least seven segments are dorsally fused to the shield (ch44:4) (Pl. 33A). Hence, the shield represents a cephalothoracic shield, including at least two limb-bearing trunk segments. The area of fusion of the body proper to the shield is very narrow, corresponding to approximately the width of the interdorsum (ch43:1). The most anterior part of the body is the hypostome/labrum complex (ch61:1). The antero-ventral part of the hypostome shows two circular bulbs (Pl. 32D) or intrusions (Pl. 33A) of about half the diameter of the basal part of the labrum, which might represent the median eye. Posterior to the hypostome/labrum complex the ventral surface of the body is a sclerotic plate, the sternum (ch62:1), which expands posteriorly just behind the insertions of the first post-mandibular limbs. It is somewhat domed with one pair of humps, the paragnaths (ch63:1) at the mandibular part (Pl. 32D). The ventral surface of the posterior body part shows segmental unpaired dome-like outgrowths (Pl. 33A). The mouth is located proximally on the posterior part of the hypostome basal to the protrusion of the labrum and anterior to the sternum (Pl. 33B, C). The mouth opening is more or less oval with a diameter of about 5 μm. It lies at the end of an atrium-like chamber. One flap of a width of about 5 μm covers the atrium from above; another two flaps, adorned with numerous fine hairs (Pl. 33C), of the same size cover the atrium from lateral. The antennula is not preserved (ch45:?, ch46:?) (Pl. 33A). The antenna and the mandible show an undivided limb stem and an endopod consisting of two portions (ch50:2, ch52:2, ch53:5, ch54:2, ch56:2, ch57:5); the condition in young stages is not known (ch51:?, ch55:?). The mandibular limb stem reaches far ventrally with a tipped elongate spatula-like lateral extension (Pl. 33B). A barrel-shaped endite is somewhat squeezed medio-distally between the limb stem and the

PLATE 31

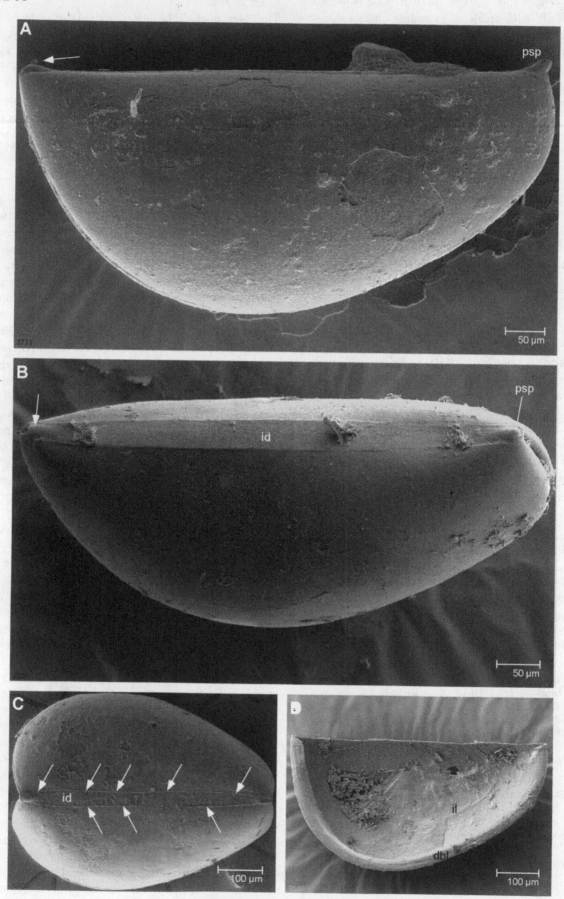

Table 28. Sample productivity of examined specimens of *Trapezilites minimus*.

Sample	Zone	Found at	Number of specimens
994	1–2	Kestad, Kinnekulle	5
5663	2a	Degerhamn, Öland	1
6404	1	West Kestad, between Haggården and Marieberg	8
6414	1	Gum (Kinnekulle), Falbygden–Billingen	1
6463	–	Degerhamn, Öland	1
6468	2a	Degerhamn, Öland	2
6473	–	Degerhamn, Öland	1
6735	1	NNE Backeborg (Kinnekulle), Falbygden–Billingen	5
6736	1	NNE Backeborg (Kinnekulle), Falbygden–Billingen	1
6750	1	Gum (Kinnekulle), Falbygden–Billingen	1

Table 29. Complete list of finds of *Hesslandona minima*, reported by different authors, with localities and horizons.

Locality	Horizon	Reference
Erratic boulders, northern Germany	Upper Cambrian?, Zone?	Kummerow 1924
Degerhamn, Öland	Upper Cambrian?, Zone?	Kummerow 1931
Merioneth Series, Great Britain	Upper Cambrian, Zone 1	Rushton 1978
Erratic boulders, northeastern Germany	Upper Cambrian, Zones 1, 2, 5?	Gründel 1981
Västergötland, southern Sweden	Upper Cambrian, Zones 1, 2	Müller 1964a (and this work)

distally inserting endopod. The post-mandibular limbs are of the same shape and morphology (ch58:1) and consist of an undivided limb stem, a medio-proximally inserting proximal endite (ch59:1) and two rami. The endopod is at least three-segmented (ch60:4) (Pl. 33A). The exopod is multi-annulated with inner setation.

Comparisons

Trapezilites minimus is similar to *Falites fala* and *H. unisulcata* in having one single prominent lobe (L₁) on the shield. It differs from these two species and from all other known phosphatocopines by the shape of the shield: the antero-posteriorly symmetrical outline of the shield plus the relatively large shield height compared with its length. *Trapezilites minimus* as well as *H. unisulcata* have a continuous interdorsum, which is absent in *Falites fala*. *Trapezilites minimus* is, again, similar to *H. unisulcata* in the hump-like thickenings on

---

Plate 31. *Hesslandona toreborgensis* n. sp.

A: Holotype (UB W 214). Image flipped horizontally. The specimen is about 680 μm in length. A lateral view of the right valve, displaying the dome-like protrusion of the anterior end of the interdorsum (arrow), the posterior spine (psp) is broken off distally.

B: UB W 215. The specimen is about 590 μm in length. A dorso-lateral view of a closed shield, displaying the interdorsum (id), its anterior hump (arrow) and the posterior spine (psp).

C: UB W 216. The specimen is about 590 μm in length. A dorsal view of an opened shield, displaying the interdorsum (id). Note the membranous furrow which separates the interdorsum from both valves (arrows).

D: UB W 217. The specimen is about 600 μm in length. A lateral view inside the right valve, displaying the doublure (dbl) and the partly preserved inner lamella (il).

both ends of the interdorsum, while other species have either a smooth interdorsum, or both interdorsal ends are drawn out into dome-like protrusions or distinct spines.

Ontogeny

The smallest stages that could be assigned to *Trapezilites minimus* are more than 400 μm in shield length. It is, however, not known how many limbs are developed in this stage (ch64:?). The material is not sufficient to describe the larval series of this species. The proportion of the length to the height of the valves does not change during development (Fig. 46).

Waldoria Gründel *in* Gründel & Buchholz, 1981

1981　*Waldoria* n. g. – Gründel *in* Gründel & Buchholz, p. 60.

1986a　*Waldoria* Gruendel, 1981 – Kempf, p. 749.

1986b　*Waldoria* Gruendel, 1981 – Kempf, p. 707.

1987　*Waldoria* Gruendel, 1981 – Kempf, p. 710.

1998　*Waldoria* Gründel *in* Gründel & Buchholz, 1981 – Hinz-Schallreuter, p. 118.

1998　*Waldoria* Gründel – Williams & Siveter, p. 31.

Derivation of name. – Combination of *Walcottella* Ulrich & Bassler, 1931 and *Bradoria* Matthew, 1899 (Gründel, 1981).

Type species. – *Waldoria buchholzi* Gründel *in* Gründel & Buchholz, 1981 by original designation (Gründel 1981).

Original diagnosis. – (Gründel 1981, translated). "Shield longer than high. Both valves of the same shape. Straight dorsal rim almost as long as maximum length or

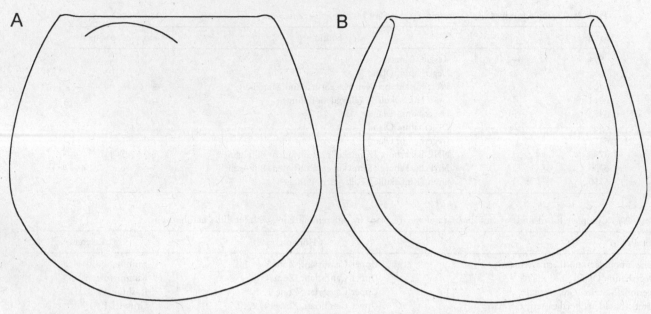

Fig. 45. Schematic drawing of the shield of *Trapezilites minimus* from outside lateral (A) and from inside lateral (B).

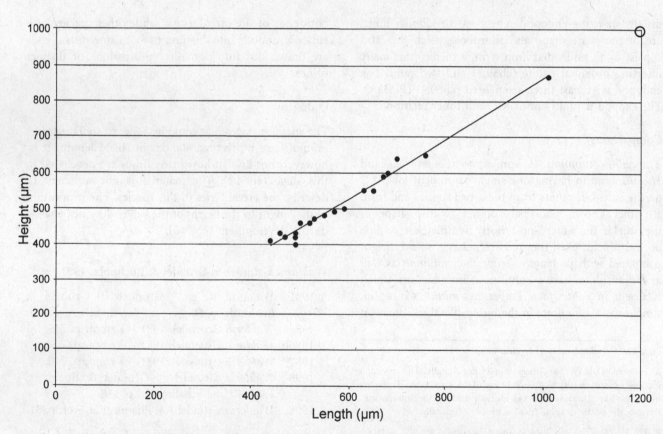

Fig. 46. Ontogeny of *Trapezilites minimus*: length versus height. Twenty-one specimens with measurable length and height plus the holotype are included. $\bigcirc$ = holotype. The trend line (the holotype is not considered in the calculation) has the polynomial function $y = 0x^2 + 0.8034x + 29.218$, y = height, x = length.

distinctly shorter than maximum length. Maximum height on antero-posterior midline [amplete] or slightly behind [post-plete]. Free shield margin strongly curved. Valves smooth apart from the lobe structure. L_1 and L_5 always present, L_2 and L_4 not always. L_3 and L_6 missing.

L_5 is the prominent structural element. It covers the main part of the valves and merges into the marginal areas. It is high and strongly convex with a rounded or spine-like pointed tip. Inner structures unknown. Hinge probably nullidont" (brackets added).

Emended diagnosis. – Valves of equal size, close tightly. Maximum length of valves between dorsal rim and midline. Surface with five lobes, which may occur more or less distinctly. Interdorsum continuous and completely smooth. Doublure rather wide.

Species referred to taxon. – The taxon *Waldoria* possibly encompasses two species, one of which, *Waldoria rotundata*, is considered in this work (cf. Appendix B):

Waldoria buchholzi Gründel *in* Gründel & Buchholz, 1981; *Waldoria rotundata* Gründel *in* Gründel & Buchholz, 1981).

Remarks. – *Waldoria* n. sp. 1 of Gründel (1981) is regarded as synonymous with *Waldoria rotundata* (see "Remarks" below).

Waldoria rotundata Gründel *in* Gründel & Buchholz, 1981

	1978	*Bradoria* sp. – Rushton, p. 275, pl. 26, figs. 13, 14 (see Williams & Siveter 1998).
p.	1978	*Falites fala* Müller, 1964 – Rushton, p. 276 (BGS BDA 1824 only), *non* BGS BDA 1820 (pl. 26, fig. 12), BDA 1844, BDA 1855, BDA 1863 (= *Falites fala*) (see Williams & Siveter 1998).
p.	1978	*Falites? minimus* (Kummerow, 1931) – Rushton, p. 277 (*partim*: BGS BDA 1771/1774, part and counterpart only), *non* BGS BDA 1167/1168, BDA 1452/1453 (= *Trapezilites minimus*) (see Williams & Siveter 1998).
	1978	*Walcottella* sp. – Rushton, p. 276, pl. 26, figs. 6, 7.
*	1981	*Waldoria rotundata* n. sp. – Gründel *in* Gründel & Buchholz, p. 61, pl. II, fig. 8; text-fig. 4.
.	1981	*Waldoria* n. sp. 1 – Gründel *in* Gründel & Buchholz, p. 62, pl. II, figs. 9, 10; text-fig. 5.
v.	1985a	*Waldoria* sp. – Müller & Walossek, figs. 4c, 6a, b (UB 770).
.	1986a	*Waldoria rotundata* Gruendel, 1981 – Kempf, p. 749.
.	1986b	*Waldoria rotundata* Gruendel, 1981 – Kempf, p. 514.
.	1987	*Waldoria rotundata* Gruendel, 1981 – Kempf, p. 710.
	1998	*Waldoria* cf. *rotundata* Gründel, 1981 – Williams & Siveter, p. 31, pl. 6, figs. 7, 8.
.	1998	*W. rotundata* – Williams & Siveter, p. 31.

Derivation of name. – Named, according to Gründel (1981), after the curved free margin of the valves, which is almost rounded like a segment of a circle.

Holotype. – A left valve illustrated by Gründel (1981) in his pl. II, fig. 8; length 2300 µm, height 1700 µm.

Remarks. – Gründel (1981) regarded *Waldoria rotundata* and *Waldoria* n. sp. 1 as two different species, because in the one form, *Waldoria rotundata*, lobe L_5 (cf. Fig. 11) is a blunt dome and in the other, *Waldoria* n. sp. 1, the lobe is drawn out into a distinct spine. Gründel had no certain information on the stratigraphic assignment of *Waldoria rotundata*, as he found his specimens in erratic boulders of the Isle of Rügen, Germany, without associated trilobite fauna, while he found *Waldoria* n. sp. 1 from Zone 1 of the Upper Cambrian. Both morphological forms appear in the material from the Upper Cambrian "Orsten" of Sweden in the *Agnostus pisiformis* Zone (Zone 1). They do not show additional differences apart from the different appearance of lobe L_5 to validate the discrimination of the two forms. We therefore regard both forms as representing the same species, namely *Waldoria rotundata*. Williams & Siveter (1998) described specimens from Warwickshire that were partly already known and described by Rushton (1978). Due to the new knowledge of the Upper Cambrian specimens from Sweden investigated herein, this material can also be assigned to *Waldoria rotundata*.

Type locality. – Erratic boulders from the Isle of Rügen, northeastern Germany.

Type horizon. – Probably Upper Cambrian, *Agnostus pisiformis* Zone (Zone 1).

Material examined. – One hundred and sixty specimens of different stages and from different areas, all from the *Agnostus pisiformis* Zone (Zone 1) (Table 30).

Dimensions. – Smallest specimen: valves about 300 µm long and about 140 µm high. Largest specimen: valves about 1600 µm long and about 1050 µm high.

Additional material. – Apart from the original material of Gründel (1981) and the material of this study, Rushton (1978) and Williams & Siveter (1998) have reported material from the Outwoods Shale Formation, *Olenus* (may be incorrect) and *Agnostus pisiformis* Zone, Merioneth Series, Merevale, Warwickshire, UK (see Table 31), which Rushton (1978), however, assigned to different species (see synonymy list above).

Distribution. – Zone 1 (*Agnostus pisiformis* Zone) of the Late Cambrian, northern middle Europe area.

Original diagnosis. – (Gründel 1981, translated). "Species of *Waldoria* with distinctly shortened dorsal rim, without spine-like outgrowth of L_5, missing individual L_5 and almost symmetrical longitudinal and cross-section at the area of maximum width, in the posterior area of the valves a pad originating from the dorsal margin."

PLATE 32

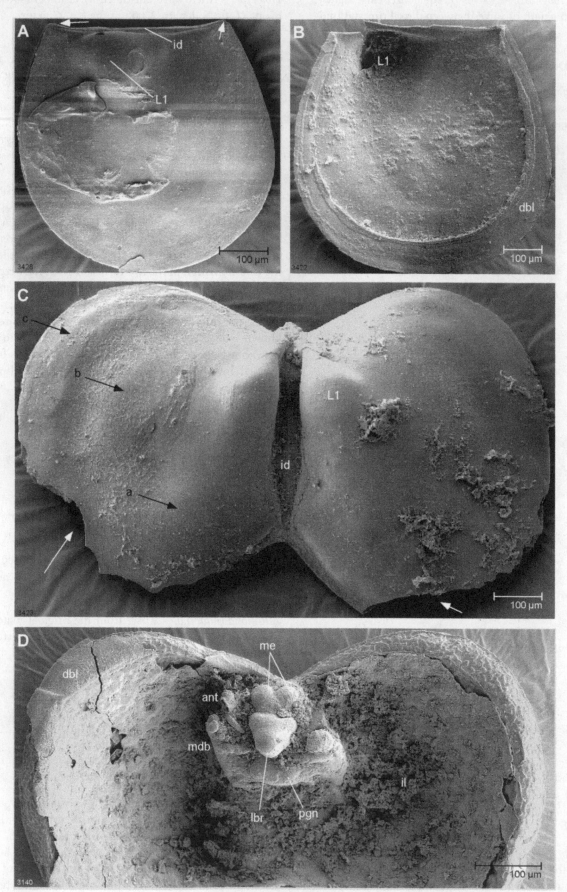

Gründel (1981) gave no diagnosis for his *Waldoria* n. sp. 1.

Emended diagnosis. – Species of *Waldoria* with prominent L_5, individually drawn out into a spine medially. Antero-dorsal lobe obliquely elongate, long axis pointing postero-ventrally. Posterior lobe small. Maximum length of valves ventral to dorsal rim. Valves slightly post-plete. Interdorsum narrow, without any structures. Doublure wide, maximum width postero-ventrally, about one sixth the length of valves. Valves close tightly.

Description. –

Shield (Fig. 47)

The bivalved shield (ch1:2) has a long and straight dorsal rim (ch2:2) (Pl. 34A, B; Fig. 47A). The maximum length of the head shield is dorsal to the dorso-ventral midline (ch3:3). The valves are of equal size and the right and left valves are symmetrical. The valves are slightly post-plete (ch4:3), about 1.5 times longer than high. The anterior free part of the shield margin, starting from the dorsal rim, curves antero-ventrally (antero-dorsal angle about 40 degrees) and equally postero-ventrally towards the ventral maximum (ch6:2). The ventral part of the margin is curved (ch7:2). Thereafter, the margin curves gently upwards – almost symmetrical to its anterior curve – towards the posterior (ch8:2, angle about 45 degrees) and curves back anteriorly – also almost symmetrical to its anterior curve – to meet the dorsal rim (ch5:3). The free margin of the shield forms almost the outline of a segment of a circle. The maximum width is slightly dorsal to the dorso-ventral midline. The margins of the valves are without outgrowths throughout (ch19:0, ch20:0), and the shield valves close tightly without any gaps (ch9:1).

Plate 32. *Trapezilites minimus* (Kummerow, 1931) Hinz-Schallreuter, 1993

A: UB W 218. Image flipped horizontally. A specimen representing a young growth stage, about 520 µm in shield length. A lateral view of the right valve. The arrows indicate the small hump-like thickenings at the anterior and posterior parts of the interdorsum (id).

B: UB W 219. Image flipped horizontally. A specimen representing an advanced growth stage, about 680 µm in shield length. A lateral view of the inside of the left valve, displaying the doublure (dbl) and the lobe (L1) from the inside. The inner lamella is missing in this specimen.

C: UB W 220. A specimen representing an advanced growth stage, about 670 µm in shield length. A dorsal view of the shield in the so-called "butterfly" position. The shield is posteriorly damaged (white arrows). The black arrows point to the additional nodes on the valves (see text pp. 109 and 111 for more details).

D: UB W 221. A specimen representing a young growth stage, about 490 µm in shield length. Ventral view. The posterior part of the shield is not shown. The median eye (me) is preserved as a pair of globe-like outgrowths (cf. Pl. 33A). The inner lamella (il) is partly covered with dirt.

The surface of the valves is smooth, but individually drawn out into five lobes, not distinctly separated from each other. The tiny lobe L_1 is located antero-dorsally (ch10:1) (Pl. 34A). Antero-laterally underneath L_1 is another, larger lobe, L_4 (ch13:1) (Pl. 34A). A further inconspicuous lobe, L_3, is located postero-dorsally (ch12:1) (Pl. 34A). A prominent lobe, L_2, is located centro-dorsally behind L_1 (ch11:1). This lobe merges ventrally into the large L_5, which is located centrally on the valves (ch14:1) (Pl. 34A). In some individuals, L_5 is drawn out into a prominent spine (ch14:1; ch16:1, ch17:1) (Pl. 35A, B), directed laterally (ch18:1) (Pl. 35A, B), called the central spine. The width of the valves with spines is about the same as the length of the valves. The dorsal and postero-dorsal area around the lobe or spine is slightly depressed. A sixth lobe, L_6, and other structures on the valves were not observed (ch15:0, ch21:0).

Interdorsum

The interdorsum is continuous from the anterior to the posterior (ch22:3), being bordered by narrow membranous furrows on both sides (Pl. 35C). The width of the interdorsum remains the same, apart from its anterior and posterior ends, where it tapers. The maximum width is about 1/30 the length of the valves (ch23:2). The interdorsum is flat in antero-posterior aspect (ch24:1), without lobes or any ornamentation (ch25:1) (Pl. 35C). Its anterior and posterior ends taper without forming any outgrowths (chs26–33:0).

Doublure

A doublure is present (ch34:1), being narrowest anteriorly (ch35:1), ventrally and posteriorly one third wider. The maximum width is postero-ventrally (ch36:4) (Pl. 35D; Fig. 47B). Approximating the dorsal rim, the doublures of both sides rapidly narrow strongly to fade out into a membranous area antero- and postero-dorsally. The maximum width of the doublure is about one seventh the length of the valves (ch37:2). On the antero-ventral and ventral margins of the doublure are small conical, dome-like outgrowths of about 5 µm in diameter irregularly arranged and curved to outside the domicilium, not arising from depressions but directly on the surface (ch38:1, ch39:2, ch40:1) (Pls. 35E, F, 36A). They are not developed on the anterior margin (Pl. 35G) or on the postero-ventral (Pl. 35H) to posterior margin. They are almost the same distance from each other. Additionally, pores with a diameter of less than 1 µm are located on the outer rim of the doublure (ch41:2) (Pl. 36A). The outer rim of the doublure shows a crest (Pl. 36A); no other structures on the doublure were observed (ch42:1).

Body

The body proper, which is completely enveloped by the bivalved shield, includes at least nine segments. Trunk

Table 30.   Sample productivity of examined specimens of *Waldoria rotundata*.

Sample	Zone	Found at	Number of specimens
6250	1	Near Ekedalen	1
6364	1	St. Stolan, Falbygden–Billingen	2
6404	1–2	West Kestad, between Haggården and Marieberg	3
6409	1	Gum (Kinnekulle), Falbygden–Billingen	6
6410	1	Gum (Kinnekulle), Falbygden–Billingen	6
6411	1	Gum (Kinnekulle), Falbygden–Billingen	1
6413	1	Gum (Kinnekulle), Falbygden–Billingen	1
6414	1	Gum (Kinnekulle), Falbygden–Billingen	5
6416	1	Gum (Kinnekulle), Falbygden–Billingen	3
6417	1	Gum (Kinnekulle), Falbygden–Billingen	20
6455	1	Degerhamn, Öland	1
6470	1	Degerhamn, Öland	1
6711	1	Stenstorp–Dala	1
6729	1	NNE Backeborg (Kinnekulle), Falbygden–Billingen	1
6730	1	NNE Backeborg (Kinnekulle), Falbygden–Billingen	4
6731	1	NNE Backeborg (Kinnekulle), Falbygden–Billingen	1
6732	1	NNE Backeborg (Kinnekulle), Falbygden–Billingen	1
6733	1	NNE Backeborg (Kinnekulle), Falbygden–Billingen	1
6734	1	NNE Backeborg (Kinnekulle), Falbygden–Billingen	7
6735	1	NNE Backeborg (Kinnekulle), Falbygden–Billingen	2
6743	1	Blomberg (Kinnekulle), Falbygden–Billingen	1
6749	1	Gum (Kinnekulle), Falbygden–Billingen	3
6750	1	Gum (Kinnekulle), Falbygden–Billingen	7
6751	1	Gum (Kinnekulle), Falbygden–Billingen	2
6755	1	Gum (Kinnekulle), Falbygden–Billingen	9
6758	1	Gum (Kinnekulle), Falbygden–Billingen	1
6760	1	Gum (Kinnekulle), Falbygden–Billingen	12
6761	1	Gum (Kinnekulle), Falbygden–Billingen	14
6762	1	Gum (Kinnekulle), Falbygden–Billingen	2
6763	1	Gum (Kinnekulle), Falbygden–Billingen	2
6764	1	Gum (Kinnekulle), Falbygden–Billingen	1
6765	1	Gum (Kinnekulle), Falbygden–Billingen	2
6769	1	Gum (Kinnekulle), Falbygden–Billingen	1
6771	1	Gum (Kinnekulle), Falbygden–Billingen	3
6772	1	Gum (Kinnekulle), Falbygden–Billingen	2
6774	1	Gum (Kinnekulle), Falbygden–Billingen	5
6776	1	Gum (Kinnekulle), Falbygden–Billingen	5
6777	1	Gum (Kinnekulle), Falbygden–Billingen	1
6779	1	Gum (Kinnekulle), Falbygden–Billingen	1
6780	1	Gum (Kinnekulle), Falbygden–Billingen	3
6781	1	Gum (Kinnekulle), Falbygden–Billingen	1
6783	1	Gum (Kinnekulle), Falbygden–Billingen	9
6784	1	Gum (Kinnekulle), Falbygden–Billingen	5

Table 31.   Complete list of finds of *Waldoria rotundata*, reported by different authors, with localities and horizons.

Locality	Horizon	Reference
Outwoods Shale, Warwickshire, UK	Upper Cambrian, Zone 1	Rushton 1978
Mecklenburg-Vorpommern, Germany	Upper Cambrian, Zone 1	Gründel 1981
Outwoods Shale, Warwickshire, UK	Upper Cambrian, Zones 1, 2	Williams & Siveter 1998
Västergötland, southern Sweden	Upper Cambrian, Zone 1	This work

segmentation of the preserved portion is retained in the insertions of the limbs and distinct segment boundaries or sternites are weakly defined. At least nine segments are dorsally fused to the shield (ch44:6) (Pls. 36D, 37A). Hence, the shield represents a cephalothoracic shield, including at least four limb-bearing trunk segments. The area of fusion of the body proper to the shield is very narrow, being not much wider than the width of the interdorsum (ch43:1) (Pl. 36D). The maximum width of the body proper is located at the mandibles (ch43:1) (Pl. 36B). Thereafter, the dorsal connection of the body proper, which is about trapezoidal in cross-section, with the inner lamella, becomes narrower posteriorly and fades out in the last possibly one sixth to one eighth.

A

B

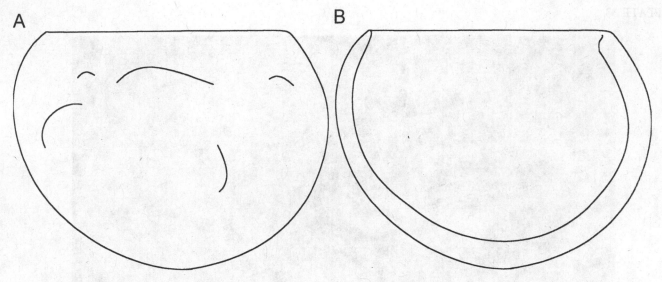

Fig. 47. Schematic drawing of the shield of *Waldoria rotundata* from outside lateral (A) and from inside lateral (B)

The posterior part of the hind body extends free from the shield into the domicilium. The limbs are arranged in a more or less regular row lateral to the body, becoming very close together at the hind body (Pl. 36B). The limbs are very small compared with the size of the shield (Pl. 37A). The anterior part of the body proper is the hypostome/labrum complex (ch61:1). The antennulae insert antero-laterally in a circular joint area (Pl. 36B). The antennae insert postero-lateral to the hypostome in a more spindle-shaped joint area, the posterior edge touching the labral part of the hypostome/labrum complex. The median eye is located antero-distally on the hypostome (Pl. 36B). The posterior end of the hypostome is drawn out distally into a lobe-like protrusion, the labrum (ch61:1) (Pl. 36B). Posterior to the hypostome/labrum complex, the ventral surface between the first to third pairs of post-antennular limbs is marked by a sclerotic plate, the sternum (ch62:1) (Pls. 36B, 37A). The sternum is slightly domed and bears one pair of humps anteriorly, the paragnaths (ch63:1), between the mandibles. The hind body is probably very soft (Pl. 36D).

A paired outgrowth of the very rear of the body is regarded as furcal rami (ch47:2). The furca is composed of a pair of flattened paddle-shaped plates of about 100 µm in length with the pointed end terminally (Pl. 36D). The furcal rami insert in a 45 degree angle to each other, i.e. they look roof-like from dorsal. Their margins are armed with a row of setae on each side plus one terminal seta, the setae being about as long as the furcal rami (ch48:2, ch49:4).

Inner lamella

The inner lamella extends along the whole doublure and medially to the dorso-lateral side of the body. The inner lamella is weakly sclerotised and frequently shows a wrinkled texture. No scar of a possibly present closing muscle is preserved (Pl. 37A).

Soft parts

The limbs are very similar to those of *H. ventrospinata*. Only some notes are made that are important for the phylogenetic analysis. The antennula is similar to that of *H. unisulcata*. It is a short appendage of about 100 µm in length in specimens with 1000 µm shield length and 300 µm head length (ch46:1). It inserts on the antero-lateral margin of the hypostome. It is uniramus and consists of about 13 irregular annuli in an individual of 1000 µm in length (ch45:1) (Pl. 37B). The last annulus bears three terminal setae, the longest is about 50 µm. The penultimate annulus bears one seta medio-terminally, about 30 µm long. The setae are adorned with fine setulae.

The antenna and the mandible consist of an undivided limb stem at least in larger specimens, a two-divided endopod and a multi-annulated exopod (ch50:2, ch51:?, ch52:2, ch53:5, ch54:2, ch55:?, ch56:2, ch57:5). The mandible has an enditic protrusion located medially between the limb stem and the endopod, recognised as the remains of the basipod (see p. 27). The post-mandibular limbs are similar to those of *H. ventrospinata*. They are separated into two groups of different shape (Pl. 37A). The first three post-mandibular pairs of limbs are serially similar to each other (ch58:1), consisting of a basipod, a setae-bearing proximal endite medio-proximally (ch59:1), a three-part endopod (ch60:4) and an annulated exopod (Pl. 36C). The endopod slightly projects medially. The first two portions are conical in posterior view with a blunt, setose median tip. The third portion is small and hump-like and extends into a single distal seta (Pl. 37C). The exopods are triangular plates with marginal setation

PLATE 33

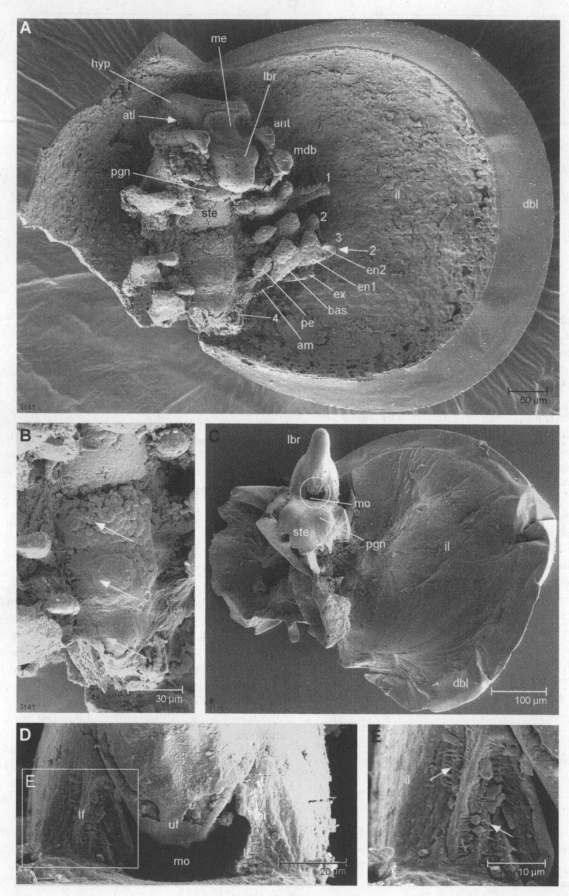

(Pl. 37C); their tip is annulated – very much like the exopods in *H. unisulcata* (cf. Pls. 11F, 12G), but the whole limb is distinctly more stout. The fourth to sixth post-mandibular pairs of limbs consist of a single limb stem and two rami. The endopod is located medio-distally on the limb stem. The exopod forms more or less the distal extension of the limb stem, without being distinctly separated from the limb stem. It is a flat plate of triangular shape with a pointed terminal tip and with marginal setation.

Comparisons

Waldoria rotundata is similar to *Veldotron bratteforsa* and *H. angustata* n. sp. in the smooth anterior and posterior ends of the interdorsum. It differs from both species in the general shape of the shield and the doublure. *Waldoria rotundata* is similar to *Veldotron bratteforsa*, *H. ventrospinata* and species of *Cyclotron* in the high number of lobes on the valves. *Waldoria rotundata* and *H. unisulcata* are very similar in the outline of the shield and the outline of the doublure, but both species differ in the occurrence of lobes, the outgrowths on the doublure, the presence of anterior and posterior interdorsal thickenings in *H. unisulcata*, and in limb morphology. The limb morphology of *Waldoria rotundata* and *H. ventrospinata* is very similar, but both species differ in several aspects: *H. ventrospinata* shows postero-ventral outgrowths of its shield margin, it has long interdorsal spines, and its shield lobes are much more conspicuous and distinct. *Waldoria rotundata* differs from all other investigated species of *Hesslandona* in the absence of interdorsal outgrowths, from *Vestrogothia spinata* and *Falites fala* in the presence of an interdorsum.

Ontogeny

During ontogeny, the proportion of the length to the height of the valves of *Waldoria rotundata* remains the same (Fig. 36). The observed range of shield lengths and heights does not indicate distinct clusters of particular growth stages (Fig. 36). The smallest stages that could be measured are more than 600 µm in shield length. It is, however, not known how many limbs are developed in this stage (ch64:?). The material is not sufficient to describe the larval series of this species.

Veldotron Gründel *in* Gründel & Buchholz, 1981

1981   *Veldotron* n. g. – Gründel *in* Gründel & Buchholz, p. 66.
1986a   *Veldotron* Gruendel, 1981 – Kempf, p. 745.
1986b   *Veldotron* Gruendel, 1981 – Kempf, p. 707.
1987   *Veldotron* Gruendel, 1981 – Kempf, p. 710.
1993b   *Veldotron* Gründel & Buchholz, 1981 – Hinz-Schallreuter, p. 334.
1993c   *Veldotron* Gründel *in* Gründel & Buchholz, 1981 – Hinz-Schallreuter, pp. 395, 402, 405, 408.
1998   *Veldotron* Gründel *in* Gründel & Buchholz, 1981 – Hinz-Schallreuter, p. 118.
1998   *Veldotron* Gründel – Williams & Siveter, p. 34.

Derivation of name. – The word "Veldotron" is an arbitrary formation (Gründel 1981).

Type species. – *Veldotron kutscheri* Gründel *in* Gründel & Buchholz, 1981 by original designation (= *Vestrogothia bratteforsa* Müller 1964).

Original diagnosis. – (Gründel 1981, translated). "Dorsal rim almost corresponding to the maximum length. Lobe formation as in *Cyclotron* [L_1, L_2 and L_3 lie close to the dorsal rim, the first two close to the antero-dorsal edge, L_3 close to the postero-dorsal edge; L_4 always (?) present, L_5 only occasionally becoming inconspicuous, L_6 present or absent]. Free margin in the anterior ventral part convex, concave postero-ventral. Posterior end widely cut, bordered by a straight, slightly to posterior and ventral sloping line."

Emended diagnosis. – Valves of equal size, leave gap postero-ventrally. Maximum length of valves between dorsal rim and midline. Surface of valves with six lobes, three of which in a line close to the dorsal rim, two anteriorly, another one in the last third of the total length; fourth lobe anterior, fifth centrally on the valve, sixth directly posterior to the fifth, both merging into each other. Interdorsum smooth throughout. Doublure wide postero-ventrally.

Plate 33. *Trapezilites minimus* (Kummerow, 1931) Hinz-Schallreuter, 1993

A: UB W 222. A specimen representing a young growth stage, about 500 µm in shield length. A ventral view of an opened shield. The median eye (me) is preserved as a pair of circular depressions (cf. Pl. 32D). The inner lamella (il) is coarsely preserved. The insertion area of the antennula (atl) is exposed. The hind body is missing, the existence of the fourth post-mandibular limb (4) is indicated by its insertion scar.

B: Close-up of the same specimen as in Pl. 33A, displaying the unpaired humps on either sternite (arrows) behind the sternum (ste), probably preservational artefacts.

C: UB W 223. A specimen representing a young growth stage, about 490 µm in shield length. A ventral view of the opened shield, displaying an internal view into the mouth area (mo). The area indicated by a circle is shown magnified in Pl. 33D from a slightly different perspective.

D: Close-up of the mouth opening illustrated in Pl. 33C, displaying the trapezoid upper flap (uf) covering the upper part and the two lateral flaps (lf) covering the right and left parts of the mouth opening. The rectangle marks the area magnified in Pl. 33E.

E: Close-up of the right lateral flap illustrated in Pl. 33C, displaying the strong fold of the surface and the fine hairs (arrows).

PLATE 34

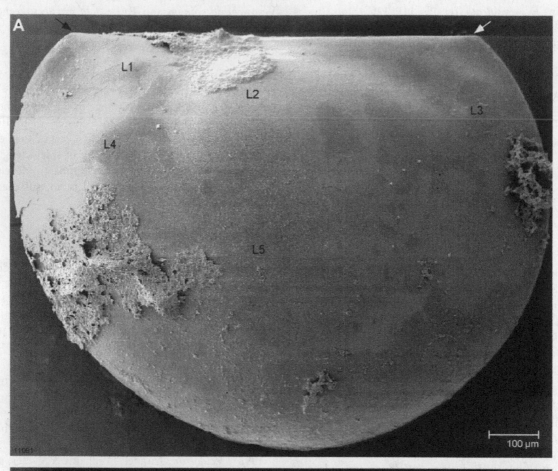

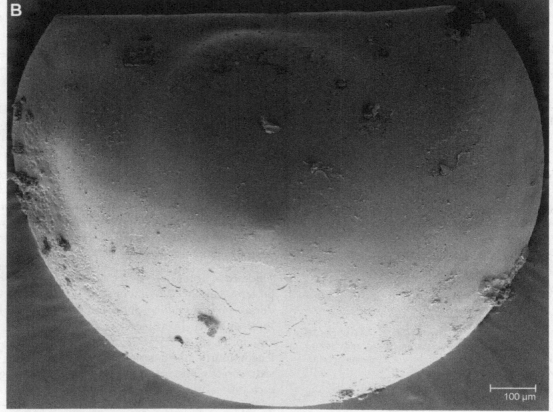

Species referred to taxon. – Two species are referred to the taxon *Veldotron*: *Vestrogothia bratteforsa* Müller, 1964 (a senior synonym of *Veldotron kutscheri* Gründel *in* Gründel & Buchholz, 1981) *Veldotron rushtoni* Williams & Siveter, 1998 (see Appendix B)

Veldotron bratteforsa (Müller, 1964) Hinz-Schallreuter, 1993

v* 1964a *Vestrogothia bratteforsa* n. sp. – Müller, p. 34, pl. 3, figs. 1, 2.

. 1965 *Vestrogothia bratteforsa* Müller – Adamczak, p. 29.

. 1972 *Vestrogothia bratteforsa* Mueller – Taylor & Rushton, p. 18.

. 1974 *F. bratteforsa* – Martinsson, p. 208 (sic!).

. 1981 *Veldotron kutscheri* n. sp. – Gründel *in* Gründel & Buchholz, p. 66, pl. III, figs. 11, 12, 15.

. 1986 *Vestrogothia bratteforsa* – Huo *et al.*, fig. 3-1 (cop. Müller 1964, pl. 3, fig. 2b).

. 1986a *Veldotron kutscheri* Gruendel, 1981 – Kempf, p. 745.

. 1986a *Vestrogothia bratteforsa* Mueller, 1964 – Kempf, p. 747.

. 1986b *Vestrogothia bratteforsa* Mueller, 1964 – Kempf, p. 101.

. 1986b *Veldotron kutscheri* Gruendel, 1981 – Kempf, p. 316.

. 1987 *Vestrogothia bratteforsa* Mueller, 1964 – Kempf, p. 436.

. 1987 *Veldotron kutscheri* Gruendel, 1981 – Kempf, p. 710.

. 1987 *Vestrogothia bratteforsa* Müller – Tong, p. 433.

. 1993b *Veldotron kutscheri* Gründel & Buchholz, 1982 – Hinz-Schallreuter, p. 334, fig. 1C.

. 1993c *Vestrogothia bratteforsa* Müller, 1964 – Hinz-Schallreuter, p. 405 (synonymised).

. 1998 *Veldotron bratteforsa* (Müller, 1964) – Hinz-Schallreuter, p. 116.

. 1998 *V. (Veldotron) bratteforsa* – Williams & Siveter, pp. 34, 35.

. 1993c *Veldotron bratteforsa* (Müller, 1964) – Hinz-Schallreuter, pp. 393, 405, figs. 9.1, 9.2.

Plate 34. *Waldoria rotundata* Gründel *in* Gründel & Buchholz, 1981

A: UB W 224. A specimen representing a late growth stage, about 1180 μm in length. A lateral view of the left valve. The arrows point to the smooth anterior and posterior dorsal ends of the shield.

B: UB W 225. Image flipped horizontally. A specimen representing a late growth stage, about 1180 μm in length. A lateral view of the right valve.

Derivation of name. – Not noted by Müller (1964) but apparently after the city of Brattefors, near where the holotype was found.

Holotype. – A complete shield illustrated by Müller (1964a) in his pl. 3, fig. 2; length 980 μm, height 5.. μm.

Remarks. – Gründel (1981) in his description of *Veldotron kutscheri* pointed out that *Vestrogothia bratteforsa* may also belong to *Veldotron* Gründel 1981. Hinz-Schallreuter (1993c) re-investigated *Vestrogothia bratteforsa* Müller, 1964 and found many similarities with *Veldotron kutscheri*. She argued that Müller (1964a) had juveniles with incomplete lobation and synonymised *Veldotron kutscheri* with *Vestrogothia bratteforsa*. Because there are no features to discriminate both forms we herein follow Hinz-Schallreuter (1993c) in regarding *Veldotron kutscheri* Gründel, 1981 as the junior synonym of *Vestrogothia bratteforsa* Müller, 1964.

Type locality. – Brattefors, Kinnekulle, Västergötland, Sweden (Müller 1964a).

Type horizon. – Upper Cambrian, Zone 2.

Material examined. – Eleven specimens of different ages and from different areas, the *Agnostus pisiformis* Zone (Zone 1) and the *Olenus gibbosus* Zone (Subzone 2) of the Upper Cambrian of Sweden (Table 32).

Dimensions. – Smallest specimen: valves about 8.. μm long and about 490 μm high.
Largest specimen: valves about 2300 μm long and about 1240 μm high.

Additional material. – Apart from the material reported by Müller (1964a) and the material studied herein, Gründel (1981) found nine left and eight right valves of up to more than 2 mm in length in glacial erratics from Mukran, Isle of Rügen, Mecklenburg-Vorpommern, Germany of unsure stratigraphic assignment. A piece of rock with more than 70 specimens (mostly valves) from glacial erratics from Damsdorf, collection of Frank Rudolph, Wankendorf, Schleswig-Holstein, Germany, is reported by Hinz-Schallreuter (1993c) of unsure stratigraphic assignment (Table 33).

Distribution. – Glacial erratics of Bralitz, Oderberg, Brandenburg, Germany, and Damsdorf, Schleswig-Holstein, Germany, and Zones 1 and 2 of the Upper Cambrian of Sweden.

Original diagnosis. – (Müller 1964a, translated). 'Dorsal rim almost as long as maximum length of valves. Valves slender, about double as long as high. Valves with five distinct lobes. L_1, L_2, and L_3 located near dorsal rim from anterior to posterior. L_1 and L_2 in the first half, L_3 in the second half of the valves. N_4 is situated below N_1 and is smaller. N_5 distinctly delimited with distinct N_6 on its posterior slope. Posterior end of valves flattened.

PLATE 35

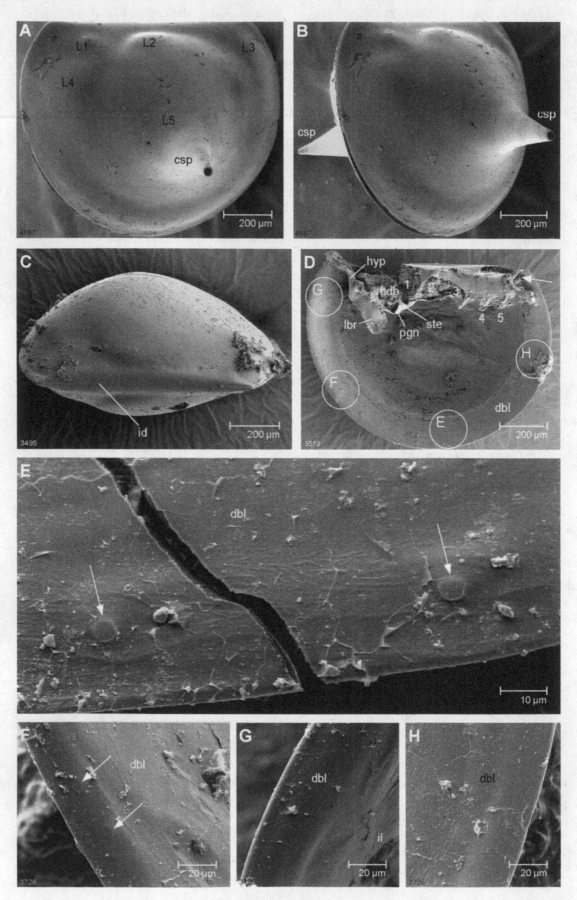

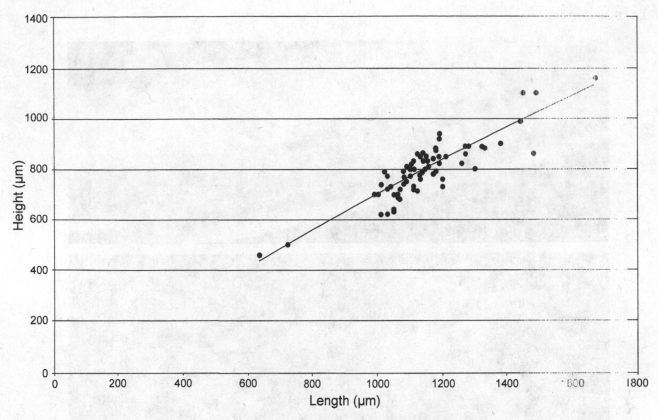

Fig. 48. Ontogeny of *Waldoria rotundata*: length versus height. Eighty-seven specimens with measurable length and height are included. T. e rend line has the polynomial function $y = 0x^2 + 0.8206x - 51.487$, y=height, x=length. Mean proportion of length to height about 1.5.

Posterior border straight and steeply descending ent- rally and posteriorly. Valves right/left symm cal, leaving a gap posteriorly."

Emended diagnosis. – Dorsal rim almost as long a: ax- imum length of valves. Valves slender, about dou e as long as high. Valves with six distinct lobes. L_1, L and L_3 located near dorsal rim from anterior to poster . L_1 and L_2 in the first half, L_3 in the second half the valves. N_4 is situated below N_1 and is smaller. dis- tinctly delimited with distinct N_6 on its posterior pe. Posterior end of valves flattened. Posterior der straight and steeply descending ventrally and poste rly. Valves right/left symmetrical, leaving gap posteric .

Description. – The material provided different g wth stages of *Veldotron bratteforsa* (Müller, 54). Preservation is almost completely restricted to is ted valves of shields.

Shield (Fig. 49)

The bivalved shield (ch1:2) has a long and straight rsal rim (ch2:2) (Pl. 38A; Fig. 49A). The maximum ler h of the head shield is slightly ventral to the dorsal of the head shield (ch3:3). The valves are of equ size and the right and left valves are symmetrical. The ax- imum height of the valves is anterior to the n line (pre-plete) (ch4:1), about 1.9 times longer than h in

---

Plate 35.   *Waldoria rotundata* Gründel *in* Gründel & Buchholz, 1981

A: UB W 226. A specimen representing a late growth stage, about 1070 µm in length. A lateral view of the left valve, displaying the lobes (L1–L5) and the central spine (see also Pl. 35B for a different view of the same specimen).

B: The same specimen as in Pl. 35A. Antero-lateral view.

C: UB W 227. A specimen representing a late growth stage, about 1050 µm in length. A dorsal view of a closed shield.

D: UB W 228. A specimen representing a late growth stage, about 1010 µm in length. A lateral view inside the right valve, displaying a partly preserved body, the terminal end of which is not preserved (arrow). The circles indicate areas from which a close-up view is documented in Pl. 35E–H, but from another specimen.

E–H: UB W 229. A specimen representing a late growth stage, about 1100 µm in length. Views of the doublure as indicated in Pl. 35D.

E: Close-up of the median part of the doublure, displaying two conical outgrowths (arrows).

F: Close-up of the antero-ventral part of the doublure, displaying the scars of two conical outgrowths (arrows).

G: Close-up of the antero-dorsal part of the doublure, displaying a smooth surface.

H: Close-up of the posterior part of the doublure, displaying a smooth surface.

PLATE 36

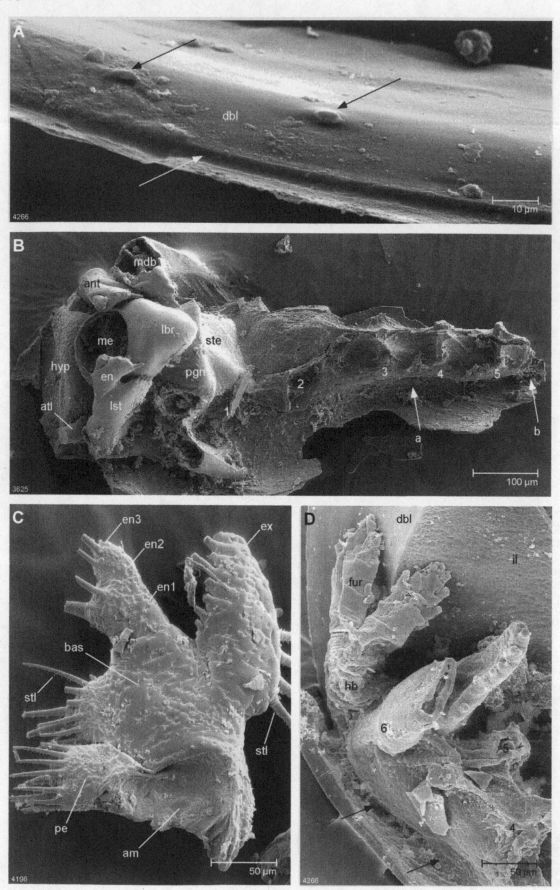

Table 32. Sample productivity of examined specimens of *Veldotron bratteforsa* including the holotype locationing (sample 955). This sample is not studied herein.

Sample	Zone	Found at	Number of specimens
955	2	Brattefors, Kinnekulle	1 (holotype)
6278	–	Between Smedsgården and Stutagården, Falbygden–Billingen	1
6404	1–2a	West Kestad, between Haggården and Marieberg	5
6470	2a	Degerhamn, Öland	2
6796	2	Between Haggården and Marieberg, Kinnekulle	1

Table 33. Complete list of records of *Veldotron bratteforsa*, reported by different authors, with localities and horizons.

Locality	Horizon	Reference
Västergötland, southern Sweden	Upper Cambrian, Zones 1, 2	Müller 1964a and this work
Mukran, Isle of Rügen, Germany	Upper Cambrian?, Zone?	Gründel 1981
Damsdorf, Schleswig-Holstein, Germany	Upper Cambrian?, Zone?	Hinz-Schallreuter 199?

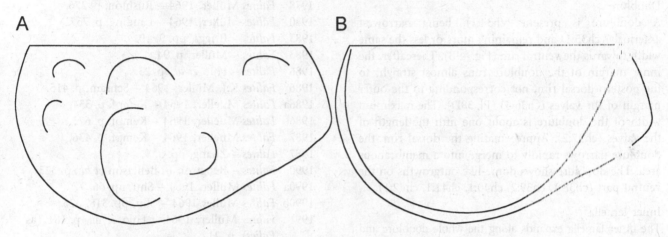

Fig. 49. Schematic drawing of the shield of *Veldotron bratteforsa* from outside lateral (A) and from inside lateral (B).

every stage. The free anterior part of the shield margin, starting from the dorsal rim, curves slightly anteriorly (ch6:2) (antero-dorsal angle about 85 degrees) and after

reaching the most anterior part of the shield, curves back postero-ventrally towards the ventral maximum (Pl. 38A). The latero-ventral outline is straight (ch1:1). Right behind the ventral maximum, the margin forks immediately upwards in a slightly excavated rim towards the most posterior part of the shield, forking anteriorly rather curved (ch8:2) to meet the postero-dorsal rim (postero-dorsal angle about 80 degrees). The margin curves back posteriorly more strongly than anteriorly (ch5:2) (Pl. 38A). The maximum width of the shield is slightly dorsal to the midline. The margins of the right and left valves are without outgrowths throughout (ch19:0, ch20:0). The shield valves leave a gap posteriorly (ch9:2) (Pl. 38B). The surface of the valves is smooth, but with six dome-like lobes, L_1–L_6 (Pl. 38A), in younger specimens lobes L_4–L_6 are much less distinct (Pl. 38C), lobes L_1–L_3 are located close to the dorsal rim. The most anterior lobe is L_1 (ch10:1), a second, slightly larger one (L_2) is located posterior to it just before the antero-posterior midline (ch11:1), a third, more inconspicuous one (L_3) is located in the last third of the valves (ch12:1). L_4 is located antero-dorsally of L_1 and is distinctly smaller (ch13:1). The medio-ventral part of the valve is more or

Plate 36. *Waldoria rotundata* Gründel *in* Gründel & Buchholz, 1981

A: UB W 230. A specimen representing a late growth stage, about 1190 μm in length (see also Pl. 36D). Close-up of the ventral part of the doublure, displaying conical outgrowths (black arrows) and a crest along the outer rim of the doublure (white arrow).

B: UB W 231. A specimen representing a large growth stage, at least 1100 μm in length. Ventro-lateral view. The shield and post-mandibular limbs are almost completely missing. Note that the body is fused to the shield along a very narrow fusion area (arrow "a"; cf. Pl. 36D). At least five post-mandibular segments are fused to the shield (arrow "b").

C: UB W 232. A post-mandibular limb of the anterior series of the specimen illustrated in Pl. 37A. Note the setulae (stl) on the median and exopodal setae. The lateral margin of the basipod (bas) extends into a curved spatula (arrow), comparable with the same structures documented in *Hesslandona unisulcata* (cf. Pl. 9C).

D: The same specimen as in Pl. 36A. Close-up of the posterior body, displaying the furcal rami (fur). The body is fused to the shield along a very narrow fusion area (arrows; cf. Pl. 36B). The hind body that is free from the shield (hb) apparently has very soft cuticle.

less well domed to form L_5 (ch14:1), which extends into a sixth lobe L_6 (ch15:1) posteriorly. All lobes are oval to circular in outline. Other structures are not developed on the valves (ch16:0, ch17:0, ch18:0, ch21:0).

Interdorsum

The interdorsum is continuous from the anterior to the posterior end of the shield (ch22:3), being laterally bordered by narrow furrows on both sides. From the anterior, the interdorsum widens first, and then remains of nearly the same width until it tapers to its posterior end. The maximum width is about 1/25 the length of the valves (ch23:1). The interdorsum is flat (ch24:0) with no ornamentation at all (ch25:1, ch26:0, ch27:0, ch28:0, ch29:0, ch30:0, ch31:0, ch32:0, ch33:0).

Doublure

A doublure is present (ch34:1), being narrowest anteriorly (ch35:1) and remaining more or less the same width towards the ventral rim (Fig. 49B). Thereafter, the inner margin of the doublure runs almost straight to the postero-dorsal rim, not corresponding to the outer margin of the valves (ch36:4) (Pl. 38D). The maximum width of the doublure is about one fifth the length of the valves (ch37:2). Approximating the dorsal rim, the doublure narrows rapidly to merge into a membranous area. The doublure shows dome-like outgrowths on the ventral part (ch38:1, ch39:2, ch40:1, ch41:1, ch42:1).

Inner lamella

The inner lamella expands along the whole doublure and extends medially to the dorso-lateral side of the body. The antero-dorsal and postero-dorsal areas of the inner lamella extend into the membranous parts of the doublure. It is concave, fitting to the inner concave surface of the valves. The inner lamella frequently shows a wrinkled texture. There are no structures on the inner lamella.

Soft parts

Soft parts are not known for this species (chs43–64:?).

Comparisons

Veldotron bratteforsa is similar to *Waldoria rotundata* and *Cyclotron lapworthi* in the high number of lobes on the valves and in the completely smooth interdorsum. It differs from these and all other species by the outline of the shield and the shape of the doublure. *Veldotron bratteforsa* is similar to *H. necopina* and *H. curvispina* n. sp. in showing dome-like structures on the ventral part of the doublure, but both species of the taxon *Hesslandona* have long spiny outgrowths on the anterior and posterior ends of their interdorsum.

Ontogeny

The smallest stages that could be assigned to *Veldotron bratteforsa* are more than 800 μm in shield length. The second largest specimen is 1070 μm long and essentially smaller than the largest one. The material is not sufficient to present a diagram of the ontogeny or even describe the larval series of this species. The proportion of the length to the height of the valves does not change during development. It remains about 1.85 throughout the known part of ontogeny.

Falites Müller, 1964

1964a *Falites* n. g. – Müller, p. 25.
1965   *Falites* Müller – Adamczak, p. 32.
1972   *Falites* Mueller – Taylor & Rushton, pp. 13, 18, 25.
1974   *Falites* Müller, 1964 – Kozur, p. 826.
1978   *Falites* Müller, 1964 – Rushton, p. 276.
1980   *Falites* Müller, 1964 – Landing, p. 757.
1983   *Falites* – Briggs, pp. 9, 10.
1983   *Falites* – Müller, p. 94.
1986   *Falites* – Huo et al., p. 23.
1986   *Falites* K.J. Müller, 1964 – Schram, p. 415.
1986a *Falites* Mueller, 1964 – Kempf, p. 354. ·
1986b *Falites* Mueller, 1964 – Kempf, p. 673.
1987   *Falites* Mueller, 1964 – Kempf, p. 436.
1987   *Falites* – Zhang, pp. 5, 9.
1990   *Falites* – Bengtson in Bengtson et al., p. 323.
1990a *Falites* Müller, 1964 – Shu, pp. 66, 77.
1990b *Falites* Müller 1964 – Shu, pp. 318, 323.
1991   *Falites* Müller, 1964 – Huo et al., p. 181 (as *Falies*), p. 212.
1993c *Falites* Müller, 1964 – Hinz-Schallreuter, pp. 386, 395, 399–403.
1995   *Falites* – Siveter et al., p. 416.
1996a *Falites* Müller, 1964 – Hinz-Schallreuter, p. 85.
1996b *Falites* Müller, 1964 – Hinz-Schallreuter, pp. 89, 91.
1998   *Falites* Müller, 1964 – Hinz-Schallreuter, pp. 104, 106–108, 112, 115.

Derivation of name. – After the province Falbygden in southern Sweden, where most of the material of Müller (1964a) was collected.

Plate 37.   *Waldoria rotundata* Gründel *in* Gründel & Buchholz, 1981

A: UB W 232. A specimen representing a late growth stage, about 1190 μm in length (see also Pl. 36A, D). A ventral view of the body, the left valve is damaged and displaced. Note the size of the shield valve in comparison with the size of the limbs. The hind body posterior to the sixth post-mandibular segment is free from the shield, possibly representing at least one additional segment (7?; cf. Pl. 36D). The rectangle marks the area of the image, displaying the antennula, which is magnified in Pl. 37B.

B: Close-up of the antennula illustrated in Pl. 37A. Two setulated setae originating from the antennula are located on the specimen (arrows) due to preservation or processing.

PLATE 37

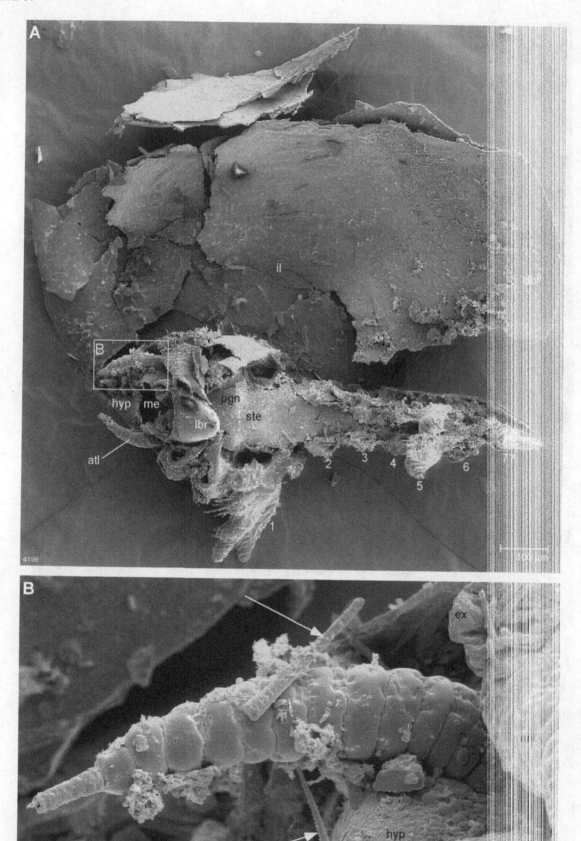

PLATE 38

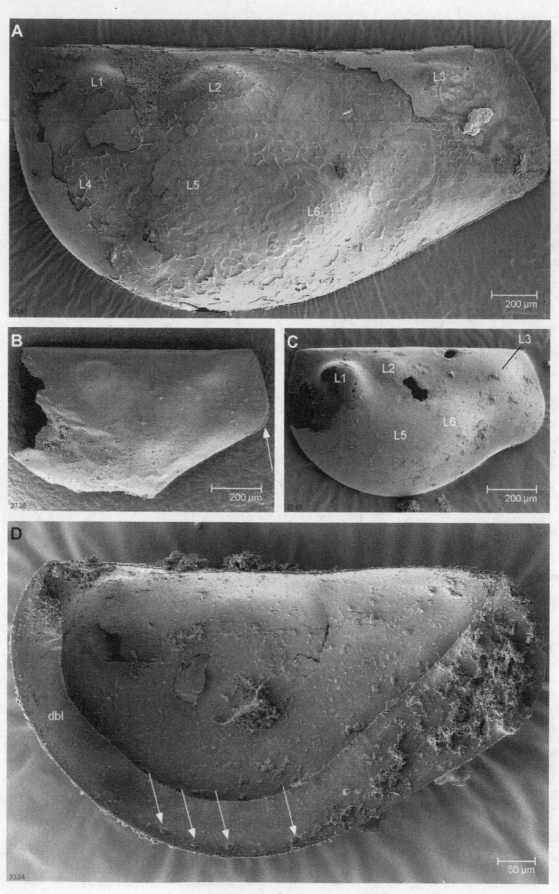

Type species. – *Falites fala* Müller, 1964 by original designation (Müller 1964a).

Original diagnosis. – (Müller 1964a, translated). "Representative of Phosphatocopina with relatively short, straight dorsal rim and one to three valve lobes, which are restricted to the dorsal part of the smooth valves. The valves have a flat marginal area and a more strongly domed central area. Doublure always distinct, but of different width."

Emended diagnosis. – Valves of equal size, close tightly. Maximum length of valves between dorsal rim and midline. Surface with three lobes in a line close to the dorsal rim, two anteriorly, another one in the last third of the total length. Interdorsum absent, only two elongated, smooth triangular plates anterior and posterior between the valves, connected by a dorsal furrow. Doublure relatively wide, especially postero-ventrally.

Species referred to taxon. – The taxon *Falites* possibly encompasses five species, of which one is considered in this work (cf. Appendix B):

Falites angustiduplicata Müller, 1964
Falites cycloides Müller, 1964
Falites fala Müller, 1964 (type species, see below)
Falites insula Hinz-Schallreuter, 1998
Falites marsupiata Cui & Wang, 1991

Remarks. – "*Falites*" *pateli* Landing (1980) probably does not belong to *Falites* (Williams & Siveter 1998; see Appendix B for references).

Falites fala Müller, 1964

v 1964a Falites fala n. sp. – Müller, p. 25, pl. 3, figs. 3–10; pl. 5, fig. 6; text-fig. 2.

Plate 38. *Veldotron bratteforsa* (Müller, 1964) Hinz-Schallreuter, 1993

A: UB W 233. The largest specimen available, about 2300 μm in length. A lateral view of the left valve, displaying lobes L_1–L_6 (L1–L6). The outer cuticle of the valve is mostly destroyed.

B: UB W 234. A specimen representing an advanced growth stage, the length was reconstructed to about 1100 μm. A lateral view of the left valve. The anterior part of the shield is destroyed and missing. The arrow points to the posterior gaping of the shield.

C: UB W 235. A specimen representing an advanced growth stage, about 1020 μm in length. A lateral view of the left valve. Note that lobes L_4–L_6 (L4–L6) are almost inconspicuous.

D: UB W 236. A specimen representing a young growth stage, about 700 μm in length. An inside view of the right valve, displaying the doublure (dbl) and its dome-like outgrowths (arrows). The inner lamella is not preserved.

. 1965   *Falites fala* Müller – Adamczak, pp. 28, 29, pl. 1, figs. 4, 5a–c; text-fig. 1.
. 1972   *Falites* sp. – Taylor & Rushton, pp. 17, 18, 25.
. 1974   *Falites fala* Müller – Martinsson, p. 21.
. 1978   *Falites fala* Müller – Rushton, p. 277 [*partim*: specimens of Monks Park Formation; non specimens from Outwoods Formation and text-fig. 2. (= *Hesslandona unisulcata*)]
. 1979a   *Falites fala* Müller, 1964 – Müller, pp. 20, figs. 1, 10, 11, 21, 25.
. 1980   *Falites fala* Müller, 1964 – Landing, p. 7.
. 1981   *Falites fala* Müller – Gründel, pp. 63, 69, pl. 2, figs. 6, 7.
. 1982c   *Falites fala* Müller – Müller, fig. 2.
. 1983   *Falites fala* Müller – McKenzie et al., p. 36, fig. 6.
. 1983   *Falites fala* Müller, 1964 – Reyment, fig. 1.
. 1986a   *Falites fala* Mueller – Kempf, p. 355.
. 1986b   *Falites fala* Mueller – Kempf, p. 216.
. 1987   *Falites fala* Mueller – Kempf, p. 436.
. 1987   *Falites fala* Müller – Tong, p. 433.
. 1987   *Falites fala* – Zhang, p. 5.
. 1989   *Falites fala* Müller, 1979 – Zhao & Tong, p. 15 [referred to Müller (1979a), fig. 10a–c]
. 1991   *Falites fala* Müller – Huo et al., p. 181.
. 1993b   *Falites fala* – Hinz-Schallreuter, p. 346.
. 1993c   *Falites fala* Müller, 1964 – Hinz-Schallreuter, p. 400.
. 1996a   *Falites fala* Müller, 1964 – Hinz-Schallreuter, p. 85.
. 1996b   *Falites fala* Müller, 1964 – Hinz-Schallreuter, pp. 89, 91, pl. 23; text-figs. 1, 2.
. 1998   *Falites fala* Müller, 1964 – Hinz-Schallreuter, pp. 112, 114, 132.
. 1998   *Falites fala* Müller, 1964 – Williams & Siveter, pp. 28, 29, pl. 5, figs. 1–4.
. 1999   *Falites fala* Müller – McKenzie et al., p. 164.
. 1999   *Falites fala* Müller, 1964 – Whatley et al., p. 344.

Derivation of name. – Not mentioned by Müller (1964a); named after the province Falbygden in southern Sweden.

Holotype. – Right valve illustrated by Müller (1964) in his pl. 3, fig. 4 (UB 29); length 1100 μm, height 8 μm (Pl. 40C).

Remarks. – The reports of specimens of *Falites* by Taylor & Rushton (1972) from the Nuneaton District, Warwickshire, UK are unsure. Probably, the material they found from Zone 1 has to be assigned to *H. unisulcata* and *Trapezilites minimus* (Zone 2). Rushton (1978) mentioned *Falites fala* from the Nuneaton District, Zone 1 of the Upper Cambrian. Later on, these specimens turned out to be *H. unisulcata* respectively *Waldoria* cf. *rotundata* (Williams & Siveter 1998).

Type locality. – Stenåsen, Falbygden, Västergötland, Sweden (see Fig. 2).

Type horizon. – Upper Cambrian, lower subzone of Zone 5c.

Dimensions. – Smallest specimen: valves about 130 μm long and about 105 μm high. Largest specimen: valves about 2300 μm long and about 1700 μm high.

Material examined. – Thirty-four specimens of different stages and from different areas, Zone 5 (Table 34).

Additional material. – Rushton (1978) found specimens in the Monks Park Formation, between the *Loptoplastus angustatus* Subzone and the *Ctenopyge postcurrens* Subzone (Zone 5) of the Nuneaton District, Warwickshire, UK. Gründel (1981) reported several specimens from Zones 1 and 5 of the Upper Cambrian of northeastern Germany (Table 35). Gründel (1981) mentioned that he only found juvenile specimens; the largest was about 930 μm long, which he illustrated on his pl. II, figs. 6, 7. Therefore, the assignment of material of Zone 1 to this species is doubtful. It has to be determined whether the material belongs to *Hesslandona unisulcata* Müller, 1982. Williams & Siveter (1998) reported material from the late St. David's Series and the Merioneth Series: Outwoods Shale and Monks Park Shale Formations of Merevale no. 1 borehole (*Olenus–Peltura minor* Zones) and Outwoods Shale Formation of Merevale no. 3 borehole (*Lejopyge laevigata–Agnostus pisiformis* Zones), Warwickshire; and the Crouch Farm borehole, Oxfordshire [for an explanation of series-level terms and for Cambrian biostratigraphy in England and Wales see Cowie *et al.* (1972) and Thomas *et al.* (1984)].

Distribution. – Zone 5 of the Late Cambrian, northern central Europe area.

Original diagnosis. – (Müller 1964a, translated). "A member of the genus with proportionally long dorsal rim and three lobes. The anterior one is the strongest, the others are missing in juveniles. The doublure is very broad and strongly widened at the postero-ventral area, which is slightly drawn out."

Emended diagnosis. – Maximum length of valves ventral to dorsal rim. Valves close tightly. Valves right/left symmetrical with three more or less rounded lobes. Antero-dorsal lobe largest, posterior to that the smallest lobe. Third lobe on the posterior third of valves. Valves post-plete. Interdorsum missing, anterior and posterior a tri-angular plate without any structures between valves. Doublure wide, greatest width postero-ventrally, about one fifth to one quarter the length of valve.

Description. –

Shield (Fig. 50)
The bivalved shield (ch1:2) has a long and straight dorsal rim (ch2:2) (Pl. 39A; Fig. 50A). The maximum length of the shield is slightly dorsal to the midline (ch3:3), the anterior-most point is located antero-dorsally, the posterior-most point of the shield is located at the dorso-ventral midline of the shield (Pl. 39A). The valves are of equal size and the right and left valves are symmetrical (Pl. 39B). The location of the maximum height of the valves is posterior (post-plete) (ch4:3). The shield is about 1.35 times longer than high. The anterior free part of the shield margin, starting from the dorsal rim, curves antero-ventrally (antero-dorsal angle about 60 degrees) and equally postero-ventrally towards the ventral maximum (ch6:2). The ventral part of the margin is curved (ch7:2). Thereafter, the margin curves gently upwards – equally to the anterior – towards the posterior at the dorso-ventral midline of the shield (ch8:2, angle about 45 degrees) and recurves anteriorly – distinctly more strongly compared with the anterior margin – to meet

Table 34. Sample productivity of examined specimens of *Falites fala*.

Sample	Zone	Found at	Number of specimens
975	5c	Stenåsen, Falbygden	24 (holotype included)
5940	5a–e	Stenåsen, Falbygden	1
5948	5a–e	Street between Stenstorp and Dala	3
5955	5d, e	Street between Stenstorp and Dala	4
5957	5d, e	Street between Stenstorp and Dala	2

Table 35. Complete list of finds of *Falites fala*, reported by different authors, with localities and horizons.

Locality	Horizon	Reference
Warwickshire, UK	Upper Cambrian, Zone 5	Rushton 1978
Northeast Germany	Upper Cambrian, Zone 5	Gründel 1981
Warwickshire, Oxfordshire, UK	Upper Cambrian, Zone 5	Williams & Siveter 1998

A

B

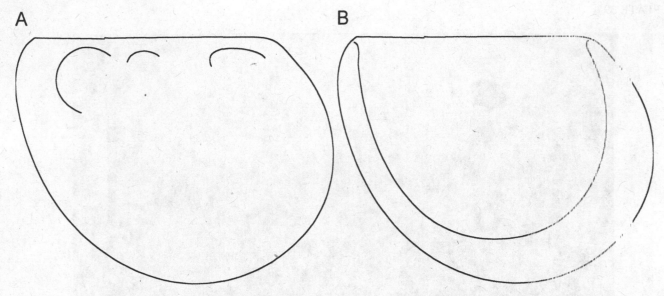

Fig. 50. Schematic drawing of the shield of *Falites fala* from outside lateral (A) and from inside lateral (B)

the dorsal rim (ch5:2). The shield appears distorted in the posterior direction. The maximum width of the shield is slightly dorsal to the midline. The margins are without outgrowths throughout (ch19:0, ch20:0), and the shield valves close tightly without any gaps (ch9:1). The surface of the valves is smooth, but with three more or less rounded dome-like lobes (Pl. 39A; cf. Fig. 9); all of them located close to the dorsal rim. The most anterior one, L_1 (ch10:1), is the largest, ellipsoidal, with its long axis expanding in the antero-dorsal/postero-ventral direction (Pl. 39A). A second distinctly smaller one, L_2, is located posterior to L_1 just before the antero-posterior midline (ch11:1), a third one of inter-mediate size, L_3, is located in the last third of the valves (ch12:1) (Pl. 39A). The outer margin of the valves, copy-ing the shape of the doublure, is distinctly depressed compared with the central area (Pl. 39A). Other lobes or any structures on the valves are not present (chs13–18:0, ch21:0). Smaller growth stages only show lobe L_1.

Dorsal rim
The interdorsum is apparently absent, the valves being separated by a narrow membranous furrow (Pl. 39B, D) from the early start of ontogeny onwards (Pl. 41D, E). This furrow continues at the anterior and posterior dorsal ends into small triangular plates, the anterior and posterior plates, which are not longer than 1/12 the shield length and about 1/80 the shield length at the anterior (ch22:2, ch23:0, ch24:0, ch25:0) (Pl. 39B, C). Both plates are without any lobes or ornamentation (chs26–33:0).

Doublure
A doublure is present (ch34:1), being narrowest anteriorly (ch35:1), ventrally and posteriorly about one

third wider. The maximum width is postero-ve ally (ch36:4). The maximum width of the doubl is between one quarter and one fifth the length the valves (ch37:3) (Pls. 40C, 41B, D; Table 40). Ap oxi-mating the dorsal rim, the doublures of both des rapidly narrow strongly to merge into a memb ous area antero- and postero-dorsally. There are no ruc-tures, such as pits or small outgrowths, develop on the doublure (ch38:1, ch39:1, ch40:0) (cf. Pl. 4 D). No pores were observed on the doublure (ch41:1 The outer edge of the doublure shows three parallel pes running along the ventral and postero-ventral ion (ch42:2). These stripes only occur in early growth ges. The more advanced, the less distinct are these pes (cf. Pl. 41A–C). The outermost margin of the do ure is slightly compressed relative to the remaining face (Pl. 40D).

Inner lamella
The inner lamella extends along the doublure and edi-ally to the dorso-lateral side of the body. It is akly sclerotised and frequently shows a wrinkled textu No scar of a possibly present closing muscle is ob ved (Pls. 40C, 41C, D).

Body
The body proper, which is completely enveloped the bivalved shield, includes at least eight segments (Pl A). Trunk segmentation of the preserved portion is re ned in the insertions of the limbs and distinct se ent boundaries or sternites are weakly defined. At leas ight segments are dorsally fused to the shield (c 4:5) (Pl. 41A). Hence, the shield represents a cephaloth acic shield, including at least three limb-bearing trun seg-ments. The area of fusion of the body proper the shield is very narrow, corresponding to approxi tely

PLATE 39

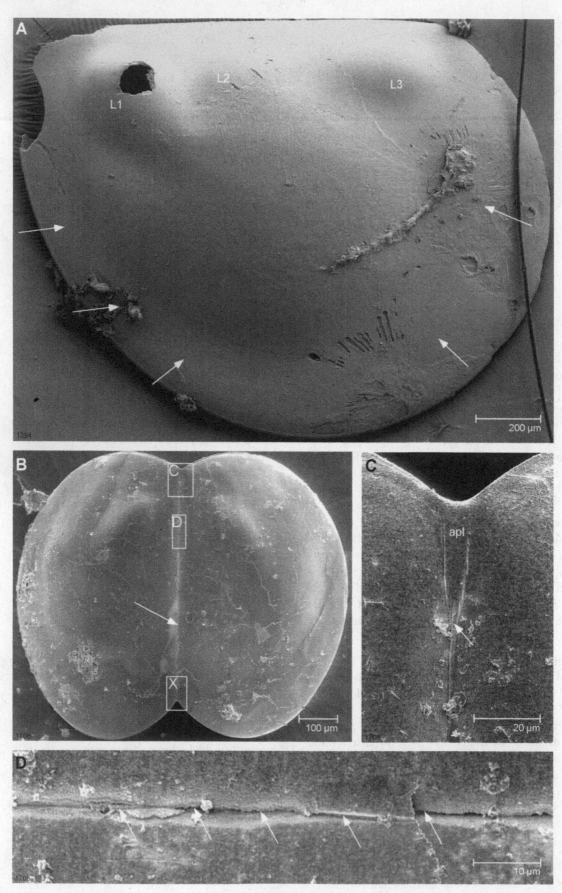

the width of the interdorsum (ch43:1) (Pl. 41A). Starting some micrometres posterior to the dorsal anterior membranous area of the doublures, the body proper extends along the inner dorsal length of the shield (Pl. 41A), but the posterior end of the body is unknown. Its posterior extension and segmental continuation remain uncertain, as well as whether there is a hind body extending free from the shield into the domicilium. The body proper, in ventral view, is oval with a blunt anterior and an elongated posterior end. The maximum width of the body proper is located at the mandibles. The cross-section of the body proper anterior to the mandibular part is almost trapezoidal with the long axis dorso-ventrally oriented. The cross-section at the mandibular segment is more or less a half-oval with the long axis in dorso-ventral aspect; the height of the body proper in this position measures less than one third the height of the valves. Posterior to it, the body proper changes its cross-section to about a half-circular shape. The height of the body decreases gradually towards the posterior. The limbs are arranged in a more or less regular row at the flanks of the body, being more or less equally set together from the anterior towards the caudal end (Pl. 41A).

The anterior part of the body proper is the hypostome/labrum complex (cf. Pl. 41A). The hypostome forms the anterior sclerotised ventral surface and has a somewhat rhomboid outline in ventral view and becomes gradually higher to the posterior in lateral view. The antennulae insert antero-laterally on the hypostome (Pl. 41A). The antennae insert postero-laterally on the hypostome in a more spindle-shaped joint area. The median eye is located antero-distally on the hypostome. The posterior end of the hypostome is drawn out distally into a lobe-like protrusion, the labrum (ch61:1). Posterior to the hypostome/labrum complex, the ventral surface between the first to third pairs of post-antennular limbs is marked by a sclerotic plate, the sternum (ch62:1) (Pl. 41C). The

sternum is slightly domed and bears one pair of umps anteriorly, the paragnaths (ch63:1), and betwee the second pair of post-antennular limbs, the mar bles. The post-mandibular limb pairs insert on a lateral lope corresponding to the cross-section of the body. The hind body, including the number of segments involv and possible furcal rami, is unknown (ch47:?, ch48:?, c 9:?).

Soft parts

The soft parts are generally similar to those descri d in detail for *H. unisulcata*. Because it is beyond the cope of this paper to describe the soft parts of *Falites* a in full detail, only some general remarks are made, i ort-ant for the phylogenetic analysis of the Phosphato ina. The detailed description is postponed for a second per. The antennula is similar to that of *H. unisulcata* eing small and consisting of less than 10 irregular nuli (ch45:1, ch46:1). The tip bears a tuft of four setae hree terminal ones and one additional subterminal seta. Again, as in *H. unisulcata*, the antenna (Pl. 4 C) consists of an undivided limb stem throughout nto-geny, but being a fusion product of the coxa and b pod (see p. 23; ch50:2, ch51:2, ch52:2), a two-part er pod (ch53:5) and a multi-annulated exopod. The s e is true for the mandible. It has an undivided lim tem throughout ontogeny and an enditic protrusion ated medially between the limb stem and the endopod eco-gnised as the remains of the basipod (see p. 27; 54:2, ch55:2, ch56:2). The endopod has two parts (7:5) and the exopod consists of several annuli, each a ulus with one medially projecting seta; the last annulu ears two setae – as in the antenna. The post-man ular limbs are serially similar to each other (ch58:1 con-sisting of a basipod, a setae-bearing proximal dite medio-proximally (ch59:1), a three-part er pod (ch60:4) and an annulated exopod (Pl. 21B, C The endopodal portions are short and slightly pr ting medially. The exopods have a circular cross- tion throughout ontogeny and never become flat plat

Comparisons

Falites fala is similar to *Vestrogothia spinata* the absence of a continuous interdorsum and the p ence of anterior and posterior plates. In these cha ters, *Falites fala* differs from all other investigated phos ato-copines. *Falites fala* is distinguished from *Vestr othia spinata* in the absence of spiny outgrowths. *Fa fala* is distinguished from *Falites angustiduplicata* the broad doublure. *Falites angustiduplicata* has a m e or less evenly developed and rather narrow doublure ilites cycloides* has a symmetrical doublure. *Falites mar iata* is only weakly known. *Falites fala* differs from ilites insula* in the shape of the doublure.

Plate 39.   *Falites fala* Müller, 1964

A: UB 55 (Müller 1964a, pl. 5, fig. 5a, b). A specimen representing a late growth stage, about 1650 µm in length. A lateral view of the left valve, displaying lobes L₁–L₃ (L1–L3). The outer rim of the shield is slightly depressed against the domed central area (arrows).

B: UB W 237. A specimen representing a young growth stage, about 720 µm in length. A dorsal view of the opened shield. An interdorsum as a dorsal bar, as documented herein for various species, is missing (arrow). The rectangles mark the areas magnified in Pl. 39C, D and 40A (X).

C: Close-up of the anterior dorsal end of the shield illustrated in Pl. 39B. The single dorsal furrow (arrow) extends into an elongated triangular plate (apl).

D: Close-up of the antero-median dorsal part of the shield illustrated in Pl. 39B. Note the single membranous dorsal furrow extending between both valves (arrows).

PLATE 40

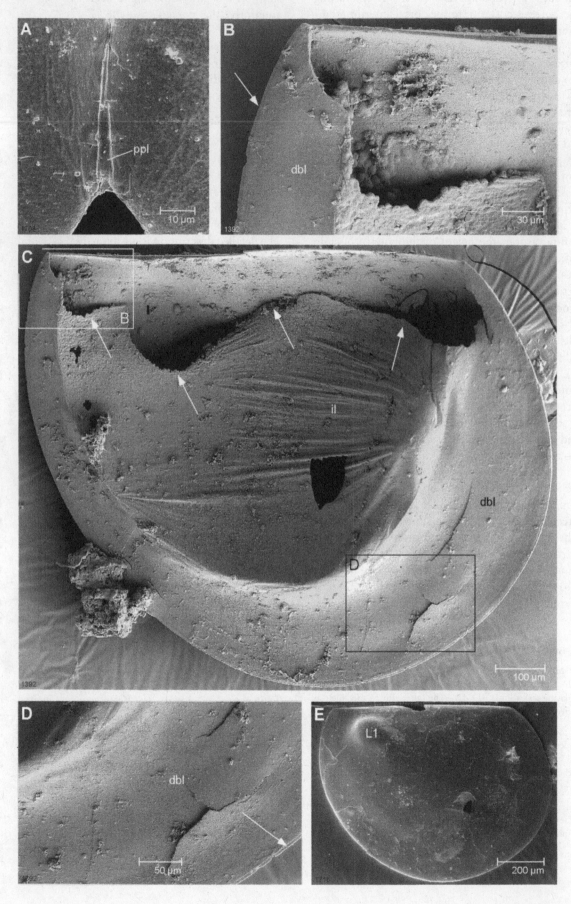

Ontogeny

During ontogeny, the proportion of the length to the height of the valves of *Falites fala* remains the same (Fig. 51). The observed range of shield lengths and heights indicates clusters, but this may be due to the low number of specimens considered. Hinz-Schallreuter (1996b) had more than 100 specimens available from the type locality, and the ontogeny of this material did not show clusters. The smallest stages that could be assigned to *Falites fala* have four pairs of functional limbs, representing a "head-larva" (ch64:1).

Vestrogothia Müller, 1964

1964a *Vestrogothia* n. g. – Müller, p. 30.
1972 *Vestrogothia* Mueller – Taylor & Rushton, p. 13.
non 1972 *Vestrogothia* Mueller – Taylor & Rushton, p. 18 (*Hesslandona*).
1974 *Vestrogothia* Müller, 1964 – Kozur, p. 827.
1980 *Vestrogothia* – Jones & McKenzie, p. 218.
1982a *Vestrogothia* – Müller, p. 287.
1983 *Vestrogothia* – Briggs, pp. 9, 10.
1983 *Vestrogothia* – Müller, p. 94.
1983 *Vestrogothia* – Reyment, p. 5.
1986 *Vestrogothia* – Huo et al., p. 23.
1986 *Vestrogothia* – Schram, p. 415.
1986a *Vestrogothia* Mueller, 1964 – Kempf, p. 747.
1986b *Vestrogothia* Mueller, 1964 – Kempf, p. 707.
1987 *Vestrogothia* Mueller, 1964 – Kempf, p. 436.
1987 *Vestrogothia* – Zhang, pp. 5, 9.
1990 *Vestrogothia* – Bengtson *in* Bengtson *et al.*, p. 323.
1990a *Vestrogothia* – Shu, pp. 61, 62, 77.
1993b *Vestrogothia* Müller, 1964 – Hinz-Schallreuter, pp. 334, 342, 344, 347.
1993c *Vestrogothia* Müller, 1964 – Hinz-Schallreuter, pp. 386, 395, 399, 402, 403, 409.

1995 *Vestrogothia* – Siveter *et al.*, p. 416.
1996b *Vestrogothia* Müller, 1964 – Hinz-Schallreuter, p. 89.
1998 *Vestrogothia* Müller, 1964 – Hinz-Schallreuter, pp. 104, 107, 116, 118, 126, 132, text-figs. 1, 3.
1998 *Vestrogothia* – Ziegler, p. 223.

Derivation of name. – From the area name Vestergötland (Latin = *Vestrogothia*) in southern Sweden, where the studied phosphatocopine species were found (K.J. Müller, pers. comm. 2002).

Type species. – *Vestrogothia spinata* Müller, 1964 by original designation (Müller 1964a).

Original diagnosis. – (Müller 1964a, translated). "Shield elongated, at least 1.5 times as long as high. Valves of equal size or almost of equal size, can gap at the anterior posterior end. With two to three small, flat, mostly circular lobes, restricted to the dorsal area. Most species are spinose and/or have lobes on their surface. Sexual dimorphism is proved."

Emended diagnosis. – Valves of equal size, close tightly. Maximum length of valves on the dorsal rim or between dorsal rim and midline. Surface without lobes, with one prominent lobe antero-dorsally or with three lobes in a line close to the dorsal rim, two anteriorly, another one in the last third of the total length. Interdorsum with small loop-like thickenings anteriorly and posteriorly or anterior and posterior end drawn out into more or less long spines. Doublure rather small or relatively wide. Valves close tightly or leave a gap.

Species referred to taxon. – The taxon *Vestrogothia* possibly encompasses six species, of which one is considered in this work (synonymy of all species is specified in Appendix B).

Vestrogothia granulata Müller, 1964
Vestrogothia hastata Müller, 1964
Vestrogothia longispinosa Kozur, 1974
Vestrogothia minilaterospinata Hinz-Schallreuter, 1998
Vestrogothia spinata Müller, 1964 (p. 168)
Vestrogothia steffenschneideri Hinz-Schallreuter, 1998

Remarks. – *Vestrogothia bratteforsa* Müller, 1964 was referred to *Veldotron* Gründel, 1981 by Hinz-Schallreuter (1993b). This is followed herein, as *Vestrogothia bratteforsa* has an interdorsum and, therefore, clearly belongs to the Hesslandonina (see p. 170). Due to missing data of all other possible species of *Vestrogothia*, only the type species stand for this study, represented by numerous specimens, was considered in the phylogenetic analysis. The phylogenetic analysis could not confirm a close relationship between *Vestrogothia* and *Falites*, both taxa traditionally assigned to

---

Plate 40. *Falites fala* Müller, 1964

A: Close-up of the posterior dorsal end of the shield illustrated in Pl. 39B. The single dorsal furrow (arrow) extends into an elongated triangular plate (ppl).

B: Close-up of the inner antero-dorsal end of the shield from the specimen illustrated in Pl. 40C, displaying a smooth doublure, apart from the slightly compressed outermost margin (arrow; cf. Pl. 40D).

C: Holotype [UB 29 (Müller 1964a, pl. 3, fig. 4)]. A specimen representing a late growth stage, about 1100 μm in length. The lateral extension of the body is retained in the dorso-median margin of the inner lamella (il, arrows). The rectangles mark the areas magnified in Pl. 40B and D.

D: Close-up of the inner antero-dorsal end of the shield illustrated in Pl. 40C, displaying a smooth doublure, apart from the slightly compressed outermost margin (arrow; cf. Pl. 40B).

E: UB W 238. A specimen representing a young growth stage, about 970 μm in length. Only lobe L_1 (L1) is developed, the more posterior ones are absent.

PLATE 41

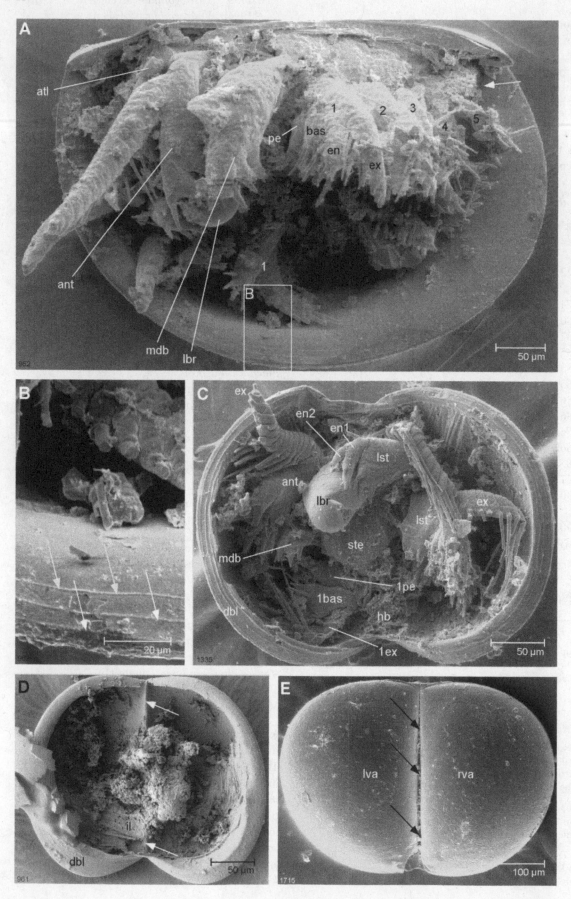

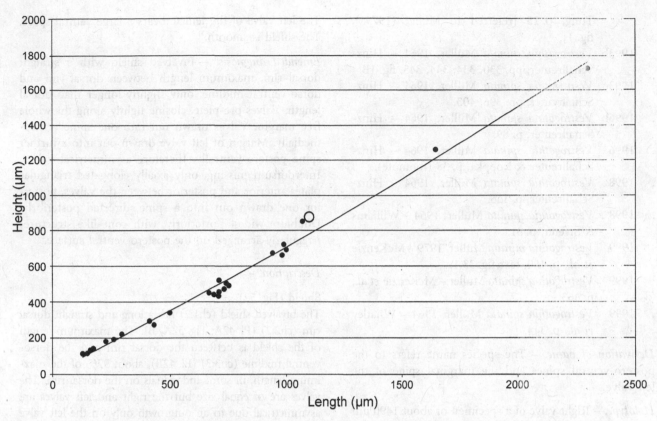

Fig. 51. Ontogeny of *Falites fala*: length versus height. Thirty-two specimens with measurable length and height are included. The trend line has the polynomial function $y = 0x^2 + 0.6486x + 10.009$, y = height, x = length. Mean proportion of length to height about 1.35. ○ = Holotype

a taxon Vestrogothiidae, respectively, Vestrogothiina (see Appendix B). Such a combination is a paraphyletic assemblage of species with plesiomorphic characters.

Vestrogothia spinata Müller, 1964

v* 1964a *Vestrogothia spinata* n. sp. – Müller, p. 30, pl. 2, figs. 4–8, 10, 11; pl. 5, figs. 1, 7–9.

. 1965 *Vestrogothica spinata* Müller – Adamczak,

Plate 41. *Falites fala* Müller, 1964

A: UB W 239. A specimen representing an early growth stage, about 300 µm in length. A lateral view inside the shield, displaying a well-preserved body. Note the similarity of the post-mandibular limbs (1–3). The rectangle marks the area magnified in Pl. 41B.

B: Close-up of the ventral part of the doublure illustrated in Pl. 41A. Note the parallel stripes on the doublure (arrows).

C: UB W 240. A specimen representing an early growth stage, about 260 µm in length. A ventral view inside the opened shield, displaying a larva with four pairs of limbs (antennulae covered). The hind body (hb) is strongly wrinkled. Note the parallel lines on the doublure (cf. Pl. 41B).

D: UB W 241. A specimen representing an early growth stage, about 430 µm in length. A ventral view of the opened shield. The inner lamella is partly missing such that the dorsal furrow is exposed (arrows).

E: UB W 242. A specimen representing an early growth stage, about 300 µm in length. A dorsal view of the opened shield, displaying the dorsal furrow (arrows).

pp. 29, 32 (*V. spinata* on p. 29, sic! on p. 32).

. 1974 *Vestrogothia spinata* Müller, 1964 – zur, pp. 827, 828.

. 1979a *Vestrogothia spinata* Müller – Müller, 23, 24, figs. 3, 4, 13, 14, 16, 18, 29, 30, 31, D.

. 1979b *Vestrogothia spinata* Müller, 1964 – ller, p. 92, fig. 1.

. 1982b *Vestrogothia spinata* Müller – Müller, 1.

. 1982c *Vestrogothia spinata* – Müller, fig. 4.

. 1982 *Vestrogothia spinata* – Schram, p. 111.

. 1983 *Vestrogothia spinata* – Briggs, p. 9.

. 1983 *Vestrogothia spinata* Müller, 1964 – Mc nzie et al., fig. 5.

. 1983 *Vestrogothia spinata* Müller, 1964 – Re ent, fig. 3.

. 1985a *Vestrogothia spinata* Müller, 1964 – M er & Walossek, fig. 2f.

. 1986 *Vestrogothia spinata* – Schram, fig. 33- –H.

. 1986a *Vestrogothia spinata* Mueller, 1964 – npf, p. 747.

. 1986b *Vestrogothia spinata* Mueller, 1964 – npf, p. 555.

. 1987 *Vestrogothia spinata* Mueller, 1964 – npf, p. 436.

. 1987 *Vestrogothia spinata* Müller – Tong, p 33.

. 1989b *Vestrogothia spinata* – Zhao, p. 471.

. 1989 *Vestrogothia spinata* Müller, 1979 – 2 o &

Tong, p. 15 [referred to Müller (1979a), fig. 1].

- . 1993b *Vestrogothia spinata* Müller, 1964 – Hinz-Schallreuter, pp. 330, 334, 344, 345, fig. 1B.
- . 1993c *Vestrogothia spinata* Müller, 1964 – Hinz-Schallreuter, pp. 396, 403.
- . 1996b *Vestrogothia spinata* Müller, 1964 – Hinz-Schallreuter, p. 89.
- . 1996 *Vestrogothia spinata* Müller, 1964 – Hinz-Schallreuter & Koppka, p. 38 (footnote).
- . 1998 *Vestrogothia spinata* Müller, 1964 – Hinz-Schallreuter, p. 126.
- . 1998 *Vestrogothia spinata* Müller, 1964 – Williams & Siveter, p. 31.
- . 1999 *Vestrogothia minuta* Müller, 1979 – McKenzie *et al.*, p. 460, text-fig. 33.1.
- . 1999 *Vestrogothia spinata* Müller – McKenzie *et al.*, p. 463.
- . 1999 *Vestrogothia spinata* Müller, 1964 – Whatley *et al.*, p. 344.

Derivation of name. – The species name refers to the postero-ventral spines and the marginal spine of the left valve.

Holotype. – Right valve of a specimen of about 1490 µm in length and 800 µm in height illustrated by Müller (1964a) in his pl. 5, fig. 1a, b (UB 23), specimen in the meanwhile slightly damaged. Posterior spine is missing due to preservation.

Remarks. – McKenzie *et al.* (1999) mentioned a species named *Vestrogothia minuta* and referred it to Müller (1979a). However, Müller (1979a) did not mention or figure a species with such a name, possibly McKenzie *et al.* (1999) confused it with "*Falites*" *minima*, a species that is treated herein as *Trapezilites minimus* (see p. 173), but this is not clear.

Type locality. – Near Stenåsen, Falbygden, Västergötland, Sweden (see Fig. 2).

Type horizon. – Upper Cambrian, *Peltura* Zone (Zone 5).

Material examined. – Seventy-seven specimens of different growth stages and from different areas of Zone 5 of the Upper Cambrian of southern Sweden (Table 36).

Dimensions. – Smallest specimen: valves about 170 µm long and about 120 µm high. Largest specimen (holotype): valves about 1490 µm long and about 800 µm high.

Additional material. – No additional material of *Vestrogothia spinata* has been reported so far.

Original diagnosis. – (Müller 1964a, translated). "A member of the genus *Vestrogothia* with a large lateral spine and a posterior spine, which developed from the doublure of the left valve. Sexual dimorphism proved.

The left valve of the female bears a large ventral spine. The shield is smooth."

Emended diagnosis. – Bivalved shield with a straight dorsal rim, maximum length between dorsal rim and dorso-ventral midline, only slightly longer than dorsal length. Valves pre-plete, closing tightly along the whole free margin. Valves drawn out into one spine ventro-medially. Margin of left valve drawn out into a further spine postero-ventrally, therefore asymmetrical valves. Interdorsum missing, only axially elongated triangular plates anterior and posterior between the valves, posterior one drawn out into a spine, directed posteriorly. Doublure widest posteriorly, with cone-like structures irregularly arranged on the postero-ventral surface.

Description. –

Shield (Fig. 52)

The bivalved shield (ch1:2) has a long and straight dorsal rim (ch2:2) (Pl. 42A; Fig. 52A, B). The maximum length of the shield is between the dorsal rim and the dorso-ventral midline (ch3:3) (Pl. 42B), about 97% of the maximum length, in some individuals on the dorsal rim. The valves are of equal size but the right and left valves are asymmetrical due to an outgrowth only on the left valve (Pl. 42A). The maximum height of the valves is anterior to the antero-posterior midline, i.e. the shield is pre-plete (cf. Fig. 9; ch4:1). The shield is about 1.8 times longer than high. The free anterior part of the shield margin starts from the relatively straight dorsal rim (antero-dorsal angle nearly 90 degrees, ch6:1), begins to curve slightly to the postero-ventral before reaching the dorso-ventral midline. After curving ventrally, the margin of both valves runs almost straight towards the ventral maximum (Pl. 42A). The latero-ventral outline of the right valve curves (ch7:2) into an almost straight margin (Pl. 42A; Fig. 52A). Thereafter, the margin of the right valve curves gently upwards into a long and straight postero-ventral margin. After reaching the most posterior edge, the margin curves slightly back (Pl. 42B; ch5:2) and becomes straight, shortly before meeting the postero-dorsal rim (ch8:1, postero-dorsal angle about 90 degrees). The margin of the right valve is without outgrowths throughout (ch19:0). The latero-ventral outline of the margin of the left valve is almost straight, postero-ventrally drawn out into a long spine of about one third to one half the length of the valves with a broad, triangular, flattened base (ch20:2) (Pl. 42A; Fig. 52B). Thereafter, the left margin runs parallel to the right one until it meets the postero-dorsal rim.

The maximum width of the shield is slightly ventral to the midline. The shield valves close tightly without any gaps in all stages along the entire free margin (ch9:1). The surface of the valves is smooth, but with one antero-central outgrowth forming a spine on both sides (ch16:1; ch17:2), the antero-central spine, of about one quarter

Table 36. Sample productivity of examined specimens of *Vestrogothia spinata*.

Sample	Zone	Found at	Number of specimens
975	5c–d	Stenåsen, Falbygden	27 (including holotype)
5940	5a–e	Stenåsen, Falbygden	3
5942	5a–e	Road cut at street between Stenstorp and Dala	1
5948	5a–e	Road cut at street between Stenstorp and Dala	14
5955	5d–e	Road cut at street between Stenstorp and Dala	15
5957	5d–e	Road cut at street between Stenstorp and Dala	8
6206	5c–d	Outcrop near Stenåsen, south of road cut between Stenstorp and Dala	2
6369	5a–e	Brattefors, Kinnekulle, Falbygden–Billingen	3
6391	5c–d	Between Haggården and Marieberg (Kinnekulle), Falbygden–Billingen	3
6392	5c–d	Between Haggården and Marieberg (Kinnekulle), Falbygden–Billingen	

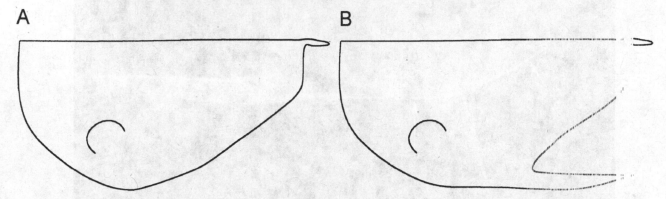

A   **B**

Fig. 52. Schematic drawing of the shield of *Vestrogothia spinata* from outside lateral. A: Right valve. B: Left valve (drawing flipped horizontally).

the length of the valves and directed laterally with a slight curve towards postero-ventral (ch18:2). Other structures are not present (chs10–15:0; ch21:0) (Pl. 42A, B).

Dorsal rim

An interdorsum, as described for the species of *Hesslandona*, is absent. As in *Falites fala*, both valves are separated from each other by a membranous furrow and axially elongated triangular plates at the anterior and posterior dorsal ends (Pls. 42B, 43A–C) (ch22:2; ch23:0; ch24:0; ch25:0). The dorsal membranous furrow branches anteriorly and posteriorly into two lines bordering the plates laterally on both sides (Pl. 43C). The anterior plate is without any lobes or ornamentation (ch26:0, ch27:0, ch28:0; ch32:0). The posterior plate is much wider than the anterior one and completely drawn out into a spine (ch29:1) of about one ninth the length of the valves (ch30:2), basally as thick as the plate, and directed posteriorly (ch31:3; Pls. 42A, 43C; Table 39). The spine is smooth (ch33:1).

Doublure

A doublure is present (ch34:1), narrowest ventrally (ch35:2), anteriorly slightly wider, and posteriorly again about double as wide as anteriorly (ch36:3; Pls. 42B, 43D; Fig. 53A, B; Table 37). Approximating the dorsal rim, the doublures of both sides rapidly narrow slightly to merge into a membranous area antero- and postero-

dorsally (Pl. 43E), which is slightly less sclerotised than the doublure itself. The maximum width of the doublure is about one eighth the length of the valves (ch37:0).

Small conical outgrowths are arranged in an irregular row on the postero-ventral margin of the doublure of more advanced growth stages (ch39:2, ch40:1). These outgrowths are exposed on the surface (ch38:1). No pores on the doublure were observed (ch41:1, ch42:1). The doublure of the left valve is drawn out into a long spine postero-ventrally (Fig. 53B).

Inner lamella

The inner lamella extends along the whole doublure and medially to the dorso-lateral side of the body. The inner lamella is weakly sclerotised (Pl. 43E) and frequently shows a wrinkled texture. No scar of a possibly present closing muscle is preserved.

Body

The body proper, which is completely enveloped by the bivalved shield, includes at least seven segments (Pl. 44B). Trunk segmentation of the preserved portion is retained in the insertions of the limbs and distinct segment boundaries or sternites are weakly defined. Minimally, seven segments are dorsally fused in the shield (ch44:4; Pl. 44B), being a cephalothoracic shield including at least four limb-bearing trunk segments. The area of fusion of the body proper to the shield is very narrow, not reaching much to the lateral sides (ch45:1).

PLATE 42

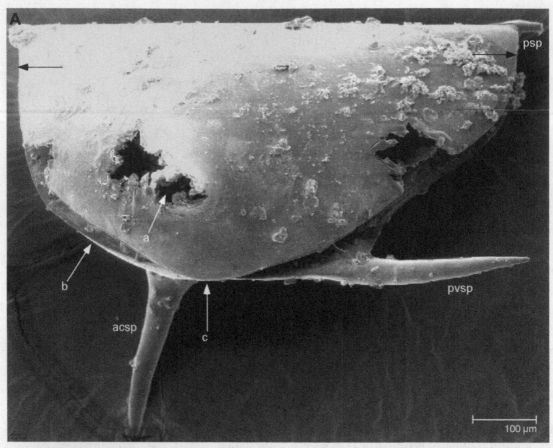

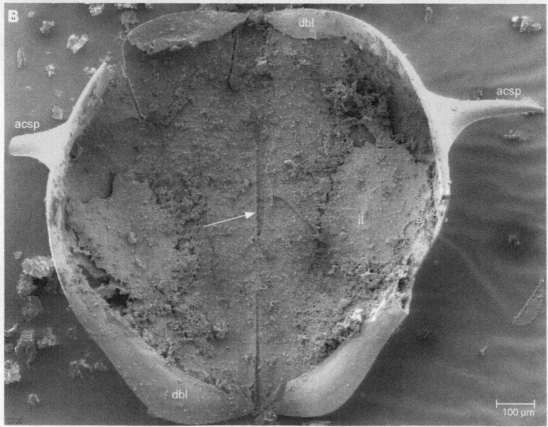

Starting a few micrometres posterior to the dorsal anterior membranous area of the doublures, the body proper extends along the inner dorsal length of the shield; the posterior end of the body is unknown (Pl. 44B). Its posterior extension and segmental continuation are uncertain, as well as if there is a hind body extending free from the shield into the domicilium. The body proper in ventral view is oval with a blunt anterior end and an elongated posterior end. The maximum width of the body proper is at the mandibles. The cross-section of the body proper anterior to the mandibular part is almost trapezoidal with the long axis dorso-ventrally oriented. The cross-section at the mandibular segment is more or less about half-oval with the long axis in dorso-ventral aspect; the height of the body proper in this position is more than one third the height of the valves. Posterior to it the body proper changes its cross-section to about half-circular. The height of the body decreases gradually towards the posterior to about one sixth the height of the valves at the posterior end of the body proper. The limbs are arranged in a more or less regular row at the flanks of the body, being more or less equally set together from the anterior towards the caudal end.

The anterior part of the body proper is the hypostome/labrum complex. The hypostome forms the anterior sclerotised ventral surface that is somewhat rhomboid and becomes gradually higher to the posterior. The antennae (Pl. 44B) insert postero-laterally to it in a spindle-shaped insertion area located on a lateral slope. The posterior end of the hypostome is drawn out into a lobe-like protrusion, the labrum (ch61:1) (Pl. 44B, D). Posterior to the hypostome, the ventral surface between the first to third pairs of post-antennular limbs is marked by a sclerotic plate, the sternum (ch62:1; Pl. 44B, C, D, F), which is covered by several rows of fine hairs (Pl. 44C). It is slightly domed and bears one pair of humps anteriorly, the paragnaths (ch63:1), between the mandibles (Pl. 44D), which are much less distinctly domed in early larvae (Pl. 45A). Posterior to the sternum, the ventral body surface becomes progressively

softer. The post-mandibular pairs of limbs insert on a lateral slope corresponding to the cross-section of the body. The hind body, including the number of segments involved and possible furcal rami, is unknown (ch47:?, ch48:?, ch49:?) (cf. Pl. 44E).

Soft parts

The soft parts are similar to those described in detail for *H. unisulcata*. Because it is outside the scope of this paper to describe the soft parts of *Vestrogothia spinata* in full detail, only some general remarks are made, important for the phylogenetic analysis of the Phosphatocopina. The antennula was not observed (ch45:?, ch46:?). The antenna (Pls. 44B, D, F, 45A, B) consists of an undivided limb stem throughout ontogeny, being a fusion product of the coxa and basipod (see p. 23; ch50:2, ch51:2, ch52:2), a two-divided endopod (ch53:5; Pl. 45A) and a multi-annulated exopod. The mandible of later growth stages of *Vestrogothia spinata* has an undivided limb stem and an enditic protrusion located medially between the limb stem and the endopod (Pl. 44D, F), recognised as the remains of the basipod (see p. 27; ch54:2, ch56:2). It is medially drawn out into an elongated conical spine and bears a seta proximal to the spine (Pl. 44D), two additional setae are located antero- and postero-distally (Pl. 44F). The endopod is two-divided (ch57:5; Pl. 44D) and the exopod consists of several annuli, each annulus with one medially projecting seta; the last annulus bears two setae — as in the antenna. In early stages the limb stem is distinctly two-divided (ch55:1) into a proximal coxa and a distal basipod; an enditic protrusion between the limb stem and the two-part endopod is absent (Pl. 45A, B; Fig. 59A). The coxa and basipod are medially drawn out into a spine-bearing gnathobase. During further ontogeny, the lateral part of the basipod fuses with the coxa, while the median enditic gnathobase slightly reduces in size and gets squeezed between the limb stem and the endopod (cf. Pl. 44D, F; Fig. 59C). The post-mandibular limbs are similar to each other (ch58:1), consisting of a basipod, a setae-bearing proximal endite medioproximally (ch59:1), a three-divided endopod (ch60:1) and an annulated exopod (cf. Pls. 44B, 45A). The endopodal portions are short and slightly projecting medially. They are drop-shaped with a blunt end medially. The exopods of the post-mandibular pairs of limbs are multi-annulated as in the antenna and the mandible but with fewer annuli; they have no lateral setation.

Comparisons

Vestrogothia spinata is similar to *Falites fala* in the absence of a continuous interdorsum and the presence of anterior and posterior plates. In these characters, *Vestrogothia spinata* differs from all other investigated phosphatocopines. *Vestrogothia spinata* is distinguished

---

Plate 42. *Vestrogothia spinata* Müller, 1964

A: UB W 243. Image flipped horizontally. A specimen representing an advanced growth stage, about 800 μm in length. A lateral view of the right valve and the left valve in the background. The antero-central spine (acsp) of the right valve is broken off at its base (white arrow "a"). Note the straight margins of both valves antero-ventrally (white arrow "b") and that the postero-ventral spine (pvsp) is only developed at the margin of the left valve, while the right valve is distinctly curved ventrally (white arrow "c"). The black arrows point to the straight antero-dorsal and postero-dorsal margins of the shield.

B: UB W 244. A specimen representing a late growth stage, about 1070 μm in length. A ventral view of both opened valves. Note that the doublure is widest posteriorly (dbl). The inner lamella is only partly preserved such that the dorsal furrow is exposed (arrow).

PLATE 43

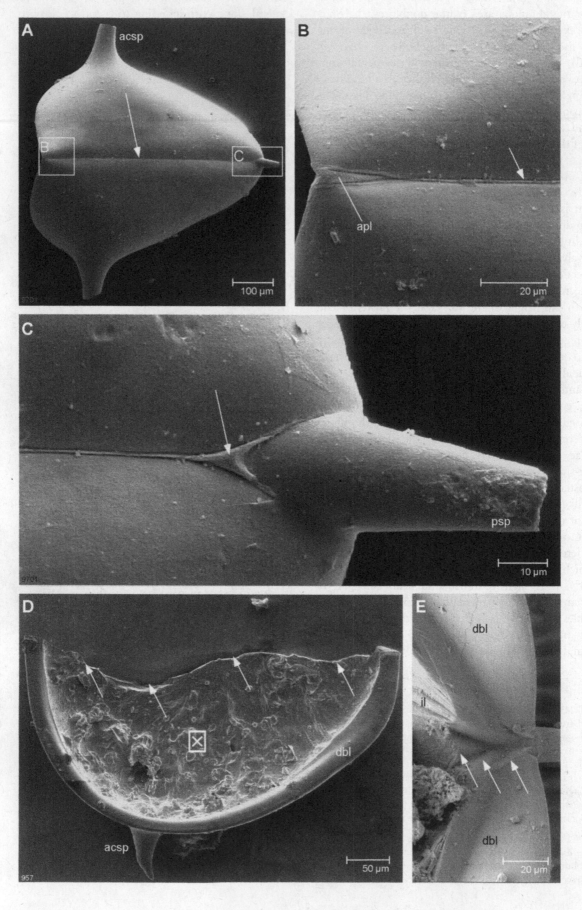

A

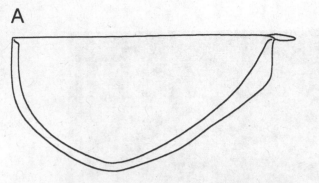

B

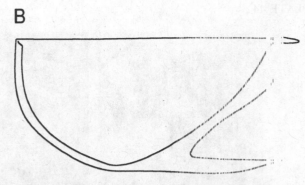

Fig. 53. Schematic drawing of the shield of *Vestrogothia spinata* from inside lateral. A: Right valve. B: Left valve (drawing flipped horizon ꞏ ꞏ).

from *Falites fala* in the occurrence of spiny outgrowths on the valves and the margin of the left valve. The posterior dorsal plate is not drawn out into a spine in *Falites fala*, and the shape of the shield and the doublure in both species is much different. *Vestrogothia spinata* differs from *Vestrogothia granulata* in the outline of the shield and the presence of outgrowths on the shield. In *Vestrogothia hastata*, the maximum length is distinctly larger than the dorsal length. *Vestrogothia longispinosa* is missing the postero-ventral spines on the left valve. *Vestrogothia minilaterospinata* has very tiny antero-central spines compared with those of *Vestrogothia spinata*, and *Vestrogothia steffenschneideri* is missing any spines on the valve and the left shield margin.

Ontogeny

During the ontogeny of *Vestrogothia spinata*, the proportion of the length to the height of the valves remains

the same (Fig. 54). The antero-central spines grow ꞏom short pointed outgrowths to distinctly elongated s ꞏꞏnes. The postero-ventral spines occur relatively late ꞏ the ontogeny. They are completely missing in individu ꞏ s of less than 800 μm in shield length, such that the ꞏ di-viduals are divided into young stages with a sr ꞏꞏoth shield margin (A in Fig. 54) and later stages ꞏvith postero-ventral spines present (B in Fig. 54). With ꞏ the stages with a smooth shield margin, two more c ꞏ less distinct gaps can be observed (arrows in Fig. 54), ꞏ ꞏich may indicate at least three possible growth stages ꞏ -III in Fig. 54). The first larval stage recognised consis ꞏ of a body with at least four limb-bearing segments d ꞏ ally fused to an all-enclosing bivalved shield (ch64ꞏ1).

Discussion

Comparative morphology

The species of Phosphatocopina described herein ꞏ ffer strongly in the maximum size of their shields. Mo ꞏ ver, they show a large number of morphological diffe ꞏ ces in their shields, which are considered as specific, r ꞏ rd-less of individual variability or differences be ꞏ een growth stages. Comparison of the investigated s ꞏ cies with each other and with other phosphatocopine ꞏ om the literature aimed to find similarities and diffe ꞏ ces in morphological structures. Of particular importa ꞏ e is the comparison of the investigated species with a re ꞏ ntly found supposed juvenile growth stage (represen ꞏ by two specimens 330 and 340 μm in maximum leng ꞏ of a phosphatocopine from the *Protolenus* lime ꞏ ne, Comley "Series" of the Lower Cambrian of Shro ꞏ ire, England (Siveter *et al.* 2001). This species, descri ꞏ l in open nomenclature as "Phosphatocopida sp." by ꞏ i ꞏ eter *et al.* (2001), is used here as an ingroup member ꞏ the phylogenetic analysis and the coding of its charac ꞏ s is also given in the following text in the form chX:Y, ꞏ ere X is the character and Y is the respective characte ꞏ ꞏate (see Appendix B for a list of all characters code ꞏ and the character matrix).

PLATE 44

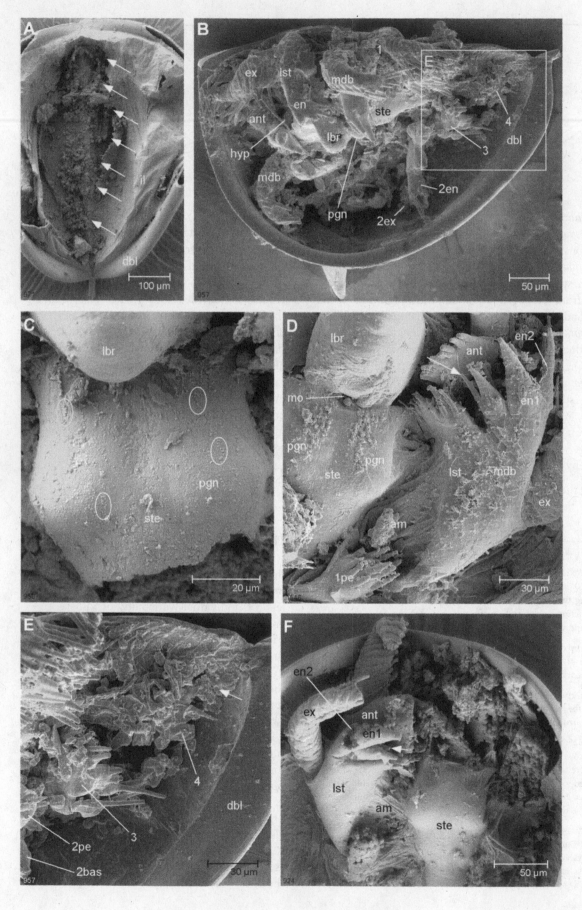

Bivalved shield. – The shield of the phosphatocopines investigated herein is bivalved and encloses the whole body. It has a dorsal furrow that might allow some movement of both valves, although a closing muscle, as in ostracodes, seems not to exist – at least it is not recognisable in the special fusion condition of the body proper to the shield. A bivalved all-embracing shield is also present in the Lower Cambrian species from England (Siveter *et al.* 2001) and the Middle Cambrian representatives described by Walossek *et al.* (1993), but the shield lacks the hinge furrow (Siveter *et al.* 2001) (ch1:2). All investigated phosphatocopines as well as any other individual representing an advanced growth stage have a straight dorsal rim. The condition in later growth stages of the Lower Cambrian species is not known (ch2:?).

The anterior free shield margin is straight in *H. necopina*, *H. curvispina*, and *Vestrogothia spinata*. It is curved in all other species. The ventral free shield margin is straight in *H. curvispina* and *Veldotron bratteforsa*. The posterior free shield margin is straight in *H. necopina*, *H. curvispina*, *H. trituberculata*, and *Vestrogothia spinata*. It is curved in all other species. The shield margin in the Lower Cambrian species is curved along its whole edge, but as the species is known from

---

Plate 44. *Vestrogothia spinata* Müller, 1964

A: The same specimen as illustrated in Pl. 43E. An inside view of both opened valves, displaying the doublure (dbl), the inner lamella (il) and the fusion area of the body proper to the shield, indicated by the margin where the body proper merges into the inner lamella (arrows).

B: UB 625 (Müller 1979a, fig. 29A–C). Image flipped horizontally. The same specimen as illustrated in Pl. 43D (here the left valve). A specimen representing an advanced growth stage, about 460 μm in length. A lateral view inside the left valve, displaying the partly preserved body. The anterior body is somewhat distorted, so the antenna of the left side (ant) is shifted to the other side. The rectangle marks the area magnified in Pl. 44E.

C: UB 600 (Müller 1979a, fig. 3A–D; Müller & Walossek 1985a, fig. 2f). A specimen representing a young growth stage, about 350 μm in length. Close-up of the sternum (ste), displaying several groups of fine hairs on the sternum and paragnaths (circles).

D: UB W 247. A specimen representing a young growth stage, about 310 μm in length. Close-up of the sternum (ste), labrum (lbr) and mandible (mdb). Note the undivided mandibular limb stem (lst), the enditic protrusion medio-distally of the limb stem (arrow) and the endopod divided into two portions (en1, en2).

E: Close-up of the hind body illustrated in Pl. 44B, displaying the posterior end of the body. The hind body is hardly preserved and does not provide any details.

F: UB W 248. A specimen of a young growth stage, about 460 μm in length. A ventral close-up of the sternum (ste) and the mandible with the limb stem (lst), the first endopodal portion (en1) and the exopod (ex) covering the antenna (ant). The second endopodal portion of the mandible (en2) is broken off. Note the enditic protrusion between the limb stem and the endopod (arrow). The remains of the soft parts are very coarsely preserved.

---

larval specimens only, the character cannot be coded (ch6:?, ch7:?, ch8:?).

The valves of *H. trituberculata* and *Veldotron bratteforsa* leave a gap postero-ventrally, whereas the valves of all other species, as well as those of younger growth stages including the Lower Cambrian one (ch9:?) close tightly along the free margin.

Position of the maximum length of the shield. – The maximum length of the shield of the species investigated herein is located either ventral to the midline, respectively, on the midline, as in *Trapezilites minimus* and *Waldoria rotundata*, or dorsal to the dorso-ventral midline of the shield, as in all other species. The maximum length of the shield corresponds to the length of the dorsal rim in *H. necopina* and *H. curvispina* as well as in the Lower Cambrian species from England (ch3:2, ch5:0).

In *H. trituberculata*, *H. kinnekullensis*, *H. toreborgensis* n. sp., and *Vestrogothia spinata*, the dorsal length is almost as large as the maximum length (Table 37). In all other species, the dorsal rim is significantly shorter than the maximum length of the valves (Table 37).

A dorsal rim being shorter than the maximum length of a bivalved shield occurs in three theoretically possible variations:

- only the anterior margin curves back more strongly than the posterior margin (Fig. 55A);
- both margins curve back equally strongly (Fig. 55B);
- only the posterior margin curves back more strongly than the anterior margin (Fig. 55C).

The first possibility (Fig. 55A) does not occur among Phosphatocopina, the second (Fig. 55B) is found, e.g. in *Trapezilites minimus* and more or less in *Waldoria rotundata*, *H. trituberculata*, *H. suecica* n. sp. and *H. angustata* n. sp. (Table 38). The third alternative is found in all other phosphatocopine species investigated (Table 38) as well as in other bivalved arthropod such as, e.g. †*Kunmingella maotianshanensis* Hou, 1987 (cf. Hou *et al.* 1996). Moreover, the third alternative is even found in non-bivalved shields of many arthropods such as †*Fuxianhuia protensa* Hou, 1987 (cf. Hou & Bergström 1997), †*Waptia fieldensis* Walcott, 1912 (cf. Bergström 1992), †*Isoxys aurius* (Jiang, 1982) (cf. Hou 1987a; Shu *et al.* 1995), and crustacean such as †*Bredocaris admirabilis* Müller, 1983 (cf. Müller & Walossek 1988).

Position of the maximum height of the shield. – The maximum height of the shield in relation to the antero-posterior midline in phosphatocopines can be located anterior to the midline (pre-plete), on the midline (amplete) or posterior to the midline (post-plete; cf. Fig. 8), i.e. three possible conditions. These three conditions, originally introduced for the description of the

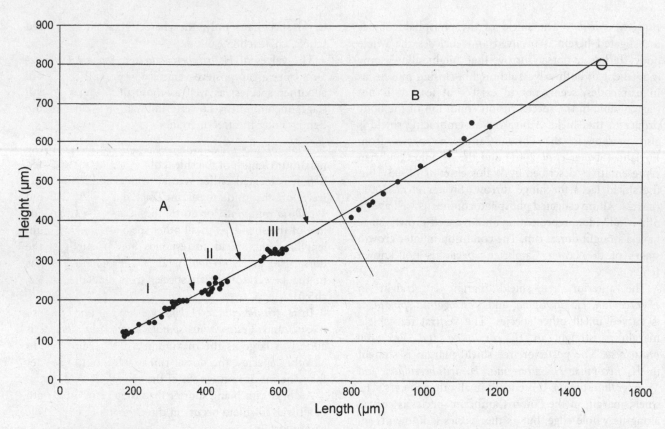

Fig. 54. Ontogeny of *Vestrogothia spinata* Müller, 1964: length versus height. Sixty-five specimens with measurable length and height were considered. ◯ = holotype. The trend line has the polynomial function $y = 0x^2 + 0.459x + 33.598$, y = height, x = length. For explanation of additional signs, see text.

Table 37. The minimum and maximum length of the shield, the maximum length of the dorsal rim and the percentage of the dorsal length of the maximum length plus the location of the maximum length in correspondence to the dorsal rim, respectively the midline of the shields in the phosphatocopine species studied herein. Data for "Phosphatocopida sp." from Siveter *et al.* (2001).

Species	Minimum	Maximum	Dorsal	%	Location of maximum length
Hesslandona unisulcata	240	1650	1390	84	Between the dorsal rim and the midline
Hesslandona necopina	260	880	880	100	On the dorsal rim
Hesslandona kinnekullensis	175	950	925	97	Between the dorsal rim and the midline
Hesslandona trituberculata	810	1460	1400	96	On the dorsal rim
Hesslandona ventrospinata	790	1730	1560	90	Between the dorsal rim and the midline
Hesslandona suecica n. sp.	230	790	730	92	Between the dorsal rim and the midline
Hesslandona angustata n. sp.	330	700	610	87	Between the dorsal rim and the midline
Hesslandona curvispina n. sp.	850	2300	2300	100	Between the dorsal rim and the midline
Hesslandona toreborgensis n. sp.	320	900	880	97	Between the dorsal rim and the midline
Trapezilites minimus	440	1010	680	67	On or ventral to the midline
Waldoria rotundata	630	1670	1380	83	On or ventral to the midline
Veldotron bratteforsa	850	2300	2060	90	Between the dorsal rim and the midline
Falites fala	130	2300	1810	79	Between the dorsal rim and the midline
Vestrogothia spinata	170	1490	1450	97	Between the dorsal rim and the midline
Phosphatocopida sp.	330	340	340	100	On the dorsal rim

ostracode shell, can be applied to any bivalved arthropod as opposed to really reflecting systematic relationships *per se*. *Falites fala* and *H. unisulcata* are the only phosphatocopine species investigated having a post-plete shield (Table 38). *Trapezilites minimus* and *Waldoria rotundata* have amplete valves; the outlines of their valves are antero-posteriorly symmetrical. The shield of the Lower Cambrian species is antero-posteriorly symmet-

rical and therefore amplete, but as it represents a young growth stage, no clear assumptions can be made (ch4:?). All other species dealt with herein have pre-plete valves (cf. Table 38). The outline of different species having a valve belonging to the same type is rather similar, e.g. the lateral outlines of the shield in the pre-plete valves of *Vestrogothia spinata* and *H. necopina* are almost equal (but see their relationships below). The outlines of the

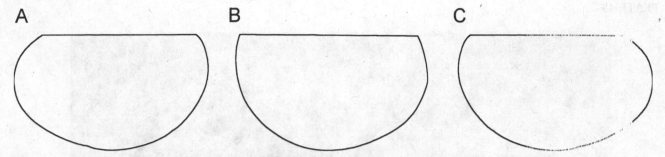

Fig. 55. Three hypothetical variations for having a maximum shield length which is shorter than the dorsal rim. Anterior left. Valves m.. not be antero-posteriorly symmetrical as the images may imply. A: The anterior margin curves back more strongly than the poster or margin. I The anterior and posterior margins curve back equally strongly. C: The posterior margin curves back more strongly than the anterior margin.

post-plete shields of *Falites fala* and *H. unisulcata* are very similar to each other, which may have led Hinz-Schallreuter (1993c) to assume that both species are closely related (cf. pp. 16–17). Variations in shields of different phosphatocopine species having a shield of the same type concern, e.g. the lobation of the valves rather than the variation of this particular type.

Ornamentation of the shield surface. – The valves of the phosphatocopine species investigated herein are either smooth or possess one or more lobes (Table 38). A smooth shield without any lobes occurs in *H. necopina*, *H. kinnekullensis*, *H. suecica* n. sp., *H. angustata* n. sp., *H. toreborgensis* n. sp., *Vestrogothia spinata*, and the Lower Cambrian member (chs10–18:0). Lobes, if present, occur symmetrically on both valves in all species. The position of the lobes seems to be stable even in different species (cf. Fig. 11). The terminology of Gründel (1981) and Williams & Siveter (1998) in labelling the different lobes in different phosphatocopine species (Fig. 11) is, hence, applied and is also used for the phylogenetic study in this investigation. A scheme for the occurrence of lobes can be presented:

- In species having one lobe (Table 38), this lo.. L_1 (cf. Fig. 11), is fairly large and is located in an a.. ro-dorsal position, as in *Trapezilites minimu* and *H. unisulcata* (Pls. 1A, 32A, C).

- If three lobes are present (Table 38), in *H. trituberculata* (sic!), *H. curvispina* n. sp., and *lites fala*, the anterior-most one is found in the sam.. osition as the single lobe L_1 in *H. unisulcat* and *Trapezilites minimus*. The second lobe, L_2, is l..ted in the first half of the length of the valves imme..tely behind the anterior one; the third lobe, L_3, lies p..terior to the midline. All three lobes are arranged p..llel to the dorsal rim (cf. Pls. 17A, 28B, C, 39A).

- If more than three lobes are present (Table 38) ..ch as in *H. ventrospinata*, *Waldoria rotundata*, *Vel..ron bratteforsa*, or in *Cyclotron lapworthi*, a species ..om the Upper Cambrian of England with six ..bes (Williams *et al.* 1994b; Williams & Siveter ..98; cf. Fig. 11), the pattern of three subdorsally ar..ged lobes L_1–L_3 is similar to the pattern of specie ..ith three lobes (cf. Pl. 38A). The fourth lobe is l..ted antero-medially, the large fifth lobe, L_5, is l..ted

Table 38. Distribution of variation in outlines as illustrated in Fig. 55 (*Hesslandona necopina* and *Hesslando..a curvispina* have an outline not indicated in Fig. 55, see text), the position of the maximum height of the shields in relation to the antero-posterior midline, and the number of lobes on the valves of the phosphatocopine species described herein. See text for explanation.

Species	Fig. 55	Pre-plete	Amplete	Post-plete	Lobes
Hesslandona unisulcata	C			+	1
Hesslandona necopina	–	+			0
Hesslandona kinnekullensis	B	+			0
Hesslandona trituberculata	B	+			3
Hesslandona ventrospinata	C	+			6–7
Hesslandona suecica n. sp.	B	+			0
Hesslandona angustata n. sp.	B	+			0
Hesslandona curvispina n. sp.	–	+			3
Hesslandona toreborgensis n. sp.	C	+			0
Trapezilites minimus	B		+		1
Waldoria rotundata	B		+		5
Veldotron bratteforsa	C	+			6
Falites fala	C			+	3
Vestrogothia spinata	C	+			0

PLATE 45

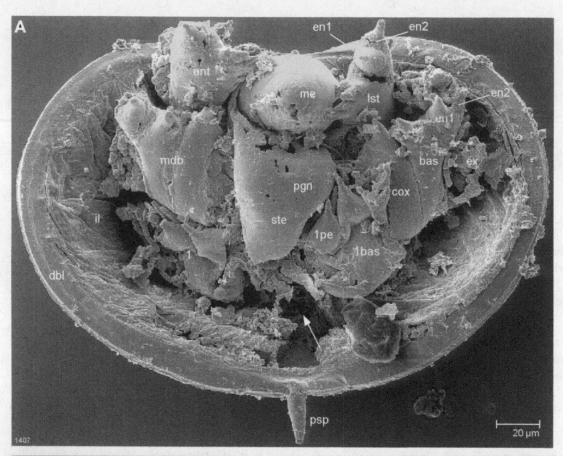

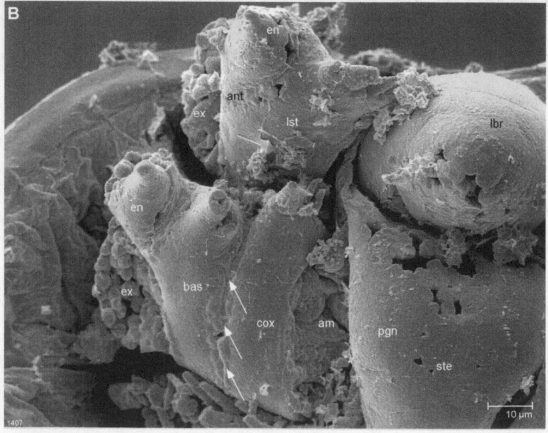

centrally on the valves, and the sixth lobe, L_6, is located postero-ventrally to lobe L_5. The antero-dorsal lobe in *Waldoria rotundata* is, however, significantly smaller compared with L_1 in, e.g. *Falites fala* or *H. unisulcata* (Pl. 34A). The shield of *H. ventrospinata* has a seventh lobe between L_4 and L_5 (Pl. 19A).

None of the phosphatocopine species from the Upper Cambrian of Sweden has two, four, or more than seven lobes.

Arthropod and crustacean shields may bear spine-like extensions on different parts of their surface. Spines also occur in phosphatocopines. Two of the Upper Cambrian Swedish phosphatocopine species show symmetrical spine-like outgrowths but on different positions on the surface of the shields. Some individuals of *Waldoria rotundata* have – instead of lobe L_5 – a spine at this position, which is, in accordance with Gründel (1981), interpreted as L_5 simply drawn out into a large central spine. The spine is only developed in larger specimens representing more advanced growth stages.

Gründel (1981) has distinguished two species, *Waldoria rotundata* with L_5 drawn out into a spine and *Waldoria* n. sp. *sensu* Gründel (1981) with a simple L_5 (see p. 115). Alternatively, the presence or absence of such a central spine might reflect sexual differences, but there is no further evidence for either of the two hypotheses. *Vestrogothia spinata* has symmetrical spine-like outgrowths on the shields antero-ventrally, in a position where no lobe occurs in any other species. The origination of the spine from a lobe, as assumed for *Waldoria rotundata*, can therefore be excluded. These spines are interpreted as completely new acquisitions and may represent an autapomorphy of *Vestrogothia*.

Other forms of ornamentation on the surface of the phosphatocopine shields besides the marginal rim (see below), such as ridges or reticulate textures, as in many ostracodes, do not occur among the species of the Upper Cambrian "Orsten" of Sweden. The valves of *Tubupestis tuber* Hinz & Jones, 1992 from the *Ptychagnostus gibbus* Zone of the Middle Cambrian of Queensland, Australia, are drawn out into numerous nodes (see Hinz & Jones 1992, cf. Appendix B) which is not shown in any other

known phosphatocopine. This condition is interpreted as an autapomorphy of *Tubupestis tuber*

The marginal rim of the valves. – The so-called free margin of the valves of the Phosphatocopina from the Upper Cambrian of Sweden extends from the antero-dorsal far towards the ventral side and backwards to the postero-dorsal corner of the shield. It is smooth except in two species. The Lower Cambrian species from England has in fact a smooth margin but this form is young and not representative of this species (chs19 ult:?). *Hesslandona ventrospinata* has flat triangular outgrowths on the margins of both valves which interlock (Pl. 19A, D), such that the valves are asymmetrical. Comparable outgrowths are present in all species of *Bidimorpha* Hinz-Schallreuter, 1993 (see Hinz-Schallreuter 1998), which, due to the presence of an interdorsum and interdorsal spines, may be part of the taxon *Hesslandona* as understood in this paper. The second case is *Vestrogothia spinata*, where the margin of the left valve of larger instars is drawn out into a long spiny outgrowth directed posteriorly, while the margin of the right valve lacks such a structure (Pl. 42A; Fig. 52A, B). Because the position of the spine is different from that of the flat triangular outgrowth of the right valve of *H. ventrospinata*, homology is ruled out and the structures are coded differently in the matrix, each are probably autapomorphies. This interpretation is supported by the appearance of similar structures in various, apparently unrelated taxa, such as *Walossekia quinquespinosa* Müller, 1983 (see Müller 1983).

Dorsal rim. – The dorsal rim of the phosphatocopine species investigated herein appears in two different conditions. The dorsal rim of *Vestrogothia spinata* and *Falites fala* shows small, elongated triangular plates anteriorly and posteriorly between the valves and a simple membranous furrow extending between the plates, thus separating the valves from each other (Pls. 39C, 40A, 41C). In all other Upper Cambrian phosphatocopines from the Swedish "Orsten", namely *Waldoria rotundata*, *Veldotron bratteforsa* and all species of *Hesslandona*, a continuous cuticular dorsal bar of specific width, the interdorsum, separates both valves and is bordered by a membranous furrow on both sides (cf. Pls. 10, 19E, 22A, 26A). The maximum width of the interdorsum of *H. angustata* n. sp. is 1/36 the length of the shield and represents the narrowest interdorsum of the phosphatocopines investigated (Table 39). *Hesslandona suecica* n. sp. has the widest interdorsum in relation to shield length, being 1/11 the length of the shield (Table 39). By contrast, *Parashergoldopsis levis* Hinz-Schallreuter, 1993 from the *Triplagnostus gibbus* Zone, Middle Cambrian of Australia, has the widest interdorsum of a phosphatocopine reported so far, being about one-seventh the length of the shield (cf. Hinz-Schallreuter

---

Plate 45. *Vestrogothia spinata* Müller, 1964

A: UB W 249. A specimen representing an early growth stage, about 170 μm in length. A ventral view of the opened shield, displaying an early larva with four limb-bearing segments (antennula covered). Note the partition of the mandibular limb stem into a coxa and a basipod. The hind body is poorly preserved. The antenna and mandible of the right side (left in the image) are magnified in Pl. 45B.

B: Close-up of the antenna and mandible illustrated in Pl. 45A. Note the undivided limb stem of the antenna and the suture between the coxa and the basipod of the mandible.

Table 39.   Dorsal rim. Width of the interdorsum in relation to the maximum length of the shield, the antero- and postero-dorsal spines and the length of the spine in relation to the maximum length of the shield of the phosphatocopine species investigated herein. Data for "Phosphatocopida sp." are from Siveter *et al.* (2001).

Species	Interdorsum width	Antero-dorsal spine	Proportional length	Postero-dorsal spine	Proportional length
Hesslandona unisulcata	1/18	○	●	○	●
Hesslandona necopina	1/14	+	1/5	+	1/4
Hesslandona kinnekullensis	1/19	+	1/6	+	1/5
Hesslandona trituberculata	1/14	+	1/11	+	1/16
Hesslandona ventrospinata	1/14	+	1/5	+	1/4
Hesslandona suecica n. sp.	1/11	+	1/15	+	1/9
Hesslandona angustata n. sp.	1/36	–	●	–	●
Hesslandona curvispina n. sp.	1/18	+	1/3–1/2	+	1/3–1/2
Hesslandona toreborgensis n. sp.	1/16	+	1/26	+	1/15
Trapezilites minimus	1/14	○	●	○	●
Waldoria rotundata	1/30	–	●	–	●
Veldotron bratteforsa	1/26	–	●	–	●
Falites fala	–	–	●	–	●
Vestrogothia spinata	–	–	●	+	1/9
Phosphatocopida sp.	–	–	●	–	●

+, spine present; –, spine absent; ○, hump-like thickening; ●, proportion not applicable as no spine present.

1993a, c). The possible phylogenetic significance of these differences could not be investigated within this paper, but will be part of a further study, including all species known to have an interdorsum.

The antero-dorsal area of the interdorsum may be smooth, such as in *H. angustata* n. sp. (Pl. 26B, D), or bear hump-like thickenings, as in *H. unisulcata* (Pl. 1A, D) and *Trapezilites minimus* (Pl. 32A), or drawn out into a more or less long spine, such as in *H. ventrospinata* (Pl. 20A; Table 39). When a spine is present, it can be directed antero-dorsally, as in *H. necopina* (Pl. 13A) and *H. curvispina* (Pl. 28A), or dorsally, as in *H. kinnekullensis* (Pl. 16A). *Hesslandona curvispina* (Pl. 28A) has the longest spines relative to shield length (Table 39). The anterior spines of all investigated species are smooth.

The Lower Cambrian species from England does not show a dorsal furrow and has neither isolated plates nor an interdorsum. This species is only represented by a young growth stage, but the young growth stages of Upper Cambrian phosphatocopines do possess such structures, so the characters can be coded for the Lower Cambrian species too (ch22:1, ch23–33:0). Putative phosphatocopine species of the taxa Dabashanellidae (e.g. Huo *et al.* 1983, 1991; Zhang 1987; Melnikova & Mambetov 1990, 1991) and Liangshanellidae (e.g. Hou 1987b; Abushik *et al.* 1990; Huo *et al.* 1991) from the Lower Cambrian of China, and the Monasteriidae (e.g. Fleming 1973; Shu 1990a) and Oepikalutidae (e.g. Jones & McKenzie 1980) from the Middle Cambrian of Australia (see Appendix B for taxonomy of taxa and for complete literature) plus unassigned Middle Cambrian forms from Australia (Walossek *et al.* 1993, fig. 4A, B) have in fact a bivalved shield with a straight dorsal rim, but no dorsal furrow. The shield of these forms is strongly curved dorsally as if there are two valves (cf. Zhang & Pratt 1993).

In the case of *Vestrogothia spinata* and *Falites fala*, which have no continuous interdorsum but elongated triangular plates, the anterior plates are smooth (Pls. 39A, 42C).

The postero-dorsal area of the interdorsum may be smooth, such as in, e.g. *Waldoria rotundata* (Pl. 34A), or bear hump-like thickenings, such as in *H. unisulcata* (Pl. 1E) and *Trapezilites minimus* (Pl. 32A), or – comparable with the antero-dorsal area – is drawn out into a more or less long spine (Table 39). When a spine is present, it can be directed posteriorly, such as in *H. trituberculata* (Pl. 17A), postero-dorsally, as in *H. necopina* (Pl. 13A), or dorsally, as in *H. kinnekullensis* (Pl. 16A). The longest spine relative to shield length occurs in *H. curvispina* (Pl. 28A); it even curves towards the postero-ventral in the latter half of its length. The posterior spines of *H. necopina* and *H. curvispina* show scalid-like outgrowths on the surface of the distal half (Pls. 13A, 29A, B); such structures are not developed in any other species and may indicate a close relationship between both species.

In the two species that do not have an interdorsum but have elongated triangular plates, the posterior plate of *Falites fala* is smooth (Pl. 40A), whereas it is drawn out into a straight spine directed posteriorly in *Vestrogothia spinata* (Pl. 42A).

Doublure. – A doublure is present in all phosphatocopine species, including the Lower Cambrian one (ch34:1). It extends more or less parallel – with variability in width – from the anterior around the ventral outline towards the postero-dorsal edge of the shield, being more or less symmetrical in both valves. The width of the doublure varies from the anterior towards the posterior in all species and even in the youngest forms. The width of the anterior part of the doublure in relation to the shield ranges from 1/23 in *H. toreborgensis* n. sp. (cf. Pl. 31D; Table 40) to 1/10 in *Veldotron bratteforsa* (cf. Pl. 38D; Table 40). The width of the median part of the doublure in relation to shield length ranges from 1/27 in *H. ventrospinata* (Table 40) to one seventh in *Falites fala*, showing much larger differences in width than the anterior part of the doublure. The width of the posterior part of the doublure in relation to shield length ranges from 1/15 in *H. toreborgensis* n. sp. (Table 40) to one quarter in *Falites fala*, being distinctly wider on average than the anterior and median parts of the doublure. The maximum width of the doublure is, therefore, located postero-ventrally to posteriorly in almost all species, including the Lower Cambrian one (ch35:1, ch36:3, ch37:1; cf. Table 40). Only in *Trapezilites minimus* is the median (=ventral) part of the doublure widest, corresponding to the antero-posterior symmetry of the outline of the valves and the relatively large height in proportion to length.

Structures on the doublure occur in several but not all species. *Hesslandona unisulcata* has bottle-like structures irregularly arranged on the surface of the median part of the doublure (Pl. 2C). *Hesslandona necopina*, *H. ventrospinata* and *H. curvispina* have conical, dome-like outgrowths which are somewhat regularly arranged close to the inner edge of the ventral part of the doublure (Pls. 15C, 20D, 30A, B), and which may indicate a close relationship. Similar structures are also present in *Waldoria rotundata* and *Veldotron bratteforsa* (Pls. 35E, 38D). All other species, including the Lower Cambrian representative, have a smooth doublure (chs38–42:0).

Soft part morphology. – Soft part morphology is described in detail for only one species, *H. unisulcata*, but other species also show at least partly preserved soft parts, such as *H. necopina*, *H. suecica* n. sp., *H. ventrospinata*, *Waldoria rotundata*, *Falites fala*, and *Vestrogothia spinata*. Their description will be presented in detail in a second paper on the "COen" Phosphatocopina. Although most phosphatocopine species are only preserved as empty shields, at least some general aspects of the soft part morphology will be discussed. Because soft part details are also known from the recently found phosphatocopine species from the Lower Cambrian of England (Siveter *et al.* 2001), they can be compared in detail with those of the phosphatocopines investigated herein. Isolated phosphatocopine limbs reported from the Middle Cambrian of the Georgina Basin, Queensland, Australia are also considered (Walossek *et al.* 1993).

The body of all phosphatocopine species is more or less similar, with a bulging anterior end including the hypostome/labrum complex, the sternum, and a conical hind body without clear segmentation. The maximum width is at the mandibular segment. The body proper is fused to the shield via a rather narrow area in all species investigated. The area of fusion of the body proper to the shield in the Lower Cambrian species is not known (ch43:?). The number of segments fused to the shield seems to be different in the particular species. The maximum number of segments observed in *H. unisulcata*, is six (Pl. 12B, D), whereas in *H. necopina* and *Waldoria rotundata*, at least nine segments are fused dorsally to the shield (Pls. 15E, 36D). In the Lower Cambrian phosphatocopine from England, four segments are fused dorsally to the shield, but as both specimens known are supposed to be early larval stages, it cannot be coded into the data matrix (ch44:?).

In all phosphatocopine species with known soft parts, including the Lower Cambrian one, the anteriormost part of the body proper is the hypostome/labrum complex (ch61:1) (Fig. 56).

Table 40. Width of the anterior, median and posterior part of the doublure as well as the maximum width of the doublure in relation to the shield length. Data for "Phosphatocopida sp." were ascertained from fig. 1A of Siveter *et al.* (2001).

Species	Anterior	Median	Postero-ventral	Posterior	Maximum width
Hesslandona unisulcata	1/16	1/22	1/11	1/15	1/11
Hesslandona necopina	1/20	1/27	1/20	1/14	1/14
Hesslandona kinnekullensis	1/17	1/23	1/17	1/14	1/14
Hesslandona trituberculata	1/15	1/15	1/8	1/12	1/8
Hesslandona ventrospinata	1/18	1/20	1/11	1/15	1/11
Hesslandona suecica n. sp.	1/15	1/21	1/13	1/15	1/13
Hesslandona angustata n. sp.	1/17	1/15	1/12	1/14	1/12
Hesslandona toreborgensis n. sp.	1/23	1/18	1/15	1/18	1/15
Hesslandona curvispina n. sp.	1/16	1/18	1/11	1/15	1/11
Trapezilites minimus	1/10	1/8	1/9	1/10	1/8
Waldoria rotundata	1/12	1/9	1/7	1/9	1/7
Veldotron bratteforsa	1/10	1/13	1/6	1/10	1/6
Falites fala	1/12	1/7	1/4	1/6	1/4
Vestrogothia spinata	1/14	1/26	1/9	1/12	1/9
Phosphatocopida sp.	1/12	1/12	1/11	1/11	1/11

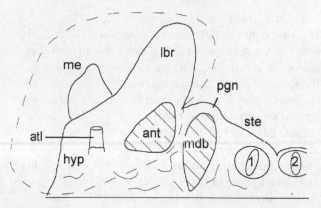

Fig. 56. Reconstruction of the head of a phosphatocopine, displaying the hypostome/labrum complex (embraced in a dashed line) with the hypostome (hyp), labrum (lbr), median eye (me), the sternum (ste) with paragnaths (pgn) and limb insertions (cf. Figs. 14B, 23).

The antero-proximal portion of the hypostome/labrum complex is the rhomboid hypostome (Fig. 56). Its posterior end is drawn out into a conical structure with a bluntly rounded end, the labrum (cf. Pls. 11C, 25A, 33C; Fig. 56) which has pores, sensilla, papilliform structures and hairs on its posterior side (cf. Pl. 3F, G). Antero-distally on the hypostome, a putative eye structure occurs medially in different states of preservation, as described for *H. unisulcata* on pp. 21–22. It consists of three cups. In species with known eyes, such as *H. unisulcata* and *H. suecica* n. sp., two of them are symmetrical anteriorly on both sides, a third inconspicuous one may be located postero-distally (cf. Pls. 3C, 33A).

The post-oral cephalic region is again a rhomboid portion with its maximum height anteriorly. It consists of the segments of the antenna, the mandible and the first post-mandibular pair of limbs, which ventrally form a single plate, the sternum (cf. Pls. 21A, 24A; Fig. 56), the fusion product of the respective sternites. The mandibular sternal part is domed into a pair of humps, the paragnaths (cf. Pls. 25A, 33C; Fig. 56). This condition is present in all species investigated, as well as in the Lower Cambrian phosphatocopine from England (ch62:1, ch63:1).

The antennulae insert on the antero-lateral side of the hypostome via a short socket (Pl. 7B; Fig. 56) and are very small compared with the head size in all species yielding soft parts, and they bear only a few terminal setae (Pls. 7B, 10A, 11A, 37B). They never reach one quarter the head length. The antennulae are more or less weakly annulated, e.g. in *H. unisulcata* (Pl. 10A), and slightly more distinctly in *Waldoria rotundata* (Pl. 37B). The morphogenesis of the antennulae of phosphatocopine species investigated herein is more or less similar between species. The antennulae are small in early instars and bear only a few setae. This condition does not change in later growth stages (Pls. 7B, 11A).

The antennula of the Lower Cambrian representative is insufficiently known (ch45:?, ch46:?), but it may also be rather small (cf. Siveter *et al.* 2001).

The antennae insert at the postero-lateral edge of the hypostome/labrum complex (Fig. 56). They consist of a prominent limb stem and two rami. The undivided limb stem carries the two-part endopod and the exopod, which is multi-annulated in all species studied (cf. Pls. 6A, 25A, 41A; Fig. 57A). The morphogenesis of the antenna shows some minor differences between the species, particularly in the development of the median gnathic edge of the limb stem and the median armature of the endopod. The limb stem is undivided in all stages of all species of Upper Cambrian phosphatocopines with preserved soft parts (cf. Pls. 5F, 10A, 41C; Figs. 57A, 65A, B) as well as in the isolated antennae from the Middle Cambrian of Australia (Walossek *et al.* 1993). The phosphatocopine species from the Lower Cambrian of England, however, has an antenna with a two-divided limb stem, consisting of a coxa and a basipod (ch50:2, ch51:1), both being drawn out medially into a spine-bearing endite (Siveter *et al.* 2001; Fig. 58A). It is unclear whether this condition changes later in ontogeny (ch52:?), but it demonstrates a condition identical to that of the mandible of the same species (see below). The endopod is similarly organised in all Upper Cambrian phosphatocopine species, consisting of two medially projecting portions.

By contrast, the antenna of the Lower Cambrian phosphatocopine has a three-divided endopod (ch53:4), so matching the condition of the post-mandibular limb (Fig. 58A, C). The exopod is similar in all known phosphatocopine species in comprising a ontogenetically varying number of setated annuli: in *H. unisulcata*, it consists of about 10 annuli in young stages and up to at least 24 annuli in late stages (Pls. 4D, 5F, 11C; Table 14). The Lower Cambrian species has an antennal exopod comprising at least eight annuli (Siveter *et al.* 2001), so matching the number in *H. unisulcata*. The Middle Cambrian isolated antennae clearly lack an exopod (Walossek *et al.* 1993).

The mandibles of later stages of the phosphatocopine species studied uniformly consist of a prominent limb stem and two rami (cf. Figs. 57B, 66A, B). The Lower Cambrian representative from England, however, has a mandible which is very similar to its antenna in having a limb stem consisting of a coxa and a basipod. Both are distinctly separated by a suture (ch54:2, ch55:1) (Fig. 58A, B). It is unclear whether this condition is the same in later stages, which are not known (ch56:?). The isolated mandibles from Australia have a two-divided limb stem as well (Walossek *et al.* 1993; medio-lateral extension of the coxa about 100 μm, cf. Table 14).

Moreover, in one species from the Upper Cambrian, *Vestrogothia spinata*, the earliest stages have a mandible

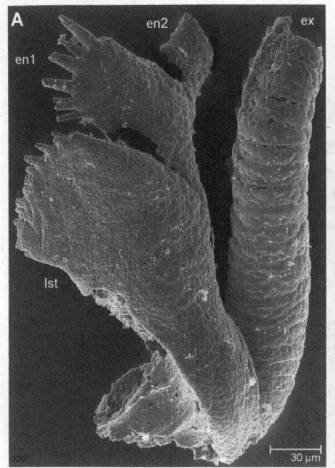
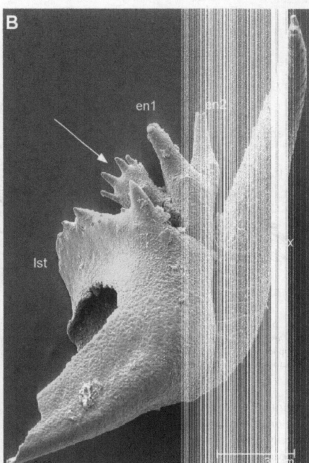

Fig. 57. A: UB W 250. Image flipped horizontally. Isolated right antenna from the anterior of *Hesslandona unisulcata*, possible growth stage III. Zone 1 of the Upper Cambrian of Gum, Kinnekulle, Sweden. The endopod is damaged and the tip of the exopod is missing. B: UB W 106 (Walossek *et al.* 1993, fig. 3F). Image flipped horizontally. Isolated right mandible from the posterior of an unknown phosphatocopine species. Zone 1 of the Upper Cambrian of Gum, Kinnekulle, Sweden. Note the high number of annuli of the exopod. The arrow points to the rest of the enditic protrusion of the basipod, which can easily be misinterpreted as the proximal portion of the endopod.

with a two-divided limb stem consisting of a coxa and a basipod (Fig. 59A). All other species have an undivided limb stem from the beginning (Fig. 59B). In *Vestrogothia spinata*, the mandibular limb stem portions gradually fuse with one another during ontogeny in a very special way: the outer edge of the basipod becomes progressively thinner, while the median portion with its seta becomes more lobate. Eventually, only a weak constriction at the basis of the exopod indicates the outer part of the basipod (Fig. 59C).

In all investigated phosphatocopines including later growth stages of *Vestrogothia spinata*, the limb stem is represented by one large portion with an oblique gnatho-base that seems to carry a three-part endopod and the multi-annulated exopod (Fig. 57B). In fact, the ontogeny of *Vestrogothia spinata* clarifies the fact that the seemingly proximal "podomere" of the endopod represents the retained median basipodal endite (arrow in Fig. 57B). Accordingly, the endopod is only divided into two medially projecting portions, which holds true for all Upper

Cambrian phosphatocopine species and the isolated mandibles from the Middle Cambrian (Walossek *et al.* 1993). The mandible of the Lower Cambrian phosphatocopine has, as in the antenna, a three-part endopod (ch57:4) (Fig. 58B).

The phosphatocopine mandibular exopod is somewhat like that of the antenna and is similar in being multi-annulated with seta-bearing and a few setaless annuli. It starts its ontogeny with about 10 annuli, in e.g. *H. suecica* (Pl. 59B), and may reach about 30 annuli, as in an isolated and unassigned limb (Fig. 57B). At least in the species investigated with antennal and mandibular exopods preserved, the number of annuli of the mandibular exopod is comparable with the number of annuli of the antennal exopod. This has also been proposed by Siveter *et al.* (2001) for the Lower Cambrian species. The Middle Cambrian isolated mandibles clearly have an exopod (Walossek *et al.* 1993).

The post-mandibular limbs of the investigated phosphatocopines are more or less serially similar, as in

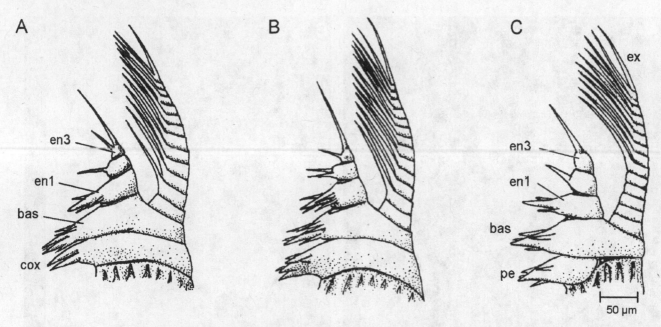

Fig. 58. Reconstruction of the limbs of the Lower Cambrian phosphatocopine species from England from the posterior [after Siveter *et al.* (2001, fig. 2)]. A: Antenna. B: Mandible. For the labelling of structures see antenna. C: First post-mandibular limb. "The length of the spines in general, the exopod of the mandible, and the exopod of the first maxilla are based on what is known from early instars of Upper Cambrian phosphatocopids ..." (Siveter *et al.* 2001, p. 481).

H. unisulcata, but may change from the anterior to the posterior, as well as differ from one species to another. Generally, they consist of a prominent basipod with a large proximal endite medio-proximally, a three-part endopod and an exopod comprising a series of setated annuli; the annuli may be partly fused (cf. Pls. 10B, 36C, 41A). This composition is also present in the fourth pair of limbs of the Lower Cambrian species from England (ch58:?, ch59:1, ch60:?; Siveter *et al.* 2001; Fig. 58C) and is comparable with that known from the isolated post-mandibular limbs from Australia, which lack the exopod as in the antenna and the mandible (Walossek *et al.* 1993).

Hesslandona unisulcata has relatively large, antero-posteriorly flattened post-mandibular limbs, which fill the whole inner space between both valves, the domicilium (Pl. 7A). The endopods consist of strongly medially elongated portions that point towards the mouth, and the exopod is a flat distally tapering plate with marginal setation (Pl. 12G). In *Waldoria rotundata*, the post-mandibular limbs are rather small and leave much space within the domocilium. They differ from those of *H. unisulcata* in that the endopod is only a small bud-like structure, while the exopod is a triangular, small and flat plate (Pl. 36C). The post-mandibular limbs of *Waldoria rotundata* are very similar to those of *H. ventrospinata*; the first three pairs of post-mandibular limbs are subequal in having a basipod with a proximal endite, a three-part endopod and a flat exopod (Pl. 36C). The succeeding pairs of limbs of both species differ strongly in consisting of an undivided limb stem without

a distinct proximal endite and two rami. The limb stems form, together with the exopods, almost medio-laterally symmetrical flat plates on which the endopods appear as small bulb-like outgrowths on the median edge (cf. Pl. 21E). The endopods of the post-mandibular limbs of *H. suecica* n. sp. and *H. necopina* are slightly medially projecting (cf. Pl. 25B), comparable with the condition found in younger stages of *H. unisulcata*. The exopods of, e.g. *H. necopina*, *H. suecica* n. sp. and *H. angustata* n. sp., are multi-annulated throughout ontogeny, and never become flat plates.

The hind body of the species investigated is usually poorly preserved and seems to be not only very soft but also weakly segmented (cf. Pls. 25B, 36D). The earliest larvae even almost completely lack a hind portion (Pl. 5A), which also holds true for the Lower Cambrian species (ch47:?, ch48:?, ch49:?; Siveter *et al.* 2001). A distinct telson portion, as is present in the ground pattern of Eucrustacea, could not be observed. An anus, supra-anal flaps, a dorso-caudal spine and latero-caudal spines could not be found; characters also represented in the ground pattern of the Eucrustacea (see Walossek 1993, 1999).

Furcal rami have long been thought to be missing in Phosphatocopina, which led Walossek (1999) to assume that a telson plus furca characterises only the Eucrustacea. Although very rarely seen, furcal rami have now been discovered in at least two of the phosphatocopines investigated, *H. unisulcata* (Pl. 12D) and *Waldoria rotundata* (Pl. 36D), and one isolated furca of unknown affinity has been discovered (Fig. 60).

In *H. unisulcata* and *Waldoria rotundata*, the furca

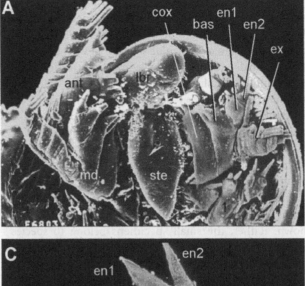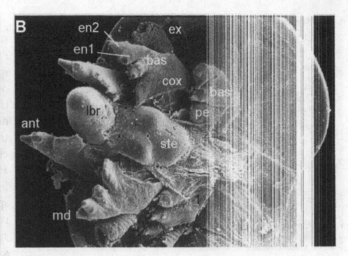

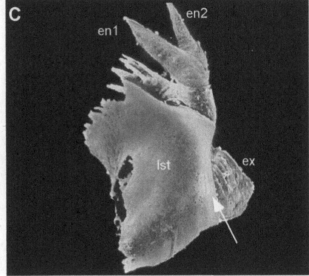

Fig. 59. Morphogenesis of the mandible. A: UB W 120 *Vestrogothia spinata*, early growth stage, having a mandible with a limb stem consisting of two portions, coxa (cox) and basipod (bas). B: UB W 140 *Hesslandona suecica* n. sp., early growth stage of about the same size as the one in (A). A suture between the coxa and the basipod of the mandibular limb stem is not present. C: UB W 139 *Vestrogothia spinata*, mandible of an advanced growth stage, limb stem now undivided. The arrow points to the rest of the disto-lateral part of the basipod.

consists of a pair of flat triangular plates with marginal setation (Pls. 12D, 36D), very similar to the exopods of the post-mandibular limbs of *Waldoria rotundata* and *H. ventrospinata*. The significance of the presence of furcal rami in Phosphatocopina will be discussed later.

Ontogeny. – The size of the adults of the species investigated herein is not clear. The largest individuals at hand with a shield length of 2300 µm are from *Falites fala*, *Veldotron bratteforsa*, and *H. curvispina* (Table 37). *Semillia pauper* Hinz, 1992 is reported with at least one specimen having a shield 2600 µm long (Hinz 1992a). Species of *Cyclotron* Rushton, 1969 in particular are reported with larger shields, such as *Cyclotron lapworthi* (Groom, 1902) from Zone 2 of the Upper Cambrian of England with shields of up to 5600 µm in length (Williams & Siveter 1998). The holotypus of *Cyclotron furcatocostatum* Gründel *in* Gründel & Buchholz, 1981

from erratic boulders of Zone 2 of the Upper Cambrian of Rügen, Germany, is 3700 µm long (Gründel 1 1).

Within the material investigated, the size of the first growth stages of the different phosphatocopine species differs significantly (Table 37; Fig. 61). The proportion of the length to the height of the shield is also variable between species. The largest individual of *H. angustata* n. sp. is only 700 µm long and has the smallest maximum size of a particular species among the material studied (Table 37; Fig. 61). The smallest individual of *H. curvispina* is 850 µm long and has the largest minimum size of any particular species (Table 37; Fig. 61). *Hesslandona curvispina* n. sp., *H. ventrospinata*, *H. trituberculata*, *Waldoria rotundata* and *Veldotron bratteforsa* probably started their development with growth stages that are significantly larger than the first instars of all other investigated species (cf. Table 37; Fig. 61).

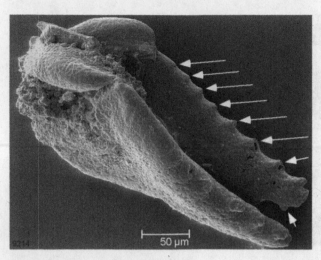

Fig. 60. UB W 251. Image flipped horizontally. Isolated furcal rami of an unknown phosphatocopine from Zone 1 of the Upper Cambrian of Gum, Kinnekulle, Sweden. The arrows point to the scars of the marginal setation.

The shields of even the smallest specimens of the five species mentioned above are very much like shields of distinctly larger growth stages; thus they can be easily discriminated.

The material investigated contains a high number of small specimens up to 400 μm in shield length, probably representing early stages of different taxa which could not be assigned to a particular species. Nevertheless, any of these small individuals are significantly different in the outline of the shield, the morphology of the interdorsum and interdorsal spines and in the shape of the doublure, as compared with instars that can be clearly assigned to *H. trituberculata, H. ventrospinata, H. curvispina* n. sp., *Waldoria rotundata* and *Veldotron bratteforsa.* Therefore, it is assumed that the unassignable specimens in the present material probably do not represent young stages of these species, i.e. from those species from which only relatively large instars are known. Rather, the small specimens belong to species from which similarly small stages are known. However, the small individuals representing early instars are all very similar and lack specific characters, characters developed only later during ontogeny. Here only ventral details could help further, which could not be studied in detail from the bulk of the small specimens.

The ontogeny of *H. unisulcata* starts with a larva having four pairs of fully developed appendages (Pl. 4D),

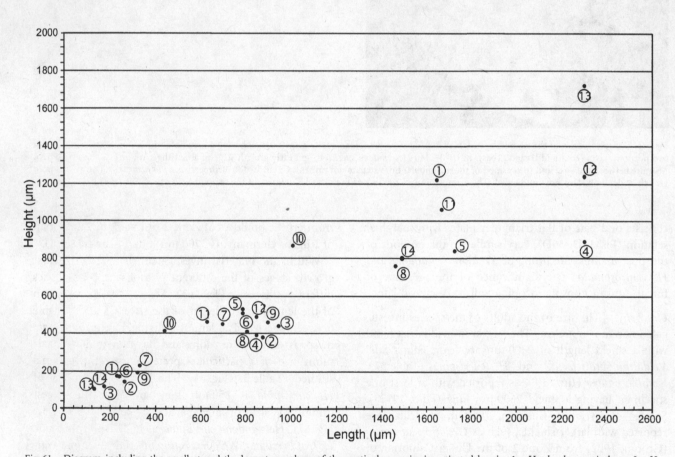

Fig. 61. Diagram including the smallest and the largest members of the particular species investigated herein. 1 = *Hesslandona unisulcata;* 2 = *H. necopina;* 3 = *H. kinnekullensis;* 4 = *H. trituberculata;* 5 = *H. ventrospinata;* 6 = *H. suecica* n. sp.; 7 = *H. angustata* n. sp.; 8 = *H. curvispina* n. sp.; 9 = *H. toreborgensis* n. sp.; 10 = *Trapezilites minimus;* 11 = *Waldoria rotundata;* 12 = *Veldotron bratteforsa;* 13 = *Falites fala;* 14 = *Vestrogothia spinata.* The shield height is strongly variable between specimens of different species of the same shield length, e.g. note the largest members of 4, 12 and 13.

a "head-larva" *sensu* Walossek & Müller (1990). A larva with less than four functional pairs of limbs has not been found in any of the remaining phosphatocopine material from the Upper Cambrian "Orsten" of Sweden or any other locality. Additionally, one of the smallest specimens, as yet of unclear affinity, present in the material studied, also has four functional pairs of limbs (Fig. 62). This is also true for the Lower Cambrian species from England (Siveter *et al.* 2001), which again represents a "head-larva" (ch64:1).

Larvae with less pairs of limbs have been described for the "Orsten" branchiopod †*Rehbachiella kinnekullensis* Müller, 1983 by Walossek (1993), representing an orthonauplius or "short-head-larva" *sensu* Walossek & Müller (1990). The first larva of the "Orsten" maxillopod †*Bredocaris admirabilis* Müller, 1983 represents a metanauplial stage, as the first pair of post-mandibular limbs is rudimentarily present (Müller & Walossek 1988, fig. 4A). Because small representatives of phosphatocopines belonging to early stages of different species are very frequent in the material, it is assumed that phosphatocopines do not have larvae comparable with a eucrustacean orthonauplius having only three pairs of functional appendages. The phosphatocopine ontogeny probably started with a "head-larva", a larva that is part

of the ground pattern of Euarthropoda (M. & Waloszek 2001a).

Phylogenetic analysis of Phosphatocopina

General remarks

There have been a few suggestions for relationships within the Phosphatocopina (e.g. Müller 82a; McKenzie *et al.* 1983; Hinz-Schallreuter 98) (Table 41), but they are not based on phylogenetic systematics *sensu* Hennig (1950). Therefore, it was necessary to establish a character-based phylogenetic analysis (Appendix A) and to reconstruct the ground pattern of Phosphatocopina.

All characters documented in the descriptive part were also used in the phylogenetic analysis. This contained mainly characters from the valves, but also a selection of soft part characters (see Appendix A for a character list and the data matrix).

As an operational outgroup, †*Agnostus pisiformis* Wahlenberg, 1822 was chosen. This is another "Orsten" arthropod, for which the body morphology of various larval stages has been described in considerable detail by Müller & Walossek (1987).

Additional ingroup taxa were (p. 14; Appendix A): the Upper Cambrian †*Martinssonia elongata* Müller & Walossek, 1986 as a representative of the derivatives of the eucrustacean stem lineage, the Upper Cambrian †*Rehbachiella kinnekullensis* Müller, 1983, the recent *Euphausia superba* Dana, 1852, as members of entomostracan and malocostracan Eucrustacea, the Lower Cambrian phosphatocopine from England, "Phosphatocopida sp." *sensu* Siveter *et al.* (2001) and the Middle Cambrian phosphatocopine *Cyclotron lapworthi* (Groom, 1902).

The phylogenetic analysis using parsimony (PAUP, see Appendix A for settings) resulted in two shortest trees, which differ only in the position of *Waldoria rotundata*. From the trees, a strict consensus tree and a 50% majority rule consensus tree (Fig. 63) were computed. Both are identical (Fig. 63), leaving *Waldoria rotundata* in an unresolved trichotomy with *Cyclotron lapworthi* and *Veldotron bratteforsa*. Because it is not possible to reconstruct the phylogeny on the basis of such a consensus tree (cf. Kitching *et al.* 1998) both originally resulting trees were used, characters compared and the unresolvable lineage collapsed to a trichotomy. This yielded a single tree (Figs. 64, 67), which is equal to both consensus trees.

Monophyletic units within Phosphatocopina (Figs 64, 67) are discussed as follows on the basis of apomorphies of particular nodes, as they resulted from the phylogenetic analysis. The nodes are numbered in

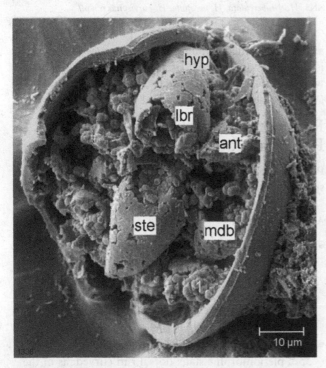

Fig. 62. UB W 252. Specimen of an early growth stage, about 125 µm in length, of an unknown species of Phosphatocopina, probably representative of the taxon *Vestrogothia*. Zone 5 of the Upper Cambrian of the area between Stenstorp and Dala, Sweden. The antennula is missing or too badly preserved. The three succeeding pairs of limbs (ant, mdb, 1) are as coarsely preserved as the whole specimen. The tip of the labrum (lbr) as well as the anterior part of the sternum (ste) are damaged.

Table 41. Comparison between the relationships within Phosphatocopina suggested by Hinz-Schallreuter (1998) and the scheme proposed in this work. Hinz-Schallreuter (1998) neither included the taxon Schallreuterinidae, which she established earlier in 1993 (Hinz-Schallreuter 1993c) nor the taxa Dabashanellidae Zhao, 1989, Liangshanellidae Huo, 1956, Monasteriidae Jones & McKenzie, 1980 and Oepikalutidae Jones & McKenzie, 1980 (see Appendix B for a complete alphabetical list of all taxa assigned to Phosphatocopina). Note that *Hesslandona unisulcata* Müller, 1982 has been regarded as a member of the taxon *Falites* by Hinz-Schallreuter (1993c). The taxa Ulopsidae, *Bidimorpha* and *Comleyopsis* are not included in the phylogenetic analysis.

Hinz-Schallreuter 1998	This work (Figs. 65, 68)
Phosphatocopida Müller, 1964	Phosphatocopina Müller, 1964
Hesslandonidae Müller, 1964	"Phosphatocopida sp." *sensu* Siveter *et al.* (2001)
Falitinae Müller, 1964	Euphosphatocopina
Falites Müller, 1964	*Vestrogothia spinata* Müller, 1964
Trapezilites Hinz-Schallreuter, 1993	NN1
Hesslandoninae Müller, 1964	*Falites fala* Müller, 1964
Hesslandona Müller, 1964	Hesslandonina Müller, 1982
Comleyopsis Hinz, 1993	Cyclotronidae Gründel, 1981
Vestrogothiinae Kozur, 1974	*Cyclotron* Rushton, 1969
Vestrogothia Müller, 1964	*Veldotron bratteforsa* (Müller, 1964)
Bidimorpha Hinz-Schallreuter, 1993	*Waldoria rotundata* Gründel, 1981*
?*Cyclotron* Rushton, 1969	Hesslandonidae Müller, 1964
?*Veldotron* Gründel, 1981	*Trapezilites minimus* (Kummerow, 1931)
?*Waldoria* Gründel, 1981	*Hesslandona* Müller, 1964
Ulopsidae Hinz, 1992	*Hesslandona unisulcata* Müller, 1982
Ulopsinae Hinz, 1992	Dorsospinata new name
Ulopsis Hinz, 1992	NN2
Tubupestinae Hinz, 1992	*Hesslandona kinnekullensis* Müller, 1964
Tubupestis Hinz, 1992	NN4
Semillia Hinz, 1992	NN3
Shergoldopsis Hinz, 1992	*Hesslandona ventrospinata* Gründel, 1981
	NN5

*The position of *Waldoria rotundata* is uncertain, see p. 177.

NN4, *Hesslandona suecica* n. sp., *H. angusta* n. sp., *H. toreborgensis* n. sp.; NN5, *H. trituberculata, H. necopina, H. curvispina* n. sp..

Figs. 64 and 67 and are referred to in the text by the form "Fig. 64:Z", whereas Z is the number of a particular node and its autapomorphies. The analysis also has bearing upon the phylogeny of Crustacea (that are identical in every tree):

- †*Martinssonia elongata* is the sister taxon to all other ingroup members;
- the two eucrustacean taxa †*Rehbachiella kinnekullensis* and *Euphausia superba* form a monophyletic unit which represents the sister group to the Phosphatocopina;
- the Lower Cambrian phosphatocopine from England (Siveter *et al.* 2001) is the sister taxon to all other phosphatocopines.

[For character sets that were assigned to earlier nodes in the tree, i.e. to the Crustacea, Labrophora (= Eucrustacea + Phosphatocopina), and Eucrustacea, see Chapter 4.3, the discussion of Crustacea and Eucrustacea]. For the monophylum embracing the Lower Cambrian "Phosphatocopida sp." plus all other phosphatocopines, the well-established name Phosphatocopina Müller, 1964 has been chosen (cf. Appendix B). The monophyletic taxon to the Lower Cambrian phosphatocopine, which comprises the remaining phosphatocopine species investigated in this

analysis, is named Euphosphatocopina new name, and is based on a set of autapomorphies (see below, with exemplary images) of its common stem species (= autapomorphies in the ground pattern of the taxon).

Reconstructed phylogeny of Phosphatocopina

Phosphatocopina ("Phosphatocopida sp." + Euphosphatocopina new name; Fig. 64:1)

Autapomorphies:

- all-embracing bivalved shield but without a dorsal axial furrow (plesiomorphic state: shield univalved, roof-shaped, anteriorly and posteriorly excavated and in profile amplete, as in the ground pattern of Labrophora and Crustacea) – already noted by Walossek (1999), extended in this paper;
- shield with a long and straight dorsal rim (Pls. 1A, 35B; plesiomorphic state: dorsal rim curved, as in the ground pattern of Labrophora and Crustacea) – this paper;
- shield with a doublure (Pls. 1C, 38D); plesiomorphic state: shield without doublure, as in the ground pattern of Labrophora and Crustacea – this paper;
- antennula smaller than one third the head length,

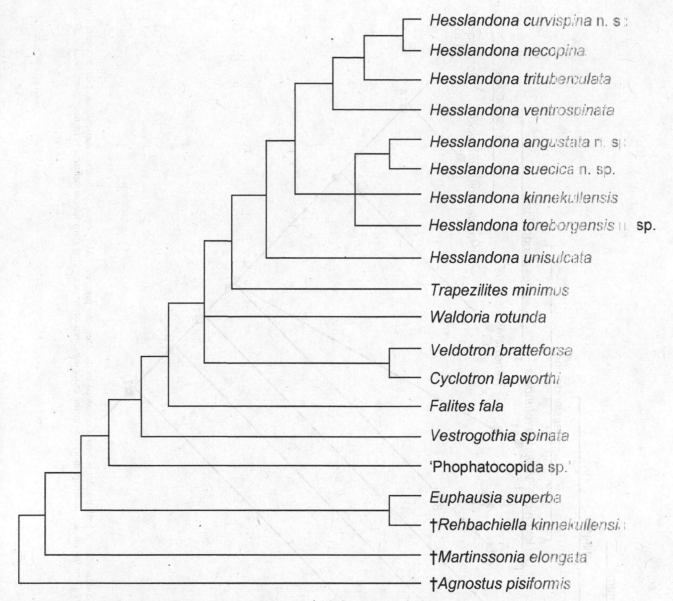

Fig. 63. Strict consensus tree and 50% majority rule tree of the two shortest trees resulting from the phylogenetic analysis including the o: :oup and all ingroup taxa from the character matrix (Appendix A) with characters coded and used as listed in Appendix A. Statistics: Nelson–: nick term information = 115; Rohlf's consistency index (1) = 0.920.

with not more than 13 annuli/portion and with no more than five distal setae (Pls. 9A, 37B); plesiomorphic state: antennula at least as long as one third the head length, with at least one seta per annulus/portion, as in the ground pattern of Labrophora and Crustacea – partly noted by Walossek (1999);

- endopodites of all post-antennular limbs with three podomeres (Fig. 58A–C); plesiomorphic state: endopodites with five podomeres, as in the ground pattern of Labrophora and Crustacea – also noted by Walossek (1999);
- number of post-cephalic limb-bearing segments less than six – noted by Walossek (1999) as "trunk largely reduced (no distinct tail)" (Pl. 37A); plesiomorphic state: number of post-cephalic limb-bearing segments

at least six, as in the ground pattern of Labro ora and Crustacea;

- no distinct trunk segmentation (Pl. 12D); j sio-morphic state: trunk distinctly segmented vith well-developed tergites, as in the ground patt of Labrophora and Crustacea – this paper.

The bivalved shield is an autapomorphy c the Phosphatocopina (Fig. 64:1). All phosphatocopine   ave a specifically shaped, bivalved shield, enclosing the ody completely (Pls. 7A, 16A). The known representat s of the stem lineage of Crustacea do not have a sir arly designed shield, and a bivalved shield does not ar-acterise the ground pattern of Eucrustacea   her (cf. Siveter *et al.* 2001). In the ground patte of

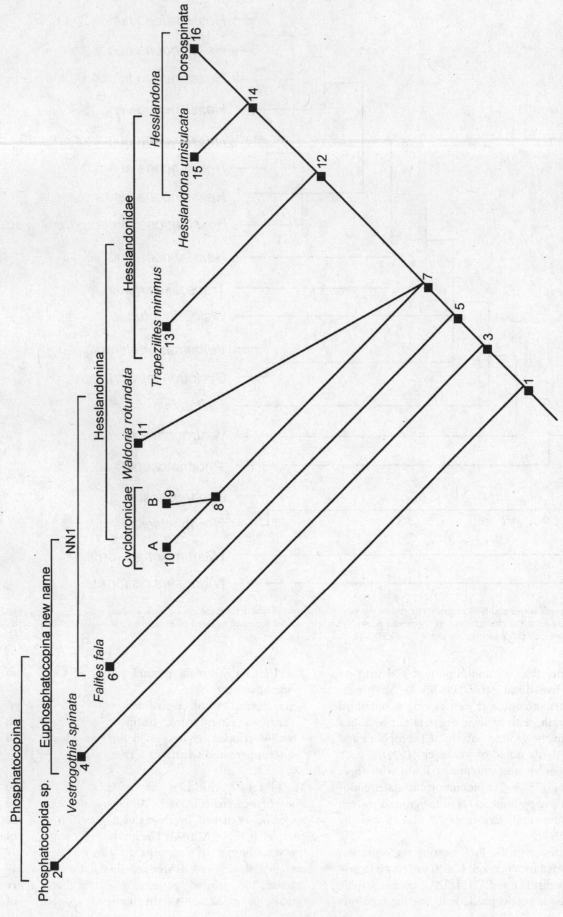

Fig. 64. Hypothesised relationships within the Phosphatocopina. A combination of both the shortest trees suggested by the computer program PAUP, equal to the strict consensus and the 50% majority rule consensus (cf. Fig. 63). See text for explanations of the numbers given for the sets of autapomorphies in the ground pattern of the according stem species before a branching event. See Fig. 67 for the relationships within Dorsospinata new name.

Eucrustacea, the shield presumably had a domed to roof-shaped design with moderate lateral extensions and a rounded, straight or slightly excavated anterior margin and an excavated posterior margin, producing two postero-lateral "wings". This design remained present in the ground patterns of the Entomostraca and Malacostraca and hence characterises the ground pattern of Labrophora and Crustacea. In fact, several euarthropods with uncertain affinities have a bivalved, unfurrowed, all-embracing shield, such as the †Bradoriida, species of †*Isoxys* Walcott, 1890, †*Tuzoia parva* (Walcott, 1912), †*Waptia fieldensis* Walcott, 1912, †*Canadaspis perfecta* (Walcott, 1912), (see, e.g. Briggs 1977; Hou *et al.* 1996; Williams *et al.* 1996; Hou & Bergström 1997), as well as the crustaceans Phyllocarida, Diplostraca, Ascothoracida and the cypris larvae of Cirripedia (Walossek *et al.* 1996); the Ostracoda (Entomostraca) even have a hinged shield. All these taxa with a bivalved shield differ in several aspects like the dorsal axis, i.e. with or without a hinge line, ventral morphology and the segmental number of the shield. Consequently, the development of a bivalved shield within Arthropoda and particularly the development of hinge lines is considered to be convergent.

All investigated phosphatocopines have a bivalved shield with a dorsal furrow from the beginning of ontogeny onwards. The bivalved shield of the Lower Cambrian phosphatocopine from England (Siveter *et al.* 2001) has no dorsal furrow. It is unknown whether a furrow develops later on, but as all Upper Cambrian earliest larval stages have a furrow, it clearly separates the Lower Cambrian representative from these. The shield of the phosphatocopine representative known from the Middle Cambrian of Siberia (Müller *et al.* 1995) and the species of the Lower Cambrian phosphatocopine taxa Dabashanellidae and Liangshanellidae from China and the Middle Cambrian taxa Monasteriidae and Oepikalutidae from Australia (see Appendix B for references) also lack a dorsal furrow. Hence, a dorsal furrow cannot belong to the ground pattern of Phosphatocopina, but represents an evolutionary novelty within this group.

Lobes on the surface of the valves are not present in *Vestrogothia spinata*, *H. necopina*, *H. suecica* n. sp., *H. angustata* n. sp., *H. kinnekullensis*, and *H. toreborgensis* n. sp. among the investigated species. All other species studied have at least one lobe, like *H. unisulcata* and *Trapezilites minimus*. Lobes are not present in the Lower Cambrian Chinese phosphatocopines of the taxa Dabashanellidae and Liangshanellidae, and they are lacking in the Middle Cambrian taxa Monasteriidae and Oepikalutidae, as well as in the specimens from the Middle Cambrian of Siberia (Müller *et al.* 1995). According to the phylogenetic analysis, and supported by the lack of lobes in Lower and Middle Cambrian

phosphatocopine species, a smooth, i.e. lobeless shield, represents part of the ground pattern of Phosphatocopina.

The maximum length of the shield in *Vestrogothia spinata*, *Falites fala*, *Cyclotron lapworthi*, *Waldoria rotundata* and many species of *Hesslandona*, lies between the dorsal rim and the dorso-ventral midline. The difference between the maximum length and the length of the dorsal midline may vary strongly between the particular investigated species (cf. Table 34). The free margin of the valves curves back dorsally (cf. Fig. 65). This back-curvature is stronger at the rear than the anterior shield end in *Falites fala*, but is more or less equal in *Waldoria rotundata* and *Trapezilites minimus*. *Hesslandona necopina* and *H. curvispina* have shields with the maximum length along the dorsal rim. It is most parsimonious to interpret the ground pattern of Phosphatocopina as the anterior and posterior free shield margin curved back, so the maximum length lay between the dorsal rim and the midline. This implies that the situation occurring in *H. necopina* and *H. curvispina*, which both have the maximum length along the dorsal rim, must have developed later in phosphatocopine phylogeny.

The anterior and posterior margins of the valves of most investigated phosphatocopine species, including the Lower Cambrian and Middle Cambrian representatives that were not considered in the phylogenetic analysis, are curved. Only in *Vestrogothia spinata*, *H. necopina*, and *H. curvispina* n. sp., is the posterior margin straight. In accordance with the phylogenetic analysis and the data from Lower and Middle Cambrian species, curved anterior and posterior margins of the valves represent a ground pattern character of Phosphatocopina.

Another autapomorphy of Phosphatocopina is the long and straight dorsal rim (Fig. 64:1). The shield of the Upper Cambrian phosphatocopine species has a long and straight dorsal rim throughout ontogeny (Fig. 4E, 12D). This is also true for the Lower Cambrian Chinese species and all Middle Cambrian representatives. The domed shield of the Eucrustacea and stem lineage representatives such as †*Martinssonia elongata*, †*Cambropachycope clarksoni*, and †*Henningsmoenicaris scutula* (cf. Walossek & Müller 1990) has a convex dorsal rim. Even the bivalved shield of Recent Ostracoda has a convex dorsal rim. There are several crustaceans which do have a straight dorsal shield midline, either bivalved shields such as †Leperditicopida (Ostracoda) or univalved, such as Euphausiacea and Decapoda. But the morphology of all of these taxa differs in so many aspects from that of Phosphatocopina to preclude any systematic affinities. Hence, in the ground pattern of Crustacea, Labrophora and Eucrustacea, the shield was probably convex.

A further autapomorphy of Phosphatocopina is the doublure (Fig. 64:1). The shield of all phosphatocopine species studied herein has a doublure along the whole free margin (Pls. 4D, 18C), and extends on both sides anteriorly and posteriorly into a membranous area (Pls. 27A, 43E). A doublure of this design is developed in the Lower Cambrian phosphatocopine from England (Siveter *et al.* 2001), in the Middle Cambrian representative from Siberia (Müller *et al.* 1995), and in one Middle Cambrian representative from Australia (Jones & McKenzie 1980). Yet, the inner side of the valves is mostly not described for other possible Middle Cambrian and Lower Cambrian representatives such as Dabashanellidae and Oepikalutidae, so the shape of the doublure is not known. The shield of at least one of the known representatives of the stem lineage of Crustacea, †*Henningsmoenicaris scutula*, does have a doublure. Among eucrustaceans, the picture is more difficult, and there is no summary literature. A doublure occurs at least in, e.g. podocopid Ostracoda (e.g. Langer 1973; Swanson 1989a, b, 1990; Becker & Adamczak 2001). In particular, taxa with weak shields or short lateral shield extensions seem to lack a doublure, but also larger forms such as euphausiaceans or decapods lack a doublure. †Trilobita (Siveter, pers. comm. 2001) and some fossil arthropods such as †*Fuxianhuia protensa* Hou, 1987 also have a doublure. If one assumes that a doublure might at least belong to the ground pattern of Euarthropoda, it must have become reduced independently in different lineages: in several euarthropods, four of the derivatives of the eucrustacean lineage, in Malacostraca, Branchiopoda, Cephalocarida and Myodocopida. The doublure remains, therefore, a weak character.

The width of the doublure varies from the anterior to the posterior. In some species, e.g. in *Falites fala* and *H. toreborgensis* n. sp., the anterior part is narrowest, in some the ventral or median part is narrowest, e.g. in *Vestrogothia spinata* and *Veldotron bratteforsa* (cf. Table 41). Either the postero-ventral part, e.g. in *Falites fala* and *H. unisulcata*, or the posterior part of the doublure, e.g. in *H. necopina*, is widest (cf. Table 41). According to the phylogenetic analysis, the anterior part was the narrowest and the postero-ventral part was the widest in the ground pattern of Phosphatocopina, ranging from one ninth to one sixth the shield length.

Another autapomorphy is the small antennula with no more than five distal setae (Fig. 64:1). The antennulae of all studied phosphatocopine species with preserved soft parts are no longer than one third the length of the head (Pls. 12B, 37B), are weakly segmented and never have more than five setae distally. In fact, the antennulae are so small that Klaus Müller missed this appendage in his earliest paper (Müller 1979a) and mislabelled the subsequent appendages accordingly, i.e. the antenna was his antennula and so on. This led to much confusion in

the understanding of the basal arthropod antennula, as this appendage was always thought to have a uniramous origin and was therefore different from the subsequent limbs that were biramous, at least in the ground pattern of Euarthropoda (Maas & Waloszek 2001a). An originally biramous antennula would therefore have considerable bearing not only on the phylogeny of the Crustacea but also of the Euarthropoda. In fact, some malacostracan taxa have "multiramous" antennulae, comprising a uniramous, three-divided peduncle and at least two multi-annulated flagella inserting terminally on the peduncle. Ontogeny demonstrates, however, origination of the malacostracan antennula from a uniramous state (e.g. Walossek 1993; Maas & Waloszek 2001b), as is true for the Entomostraca, Phosphatocopina, and stem lineage representatives of Eucrustacea (Walossek & Müller 1990).

In the fossil phosphatocopines, the antennulae, located at the flanks of the hypostome, are disguised by the prominent antennae and mandibles, and are typically not even preserved (Pl. 21A). All known derivatives of the stem lineage of Crustacea had longer, segmented, uniramous antennulae with various setae well adapted for swimming and for food gathering. Similarly, Cephalocarida, Maxillopoda, Branchiopoda, and Malacostraca probably have long, segmented and well-setated antennulae, at least in their ground pattern. Several eucrustaceans have very short antennulae with few setae, comparable with those found in the Phosphatocopina studied herein, such as Cladocera and harpacticoid copepods. But the distribution of these taxa within Eucrustacea strongly suggests that a reduction of the antennulae has occurred several times independently within the Eucrustacea, probably as an adaptation to special life habits and therefore also separately in the lineage towards the Phosphatocopina.

The antennula of the Early Cambrian phosphatocopine is poorly known. One specimen shows a slender seta that might belong to the antennula (Siveter *et al.* 2001). The size or shape is not reconstructable, but it cannot have been a prominent appendage, so a miniaturisation of the antennula had probably already occurred in the ground pattern of the Phosphatocopina.

A further autapomorphy of Phosphatocopina is the three-divided endopod in all post-antennular limbs (Fig. 64:1). None of the semaphoronts of Upper Cambrian phosphatocopines and the Middle Cambrian representative (Walossek *et al.* 1993) as well as the recently discovered specimens of phosphatocopines from the Lower Cambrian of England (Siveter *et al.* 2001) has more than three endopodal segments and/or more than two segment boundaries in all post-antennular limbs. There are crustaceans with similarly few or fewer endopodal segments, but this must be judged in relation to the ground patterns of Euarthropoda and Crustacea.

Euarthropoda have seven endopodal podomeres basally, i.e. six segment boundaries, while Crustacea have five endopodal podomeres and/or four segment boundaries in all post-antennular limbs, and this is also true for Eucrustacea *sensu* Walossek (1999). Five endopodal podomeres are conservatively retained in the thoracopods I–VIII of Malacostraca and in the post-maxillulary limbs of the Cephalocarida among the Entomostraca. All other Entomostraca have at most four-segmented endopods in all post-antennular limbs, including the antenna and the mandible [see Olesen (2001) for a mystacocarid; see Walossek (1993) for a general discussion of this feature]. The lower number of endopodal portions must, therefore, have occurred convergently within the Eucrustacea and has to be considered an apomorphy of particular ingroup taxa. The occurrence of three-divided endopods, i.e. two segment boundaries, in all post-antennular limbs is, according to the phylogenetic analysis, an autapomorphy of Phosphatocopina, resulting from a reduction from five portions, i.e. four segment boundaries, as in the ground pattern of Labrophora and Crustacea.

A fifth autapomorphy of Phosphatocopina is the low number of limb-bearing thorax segments (Fig. 64:1). The maximum size of specimens with well-preserved soft parts is usually essentially smaller than the largest isolated valves of a given species. The number of limb-bearing segments in different arthropods such as †*Fuxianhuia protensa* is about 17 (Hou & Bergström 1997); trilobites have numerous segments. The thorax in the ground pattern of Euarthropoda probably consisted of significantly more than five segments. The number of segments in the stem lineage derivatives of Eucrustacea is not clear as the known species are described on larval material only (Walossek & Müller 1990). Eucrustacean taxa have at least seven thoracic segments, as in Maxillopoda (Walossek & Müller 1998a, fig. 7), the head being composed of five segments. The observation that the last three possible growth stages of *H. unisulcata* have the same number of segments, i.e. three trunk segments (Figs. 21, 23), suggests that this segmental number of the body might be generalised for later growth stages up to the adult. The highest number found is four trunk segments, i.e. six post-mandibular segments (cf. Pl. 15D for *H. necopina*; Pl. 36D for *Waldoria rotundata*).

Another autapomorphy of Phosphatocopina is the indistinct trunk segmentation (Fig. 64:1). None of the investigated specimens with preserved soft parts shows a distinct segmentation of the trunk, and distinct tergites are lacking. The last body segments are more or less laterally fused, and only slight depressions between limb-bearing portions indicate segment boundaries (Pl. 12D). In eucrustacean taxa, apart from various parasitic taxa, the trunk is distinctly segmented and shows

well-developed tergites, as in myriapods and insects. This is also the case in fossil euarthropods, such as †*Fuxianhuia protensa*, †*Canadaspis perfecta*, and †Trilobita. Hence, in the ground pattern of Euarthropoda, Crustacea and Eucrustacea, the trunk was distinctly segmented and had well-developed tergites, while the stem species of the Phosphatocopina changed this character.

"Phosphatocopida sp."

No autapomorphies were found for this species (Fig. 64:2), preliminarily described by Siveter et al. (2001). A detailed description is in progress (Siveter & Waloszek, pers. comm. 2002).

Euphosphatocopina new taxon (*Vestrogothia spinata* + NN1; Fig. 64:3)

The new taxon Euphosphatocopina is established on the basis of the autapomorphies of the stem species:

- shield with a simple dorsal furrow (Pls. 39D, 43A; plesiomorphic state: shield without a dorsal furrow, both valves dorsally fixed) – this paper;
- anterior and posterior dorsal rim of the shield with small triangular plates between both valves (Pls. 39D, C, 40A, 43B, C; plesiomorphic state: no such plates, no dorsal furrow) – this paper;
- shield cephalothoracic, at least three post-mandibular segments dorsally fused with the shield (Pls. 11F, 42B; plesiomorphic state: head shield, fusion of body with shield restricted to head segments) – this paper;
- coxa and basipod of antenna fused to a single limb stem throughout ontogeny (Pl. 5F, Fig. 57A; plesiomorphic state: coxa and basipod separated, cf. Fig. 58A) – also noted by Walossek (1999);
- coxa and basipod of mandible fused to a single limb stem from later larvae on (Figs. 57E, 59C; plesiomorphic state: coxa and basipod separated) – also noted by Walossek (1999);
- endopod of antenna with two podomeres (Fig. 43B; plesiomorphic state: endopod with three podomeres, cf. Fig. 58A) – this paper;
- endopod of mandible with two podomeres (Fig. 43B; plesiomorphic state: endopod with three podomeres, cf. Fig. 58B) – this paper.

An autapomorphy of Euphosphatocopina is the simple dorsal furrow with two isolated dorsal plates (Fig. 64:3). As stated above, the phosphatocopine species from the Upper Cambrian investigated herein have either a simple dorsal furrow with two isolated plates at both dorsal ends (Pls. 39C, D, 40A) or a continuous dorsal bar, the interdorsum (Pl. 26A). The interdorsum is regarded as an evolutionary novelty that probably originated from two isolated plates as developed in *Vestrogothia spinata* and *Falites fala* among the studied species. All phosphatocopines without such a character, e.g. the Lower Cambrian phosphatocopine from England, the dabashanellids and liangshanellids from the Lower Cambrian of China, the monasteriids and oepikalutids from the Middle Cambrian of Australia and the Middle Cambrian members of Siberia (Walossek *et al.* 1993) can be excluded from the euphosphatocopines. A further resolution of the relationships of these taxa is not possible yet due to missing data.

Another autapomorphy of Euphosphatocopina is the cephalothoracic shield (Fig. 64:3). The species of Euphosphatocopina investigated herein have a cephalothoracic shield that is fused dorsally with the complete head and, at least, additionally up to the fourth post-mandibular segment (cf. Pl. 15D). Cephalothoracic shields may also occur in different eucrustacean groups, e.g. in the Eumalacostraca apart from Bathynellacea, in the Copepoda, in the Remipedia, and in the Tantulocarida. Cirripedia and Cephalocarida do not have a cephalothoracic shield. In the stem lineage representatives of Eucrustacea, only four limb-bearing segments were fused dorsally to the shield. The distribution of the cephalothoracic shields within Crustacea suggests that the shield in the ground pattern of Crustacea, Labrophora and Eucrustacea was cephalic. Hence, the head shield is a product of fusion of tergites back to the segment of the first post-mandibular pair of limbs, the maxillulae. Only later in the lineage within the Eucrustacea, the segment of the succeeding limb pair, the maxillae, is added to the head and the shield elongated accordingly. The cephalothoracic shield is therefore regarded as convergently developed in different eucrustacean lineages and the phosphatocopines. The Lower Cambrian phosphatocopine species is represented by two larval specimens only and it is not known whether the shield is cephalothoracic also in this species. Therefore, the fusion of several thoracomeres to the shield has to be considered an autapomorphy of at least the Euphosphatocopina.

A third autapomorphy of Euphosphatocopina is the fused antennal limb stem from early ontogeny onwards (Fig. 64:3). The limb stem or "protopod" of the antenna of all known growth stages of all Upper Cambrian "Orsten" phosphatocopines, of which soft part morphology is present, as well as in isolated Middle Cambrian phosphatocopine antennae (Walossek *et al.* 1993) is undivided and medially drawn out into one single antero-posteriorly compressed, vertically oriented endite (cf. Figs. 57A, 65A, B).

The antenna of the Lower Cambrian phosphatocopine from England has a clear subdivision of its limb stem into a proximal coxa with one endite and a distal basipod with one endite (Siveter *et al.* 2001) (Fig. 58A). For the ground pattern of the crown-group of Crustacea, it must

be assumed that the coxa and the basis of the antenna are separated portions. All post-antennular pairs of limbs from all known derivatives of the stem lineage of the crustacean crown-group (Müller & Walossek 1986a; Walossek & Müller 1990; Walossek 1999) have no coxae but a small, lobate, setae-bearing element medially underneath the basipod, the "proximal endite" *sensu* Walossek & Müller (1990). This proximal endite is, according to these authors, the phylogenetic "precursor" of a coxa. It must be added that Walossek & Müller (1990) also concluded that coxal portions evolved along the limb series at different evolutionary levels, and did not evolve on all legs in one step. The often cited "crustacean limb" (e.g. Schram & Koeneman 2001) did not have a coxa. The proximal endite is present in all post-mandibular limbs of phosphatocopines – which accordingly have no coxae. It is therefore assumed that the limb stem of the Middle Cambrian and Upper Cambrian phosphatocopines evolved as a product of fusion of the originally separated coxa and basipod. If so, the fusion of both portions to form an undivided limb stem should represent an autapomorphy of Euphosphatocopina (Fig. 64:3). The situation later in the ontogeny of the Lower Cambrian species is not known. The fusion might or might not have taken place until the adult stage. But in either case, the Lower Cambrian species differs from any other later phosphatocopine in having a separate limb stem, at least in the first growth stage. Further confirmation of this evolutionary path comes from the developmental fate of the mandible, as discussed below.

A further autapomorphy of Euphosphatocopina is that the coxa and the basipod of the mandible fused to a single limb stem from the later larvae onwards (Fig. 64:3). The limb stem of the mandibles in most early and all later stages of all Upper Cambrian phosphatocopines, for which soft part morphology is known, and in phosphatocopine material from the Middle Cambrian (Walossek *et al.* 1993) is undivided, as in the antenna, and medially drawn out into a spinose, but oblique, median edge (Figs. 57B, 66A, B). A lobate endite is present on it, apex somewhat pressed between the protopod and the two-part endopod (Pl. 5D; Fig. 66A, B). The median edges of the limb stem, the enditic protrusion and the endopod are guided by a set of setae anteriorly and posteriorly (Fig. 66A).

The exceptions are the youngest growth stages of *Vestrogothia spinata*, where the limb stem of the mandible is still subdivided in a proximal gnathobasic part, the coxa, drawn out medially into a spinose and pointed endite, and a distal part, the basipod, with one medially tipped endite (Pl. 45A, B; Fig. 59A). In later stages, both stem portions fuse during ontogeny to form a uniform stem, as is the case in all other Upper Cambrian phosphatocopine taxa from the beginning of ontogeny (cf. Pls. 21B, 44D, F; Figs. 59B, 66B).

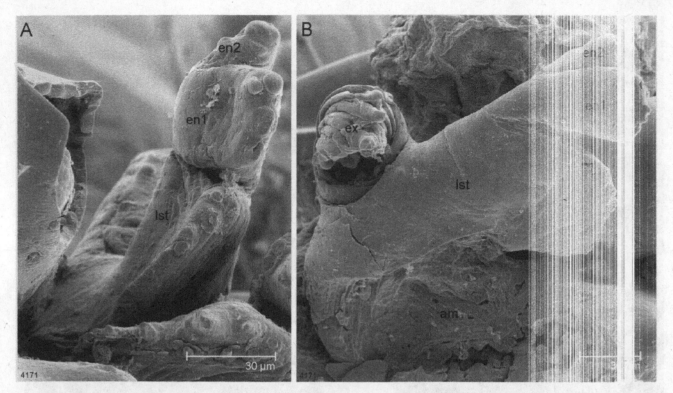

Fig. 65. UB W 253. *Hesslandona suecica* n. sp. Specimen of an advanced growth stage, about 630 µm in shield length. Zone 1 of Gum, Kir ulle. Antenna, displaying the uniform limb stem and two-part endopod. A: Median view. B: Posterior view.

The exopod-bearing part of the basipod becomes a very narrow ring around the insertion area of the exopod. Because all post-mandibular limbs have three-segmented endopods, it looks as if there was a large protopod, on which a three-segmented endopod inserted apically and an exopod on the sloping outer edge. Traditionally, this protopod would be understood as a coxa, and confusion is very likely in that a basipod was absent and the rami inserted directly on the coxa – which is not the case in any arthropod (Walossek & Müller 1990), particularly not in Crustacea.

Homologisation on traditional models would, and has, led to considerable confusion and misunderstanding of this limb subdivision. A helpful and reliable reference is always that the basipod is the element that medio-distally continues into the endopod, and laterally carries the exopod on a sloping rim (cf. Pl. 9C). In *Vestrogothia spinata*, the lobate endite that is pressed between the limb stem and the endopod in later stages clearly originates from the basipodal endite of the earliest larvae. Ontogeny thus helps to clarify the aberrant situation of Phosphatocopina. Because the morphology in other phosphatocopine species is more or less the same, it is concluded that this endite is homologous to the median part of the basipod and, hence, cannot be interpreted as the proximal-most endopodal portion. In the Lower Cambrian representative of Phosphatocopina, the coxa and the basipod of the mandible are separated portions as well (Siveter *et al.* 2001). Whether there is a fusion

in later stages of this species is not known due to m sing material. The mandibular protopod is, in conseq ice, regarded as a product of partial fusion of the co and the basipod, leaving the median enditic part of th asi-pod free, in contrast to the antenna, where the b pod is completely fused with the coxa. In the ground p ern of Eucrustacea, the coxae and basipods of the mai bles are clearly separated. The partial fusion of both po ons in the mandible to form a large limb stem with still separated basipodal endite is regarded, according the phylogenetic analysis, as a further autapomorj of Middle and Upper Cambrian Phosphatocopin The condition in the Lower Cambrian species is uncl

Another autapomorphy of Euphosphatocopina the two-part antennal endopod (Fig. 64:3). All sen ho-ronts of Upper Cambrian phosphatocopines ai the Middle Cambrian representative (Walossek *et al.* 93) have two endopodal podomeres in the antenna, i one segment boundary (Pl. 27B). As stated above, the luc-tion from five- to three-segmented endopods all post-antennular limbs including the antenna m be considered an autapomorphy of Phosphatoc ina. Hence, the further reduction of the number of do-meres from three to two in the antenna is consi ered an autapomorphy of Euphosphatocopina.

A further autapomorphy of Euphosphatocopina the two-part mandibular endopod (Fig. 64:3). All sen ho-ronts of Upper Cambrian phosphatocopines ai the Middle Cambrian representative (Walossek *et al.* 93)

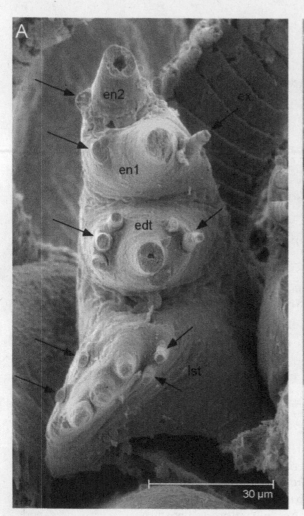 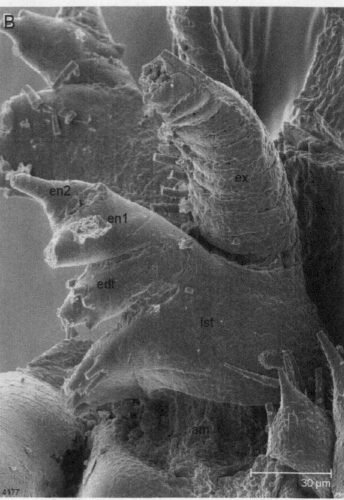

Fig. 66. UB W 255. Undetermined phosphatocopine species from Zone 1 of Backeborg (Kinnekulle), Falbygden–Billingen. Specimen of about 300 µm in shield length. Mandible, displaying the uniform limb stem and two-part endopod. A: Median view, anterior left. Note the anterior and posterior setae, which guide the median elongations of the limb stem (lst), the enditic protrusion (edt) and the endopod (arrows). B: Posterior view.

have two endopodal podomeres in the mandible (Pl. 27C). As stated above, the reduction from five- to three-segmented endopods in all post-antennular limbs including the mandible must be considered an autapomorphy of Phosphatocopina. Hence, a further reduction of the number of podomeres from three to two in the mandible is considered an autapomorphy of Euphosphatocopina.

Vestrogothia spinata

The autapomorphies of *Vestrogothia spinata* are (Fig. 64:4):

- ventro-central surface of both valves drawn out into a spine directed postero-laterally (Pl. 42B; plesiomorphic state: no such outgrowths, as in the ground pattern of Euphosphatocopina and Phosphatocopina);
- postero-ventral margin of the left valve drawn out into a spine directed posteriorly (Pl. 42A; plesiomorphic state: no such outgrowths, as in the ground pattern of Euphosphatocopina and Phosphatocopina);

- posterior margin of the valves rather straight (Pl. 42A; plesiomorphic state: curved, as in the ground pattern of Euphosphatocopina and Phosphatocopina);
- postero-dorsal plate completely drawn out into a posteriorly directed spine (Pl. 43A, C; plesiomorphic state: posterior plate smooth, not drawn out like the anterior plate, as in the ground pattern of Euphosphatocopina and Phosphatocopina and retained in *Falites fala*).

Only a few species of Phosphatocopina have a shield drawn out into spines. In *Vestrogothia spinata* and *Waldoria rotundata*, the central area of both valves is drawn out into a laterally and/or ventro-laterally to postero-laterally oriented spine (Pls. 35B, 42B), which is, in *Waldoria rotundata*, regarded as the strongly elongated lobe L_5. *Vestrogothia spinata* is a species without lobes and a complete interdorsum but with antero- and postero-dorsal plates, of which the posterior is drawn out into a spine. *Waldoria rotundata*, by contrast, has a complete interdorsum, lacking spines on either end.

According to the phylogenetic analysis, the relationships of these species are not close, implying that the outgrowths of the valves have no common origin. Accordingly, the outgrowth is an apomorphy of *Vestrogothia spinata*. Because in *Vestrogothia longispinosa* Kozur, 1974, which is not included in the analysis due to the lack of more structures, the postero-ventral area of both valves is also drawn out into a spine (Williams *et al.* 1994a), very similar to the condition in *Vestrogothia spinata*, this spine would be a synapomorphy of both species.

Another autapomorphy is the postero-ventral spine of the margin of the left valve (Fig. 64:4). Marginal outgrowths of the valves only occur in *Vestrogothia spinata* and *H. ventrospinata*, while *Vestrogothia longispinosa* does not show marginal outgrowths (Kozur 1974; Williams *et al.* 1994a). In *Vestrogothia spinata*, the marginal outgrowth occurs only on the left valve and is represented by a long spine with a flattened, triangular proximal part (Pl. 42A). In *H. ventrospinata*, both valves of late developmental stages have a pair of plate-like triangular outgrowths, which interlock (Pl. 19A). *Vestrogothia spinata* is a species without a complete interdorsum but with antero- and postero-dorsal plates, and is clearly different from *H. ventrospinata* with a complete interdorsum and anterior and posterior ends both being drawn out into spines, as is developed in many species of *Hesslandona*. Therefore, it is assumed that the outgrowths of *H. ventrospinata* and *Vestrogothia spinata* have no common origin and have to be regarded as autapomorphic features of the respective species.

A third autapomorphy is the straight posterior margin (Fig. 64:4). In *Vestrogothia spinata*, the posterior shield margin is rather straight (Pl. 42A). As stated above, a curved posterior shield margin must represent a ground pattern character of Phosphatocopina.

A further autapomorphy is the spiny outgrowth of the posterior dorsal plate (Fig. 64:4). The posterior dorsal plate of *Vestrogothia spinata* is drawn out into a slender, straight, posteriorly directed spine (Pls. 42A, 43A, C). Posterior spines are also present in various hesslandonid species, where the spine is an outgrowth of the posterior end of the interdorsum (cf. Pl. 22A, C). *Falites fala*, *Waldoria rotundata*, *Cyclotron lapworthi*, *Veldotron bratteforsa*, *Trapezilites minimus*, *H. unisulcata*, and *H. angustata* n. sp. do not have such spines. Several species of *Hesslandona*, however, show not only posterior but also anterior spines. The position of spine-bearing species within the phylogenetic tree (cf. Figs. 64, 67) implies a spineless postero-dorsal end of the shield in the ground pattern of Euphosphatocopina and Hesslandonina. Therefore, the spine of *Vestrogothia spinata* is probably of different origin compared to the spines occurring in species of *Hesslandona*.

NN1 (*Falites fala* + Hesslandonina)

The autapomorphies of the stem species of NN1 are (Fig. 64:5):

- shield valves with one prominent lobe L_1 (Pl. 1A; plesiomorphic state: shield valves without such a lobe);
- coxa and basipod of the mandible of early stages fused (Pl. 4D; plesiomorphic state: separate, as in Euphosphatocopina and Phosphatocopina).

An autapomorphy of NN1 is the prominent lobe L_1 on the shield valves (Fig. 64:5). The shields of *Trapezilites minimus* and *H. unisulcata* possess one prominent lobe L_1. In *Falites fala* and several species of *Hesslandona*, L_1 is also present together with other lobes, whereas several hesslandonid species, such as *H. necopina* and *H. kinnekullensis*, have a smooth shield. The most parsimonious explanation that is in accordance with the phylogenetic analysis, is a smooth shield being derived from a lobe-bearing shield within the taxon NN1 (cf. Fig. 67).

Another autapomorphy is the undivided limb stem of the mandible from early growth stages onwards (Fig. 64:5). In only the earliest growth stages of *Vestrogothia spinata* among the studied phosphatocopine species is the limb stem of the mandible distinctly divided into two parts, a coxa and a basipod (see above). In all other investigated species, the coxa and the basipod of the mandible are fused from the earliest growth stages on. The suture between the coxa and the basipod of the mandible in early larvae and adults is therefore constructed as a plesiomorphy occurring in the ground pattern of Phosphatocopina, while an undivided limb stem represents an apomorphic character state.

Falites fala

The autapomorphies of *Falites fala* are (Fig. 64:6):

- shield valves with prominent lobes L_2 and L_3 in addition to lobe L_1 (Pl. 39A; plesiomorphic state: shield valves only with lobe L_1, as in NN1; more plesiomorphic is a shield lacking any ornamentation, as in the ground pattern of Euphosphatocopina and Phosphatocopina);
- maximum width of the doublure more than or a fifth the length of the valves (cf. Pl. 40C; plesiomorphic state: maximum width of the doublure between one sixth and one ninth the length of the valves, as in the ground pattern of Euphosphatocopina and Phosphatocopina);
- doublure showing parallel lines (Pl. 39A, B; plesiomorphic state: doublure smooth, as in the ground pattern of Euphosphatocopina and Phosphatocopina).

An autapomorphy of *Falites fala* is the prominent lobes

L_2 and L_3 in addition to lobe L_1 (Fig. 64:6). The surface of the shield of *Falites fala* shows three prominent lobes (Pl. 39A). Three lobes also occur in *H. curvispina* n. sp. and *H. trituberculata*. All other species have either no lobes, a single lobe or six lobes. The most parsimonious solution of the evolution of lobes on the valves of Euphosphatocopina is the assumption that a single lobe developed in the ground pattern of NN1 (= *Falites fala* + Hesslandonina; Fig. 64). This lobe is retained in all lineages except two (NN3 and *H. necopina*). Additional lobes occur in other lineages such as *Falites fala*, Cyclotronidae, and NN5. The occurrence of three lobes at different nodes of the cladogram (Fig. 64) implies that three lobes is therefore an autapomorphy of *Falites fala*. The consideration of more phosphatocopine species could help to further enlighten the evolution of the lobes.

A further autapomorphy is the maximum width of the doublure being more than one fifth the length of the valves (Fig. 64:6). The doublure of *Falites fala* widens strongly in the postero-ventral region of the valves, being more than one fifth the length of the valves. Such a maximum width of the doublure does not occur in any of the phosphatocopine species studied (cf. Table 41). The maximum width of the doublure in Euphosphatocopina is generally between about one sixth and one ninth the length of the valves.

Another autapomorphy of *Falites fala* is the parallel lines on the doublure (Fig. 64:6). The doublure of *Falites fala* shows parallel lines throughout ontogeny, which are recorded in many specimens (Pl. 39A–C). Specimens of *Vestrogothia spinata* occur in the same samples as *Falites fala* but lack such lines. Therefore, it is difficult to regard the lines as simply preservational, but preferably as real structures. Because similar structures are absent in any other investigated phosphatocopine species, they are considered an autapomorphy of *Falites fala*.

Hesslandonina (Cyclotronidae + Hesslandonidae)

The name Hesslandonina (see Appendix B for synonymy) was introduced by Müller (1982a) to combine all phosphatocopines with a continuous interdorsum. Hinz-Schallreuter (1993c) argued that the interdorsum has become convergently reduced within Phosphatocopina at least twice, to produce structures like the anterior and posterior plates as described for *Vestrogothia spinata* and *Falites fala*. Consequently, she synonymised the names Hesslandonina, Hesslandonidae, Vestrogothiina (respectively Hesslandonocopina and Vestrogothicopina, see Appendix B) and Phosphatocopina and used the latter only (see also Hinz-Schallreuter 1995). Because this investigation provides evidence that the interdorsum did evolve only once in the phosphatocopine lineage and all other structures have to be inter-

preted at best as precursor structures, we re-establish the name Hesslandonina and propose to use it in the sense of Müller (1982a), i.e. defined by the autapomorphy of its stem species, the presence of an interdorsum.

A single but very distinctive autapomorphy characterises this node (Fig. 64:7):

- dorsal area of the shield with a complete dorsal bar, the interdorsum, width less than 1/25 the length of the valves (Pl. 31B; plesiomorphic state: shield valves with small triangular plates antero-dorsally and postero-dorsally, as in Euphosphatocopina, plesiomorphically retained only in *Vestrogothia spinata* and *Falites fala*).

The interdorsum is a prominent feature of phosphatocopines that requires some comments. An interdorsum is present as a dorsal bar running from the anterior to the posterior edge in several phosphatocopine species. Its width may differ between the species (Table 40). However, dorsal structures are lacking in several species described in the literature as phosphatocopines, none of which were investigated in this study: the Dabashanellidae and Liangshanellidae from the Lower Cambrian of China (see Appendix B for references), the Lower Cambrian representative from England (Siveter *et al.* 2001), the Monasteriidae and Oepikalutidae from the Middle Cambrian of Australia (see Appendix B for references), and the Middle Cambrian representatives from Siberia (Müller *et al.* 1993). A dorsal bar is also missing in *Vestrogothia spinata* and *Falites fala*, but these two species have two isolated dorsal plates anteriorly and posteriorly. The species of *Cyclotron* Rushton, 1969 were originally also thought to lack an interdorsum. In the specimens used in the early descriptions of *Cyclotron lapworthi* (Groom, 1902) the interdorsum was not preserved as mostly single valves were found (e.g. Groom 1902; Ulrich & Bassler 1931; Rushton 1969). Williams *et al.* (1994b) could, however, show that an interdorsum is present, as it is in *Waldoria rotundata*, *Veldotron bratteforsa*, *Trapezilites minimus*, and all species of *Hesslandona* among the investigated taxa.

Müller (1964a) originally distinguished two taxonomic units ("families") within Phosphatocopina on the basis of the absence or presence of an interdorsum: the Hesslandonidae with an interdorsum and the Falitidae without. Subsequently, the name Vestrogothiidae was introduced by Kozur (1974) for phosphatocopines without a continuous interdorsum. This group is still in use, embracing also the taxon *Falites*. The interdorsum was frequently suggested to be analogous to a "third valve" (Melnikova & Mambetov 1990, 1991; Hinz-Schallreuter 1994), but without arguing this "third valve" in phylogenetic aspects of euarthropod evolution. Hinz-Schallreuter (1993b) claimed that a separation of two such taxa because of the presence or absence of an interdorsum is not possible, noting without giving any reference, that "this has been proved by morphological and phylogenetic studies". Hinz-Schallreuter (1993a) also pointed out that the interdorsum should have evolved from an originally broad elongated oval bar in the Lower Cambrian to a parallel-bordered strip, and eventually, became lost in the Upper Cambrian (see also Hinz-Schallreuter 1997). The interpretation of an interdorsum in the ground pattern of the Phosphatocopina has considerable bearing on the systematic position of species within the two major lineages because the presence or absence of an interdorsum is important information. Hinz-Schallreuter also claimed in another paper (1993c) that an interdorsum might have evolved within several genera independently, but again gave no reason for such an assumption.

In fact, only one character, either the absence or presence of an interdorsum, can be the valid ground pattern character for Phosphatocopina, which subsequently represents a symplesiomorphy of ingroup taxa. Only *Vestrogothia* and *Falites* show a morphology where elongated triangular plates on the most anterior and posterior parts of the dorsal rim separate the valves. Hinz-Schallreuter (1993b) reported that the original material of 'Müller (1964a) included a species of *Vestrogothia*, without giving a species name, which shows a small, continuous interdorsum. This information is not considered herein as it is not known which species she actually meant and whether the structure she saw might have been the membrane squeezed outside the valve. Hinz-Schallreuter (1993c) even diagnosed *Vestrogothia spinata* as having a "broad to very small or missing interdorsum" (which includes all the different stages), but without giving any explanations or illustrations. Because the available material never showed such structures, *Vestrogothia spinata* is considered to have no interdorsum, and the data of Hinz-Schallreuter are probably based on misidentification. According to the phylogenetic analysis, the interdorsum characterises the phosphatocopine ingroup Hesslandonina, whose representatives all possess an interdorsum. A complete reduction, as proposed by Hinz-Schallreuter as a general tendency in various papers, does not occur. The condition present in *Vestrogothia spinata* and *Falites fala* probably represents an intermediate character state between a shield without dorsal structures and a shield with a continuous interdorsum.

Waldoria rotundata, *H. angustata*, and *Veldotron bratteforsa* as well as *Cyclotron lapworthi* (Williams *et al.* 1994b) have rather narrow interdorsa. The width of the interdorsum in all these species is less than 1/25 the shield length (cf. Table 40). In all other investigated species, the interdorsum is wider (cf. Table 40), but then always in combination with anterior and posterior outgrowths. According to the phylogenetic analysis, the most parsimonious explanation is the assumption of a

rather narrow interdorsum in the ground patt   of Hesslandonina, while the wider interdorsa of   eral hesslandonids are probably subsequently derived.

Cyclotronidae (Veldotron bratteforsa + Cyclotron lapworthi)

Gründel (1981) introduced the new name Cyclotr dae to accommodate *Cyclotron* and *Veldotron*. Becaus oth taxa were later placed within Vestrogothiidae (W ms & Siveter 1998), the name Cyclotronidae was no ubsequently used. Because our investigations confi the close relationship of *Cyclotron* and *Veldotron* as ned by Gründel (1981), we therefore re-establish the ame Cyclotronidae, comprising *Cyclotron* Rushton, 19 and *Veldotron* Gründel, 1981. *Waldoria* Gründel, 198 may be another member of this taxon, but this can be stated yet. However, the phylogenetic analysis cou not resolve the relationships of all three species rela   to each other. The autapomorphy of Cyclotroni e is (Fig. 64:8):

- shield valves with prominent lobes L_2, L_3, L_4, d L_5 (perhaps also L_6) additionally to lobe L. (F 4A; plesiomorphic state: shield valves only with lo L_1, as in NN1).

The shields of *Cyclotron lapworthi*, *Waldoria rot ata*, *Veldotron bratteforsa*, and *H. ventrospinata* have   least five lobes [cf. Pls. 19A, 34A, 38A; Gründel (198 for *Cyclotron lapworthi*]. Lobe L_6 could not be obser d in *Waldoria rotundata*, but as the shape of the lo s of *Waldoria rotundata* is slightly differen from t   in *Cyclotron lapworthi* and *V. bratteforsa*, more cies should be investigated first. All other inves ated hesslandonine species have no more than three bes. The presence of such a high number of lobes is arly apomorphic, and it is most likely that this state c racterises the taxon Cyclotronidae, but, according   the phylogenetic analysis, excluding *H. ventrospinata* e to the presence of interdorsal spines, which are ab t in the species of *Waldoria* and *Cyclotron*.

Veldotron bratteforsa

The autapomorphy of *Veldotron bratteforsa* is (Fig 4:9):

- valves gaping posteriorly (Pl. 38C; plesiom phic state: valves closing tightly along their who free length, as in Euphosphatocopina).

The early phosphatocopines could not close their lves as they had a stiff shield, as is, e.g. developed   the dabashanellids, liangshanellids and monasteriid e.g. Hou 1987b; Zhang 1987; Huo *et al.* 1991 and the wer Cambrian larva from England (Siveter *et al.* 20   . In all studied phosphatocopine species, the valves have a simple hinge furrow and could probably close the valves. In *Veldotron bratteforsa* and *H. trituberculata*, the closed valves gape posteriorly (Pls. 38C, 17F), while in all other species the valves close along their whole free margin. As a result of the phylogenetic analysis, *Veldotron bratteforsa* is a species within the taxon Cyclotronidae. *Hesslandona trituberculata*, on the other hand, is the sister taxon of *H. necopina* plus *H. curvispina* within the monophyletic taxa *Hesslandona* and Hesslandonidae (see below, Figs. 64, 67). The assumption of a gaping shield in the ground pattern of Hesslandonina would imply the loss of this character in at least seven different lineages. As favoured in the analysis, and seemingly more parsimonious, the development of a gaping shield evolved twice in *Veldotron bratteforsa* and *H. trituberculata*. Consequently, a shield with a closing free margin must be interpreted as the plesiomorphic condition within the ground pattern of Euphosphatocopina and Hesslandonina.

Cyclotron lapworthi

The autapomorphy for *Cyclotron lapworthi* is (Fig. 64:10):

- maximum length of the shield along the dorsal rim (cf. Gründel 1981; Williams & Siveter 1998; plesiomorphic state: maximum length between the dorsal rim and the midline, as in Cyclotronidae).

The location of the maximum length of the shield seems to vary among Phosphatocopina. In *Falites fala*, *Waldoria rotundata*, *H. ventrospinata*, *Veldotron bratteforsa*, and many species of *Hesslandona*, the maximum length of the shield is located distinctly between the dorsal rim and the midline (Table 40), a condition reconstructed for the ground pattern of Phosphatocopina. In *Cyclotron lapworthi*, together with the investigated phosphatocopine species *H. necopina* and *H. curvispina* n. sp., the maximum length of the shield corresponds to the dorsal rim. Both species of *Hesslandona* belong to a taxon named NN7 (cf. Fig. 67). The position of NN7 within the phylogenetic tree of the Phosphatocopina indicates the convergent development of the maximum length along the dorsal rim in *Cyclotron lapworthi* and NN7.

Waldoria rotundata

The autapomorphies of *Waldoria rotundata* are (Fig. 64:11):

- maximum height of the shield posterior to the antero-posterior midline, post-plete (Pl. 34A; plesiomorphic state: maximum height of the shield anterior to the midline, pre-plete, as in the ground pattern of Phosphatocopina);
- free shield margin curves back dorsally equally

anteriorly and posteriorly, i.e. the outline of the valves is antero-posteriorly symmetrical (Pl. 34A; plesiomorphic state: shield margin curves back posteriorly more strongly than anteriorly, as in the ground pattern of Phosphatocopina);

- lobe L_5 is drawn out into a spine directed laterally to ventro-laterally (Pl. 35A; plesiomorphic state: no L_5, no such outgrowth, as in the ground pattern of Phosphatocopina, L_5 not drawn out, as in the ground pattern of Cyclotronidae);
- doublure with small conical outgrowths (Pls. 35E, F, 36A; plesiomorphic state: doublure smooth, without outgrowths, as in the ground pattern of Phosphatocopina);
- doublure with pores (plesiomorphic state: doublure smooth, without pores, as in the ground pattern of Phosphatocopina).

A post-plete shield is an autapomorphy of *Waldoria rotundata* (Fig. 64:11). In almost all phosphatocopine species investigated, apart from *Waldoria rotundata* and *Trapezilites minimus*, the maximum height of the shield is located anterior to the antero-posterior midline; i.e. the valves are pre-plete. The position of both species in the reconstructed tree of Phosphatocopina presented herein implies a pre-plete valve as representing the plesiomorphic condition among Phosphatocopina.

The equal back-curvature of the free anterior and posterior dorsal margins of the valves is another autapomorphy of *Waldoria rotundata* (Fig. 64:11). The outline of the shield of *Waldoria rotundata* is antero-posteriorly symmetrical, i.e. the anterior free margin shows an equal back-curvature as the posterior free margin (cf. Fig. 40). The same status occurs in *Trapezilites minimus* and the representatives of the taxon NN3 (Figs. 64:14, 67:18), whereas in all other investigated phosphatocopines, the free dorsal margin curves back not at all or posteriorly more strongly than anteriorly. The condition at the ground pattern level of Euphosphatocopina is a slight back-curvature of the margins, as represented in *Vestrogothia spinata* (Pl. 42B). The position of *Waldoria rotundata* and *Trapezilites minimus* in the tree derived from the phylogenetic analysis does not give any evidence for a close relationship between *Waldoria rotundata* and *Trapezilites minimus* (Fig. 64). In the ground pattern of the Hesslandonina, the posterior free margin curves back more strongly than the anterior one (see above). Therefore, the equal curvature of the free anterior and posterior dorsal margins of the valves is regarded as convergently developed in *Waldoria rotundata*, *Trapezilites minimus* and the stem species of NN3.

A further autapomorphy of *Waldoria rotundata* is lobe L_5 drawn out into a spine (Fig. 64:11). No other phosphatocopine species studied has a lobe drawn out into a spine; the lobe itself is an autapomorphy in the ground pattern of Cyclotronidae (see above, Fig. 64:8). The spine on the valves of *Vestrogothia spinata* (Pl. 42A) is a different character that should not be confused with the spine of *Waldoria rotundata* (Pl. 35B). About half of the individuals of *Waldoria rotundata* investigated have a simple L_5 (Pl. 35C); the remaining specimens show the spine (see p. 117). It is not unlikely, therefore, that this spine reflects sexual dimorphism in this phosphatocopine species.

Another autapomorphy is the presence of outgrowths and pores on the doublure. *Waldoria rotundata* is the only species among the Cyclotronidae to have small conical to dome-like outgrowths and pores on the doublure of the shield. Such structures do not occur in *Vestrogothia spinata* or *Falites fala*. The presence of outgrowths and pores on the doublure cannot therefore be part of the ground patterns of Euphosphatocopina or NN1 (cf. Fig. 64). Similar outgrowths on the doublure occur in several hesslandonid species, such as *H. necopina* and *H. curvispina*, while they are missing in, e.g. *H. angustata* n. sp. and *Trapezilites minimus*. Regarding the position of *Waldoria rotundata* in the phylogenetic diagram (Fig. 64), such outgrowths and pores cannot characterise the ground pattern of Hesslandonina and are therefore considered as convergently developed structures among the phosphatocopines.

Hesslandonidae (*Trapezilites minimus*+ *Hesslandona*)

The autapomorphies of Hesslandonidae are (Fig. 64:12):

- anterior end of the interdorsum with a small, dome-like thickening (Pls. 1A, 32A; plesiomorphic state: interdorsum smooth, without anterior thickenings, as in Hesslandonina);
- posterior end of the interdorsum with a small dome-like thickening (Pls. 1A, 32A; plesiomorphic state: interdorsum smooth, without posterior thickenings, as in Hesslandonina);
- width of the interdorsum between 1/20 and 1/10 the shield length (plesiomorphic state: width less than 1/25 the shield length, as in the ground pattern of Hesslandonina).

An autapomorphy of the Hesslandonidae is the presence of dome-like thickenings of both ends of the interdorsum. Both characters are listed separately because the anterior end of the shield is not necessarily a symmetrical copy of the posterior end, as demonstrated by *Vestrogothia spinata*, which has a smooth anterior end and a spine-bearing posterior dorsal end (Pl. 43B, C). The anterior and posterior ends of the interdorsum of *Trapezilites minimus* and *H. unisulcata* show small hump-like thickenings (Pls. 1A, 32A). The interdorsum

of *H. angustata* and the species referred to as Cyclotronidae, *Cyclotron lapworthi*, *Waldoria rotundata*, and *Veldotron bratteforsa*, is completely smooth (e.g. Pls. 26A, B, D, 34A). In all other species, the interdorsum is drawn out into more or less long spines anteriorly and posteriorly, although the anterior protrusion of *H. toreborgensis* (Pl. 31A) is somewhat dome-like but not as inconspicuous as in *H. unisulcata* and *Trapezilites minimus*. The most parsimonious assumption with the phylogenetic analysis regarding the distribution of smooth interdorsal ends, dome-like thickening, and spines, is spines originated from dome-like thickenings. Thus, the dome-like thickenings of *Trapezilites minimus* and *H. unisulcata* represent a plesiomorphy of both species, but represent an autapomorphy of the taxon Hesslandonidae. The anterior thickening of *H. toreborgensis* n. sp., however, has to be regarded as a reduced spine (see above).

As stated above, the interdorsum in the ground pattern of Hesslandonina had a width of less than 1/25 the shield length, retained in the Cyclotronidae. *Hesslandona angustata* n. sp. is the only species among the Hesslandonidae that has such a narrow interdorsum, it even has the narrowest interdorsum among the investigated species and no outgrowths on either end. The interdorsum is 1/36 the length of the shield (Tab. 36). All other species of Hesslandonidae have distinctly wider interdorsa of at least 1/20 the shield length. If one assumes a narrow interdorsum for the ground pattern of Hesslandonidae, the relationships of *H. angustata* n. sp. imply the widening of the interdorsum five times independently, i.e. in the ground pattern of *Trapezilites minimus*, *H. unisulcata*, *H. suecica* n. sp. plus *H. toreborgensis* n. sp., *H. kinnekullensis* and NN5. The more parsimonious explanation is the assumption of a wider interdorsum in the ground pattern of Hesslandonidae, retained in all lineages except *H. angustata* n. sp. (Fig. 67:23, see below).

Trapezilites minimus

The autapomorphies of *Trapezilites minimus* are (Fig. 64:13):

- maximum height of the shield at the antero-posterior midline, amplete (Pl. 32A; plesiomorphic state: anterior to midline, pre-plete, as in the ground pattern of Phosphatocopina);
- maximum length of the shield at the dorso-ventral midline (Pl. 32A; plesiomorphic state: between dorsal rim and midline, as in the ground pattern of Phosphatocopina);
- smallest width of the doublure anteriorly and posteriorly (Pl. 32B; plesiomorphic state: only anteriorly, as in the ground pattern of Phosphatocopina);
- maximum width of the doublure ventrally, ranging

between one ninth and one sixth the shield length (Pl. 32B; plesiomorphic state: posteriorly, as in the ground pattern of Phosphatocopina).

An autapomorphy of *Trapezilites minimus* is the amplete shield (Fig. 64:13). In almost all phosphatocopine species investigated, apart from *Waldoria rotundata* and *Trapezilites minimus*, the maximum height of the shield is located anterior to the antero-posterior midline, i.e. such valves are termed pre-plete (cf. Fig. 8). In *Trapezilites minimus* and *Waldoria rotundata*, the maximum height is located on the antero-posterior midline of the shield, i.e. the shields of both species are amplete (cf. Fig. 8). The position of both species in the reconstructed tree of Phosphatocopina presented herein implies a pre-plete valve as representing the plesiomorphic condition among the Phosphatocopina.

Another autapomorphy of *Trapezilites minimus* is the location of the maximum length of the shield ventral to the dorso-ventral midline (Fig. 64:13). *Trapezilites minimus* is the only phosphatocopine species investigated in which the maximum length of the shield is located ventral to the dorso-ventral midline (cf. Pl. 32B). In all other species the maximum length is either located along the dorsal rim or between the midline and the dorsal rim, the latter probably representing the ground pattern condition of Phosphatocopina.

A further autapomorphy is the location of the smallest width of the doublure anteriorly plus posteriorly (Fig. 64:13). *Trapezilites minimus* is the only phosphatocopine species investigated in which the smallest width of the doublure is located anteriorly as well as posteriorly (Pl. 32B). In all other species the minimum width is located either anteriorly or ventrally, the latter probably representing the ground pattern condition of Phosphatocopina.

Another autapomorphy of *Trapezilites minimus* is the location of the maximum width of the doublure in the ventral part (Fig. 64:13). *Trapezilites minimus* is the only phosphatocopine species investigated in which the maximum width of the doublure is located ventrally (Pl. 32B), being about one eighth the shield length (cf. Table 40). In all other species, the maximum width is located either postero-ventrally or posteriorly, the latter probably representing the ground pattern condition of Phosphatocopina.

Hesslandona

The autapomorphy of *Hesslandona* is (Fig. 64:14):

- maximum width of the doublure less than or equal to 1/10 the shield length (plesiomorphic state: ranging between one ninth and one sixth the shield length, as in the ground pattern of Phosphatocopina).

The maximum width of the doublure is between one

sixth and one ninth the shield length in the ground pattern of Euphosphatocopina. Such a maximum width of the doublure occurs in *Vestrogothia spinata*, *Waldoria rotundata*, *Veldotron bratteforsa*, *Trapezilites minimus*, *H. unisulcata*, and *H. curvispina*. The doublure of *Cyclotron lapworthi* is unknown (cf. Williams *et al.* 1994b). *Falites fala* has autapomorphically widened the postero-ventral part of its doublure to one quarter the shield length. All other investigated phosphatocopine species have a doublure whose maximum width does not exceed 1/10 the shield length. A maximum width of the doublure of 1/10 the shield length in the ground pattern of *Hesslandona* requires a widening of the doublure only once, i.e. in *H. curvispina* n. sp. (cf. Fig. 67). This is more parsimonious than assuming a wider maximum width in the ground pattern of *Hesslandona*, implying a narrowing of the doublure in several lineages within *Hesslandona* independently.

Hesslandona unisulcata

The autapomorphy of *H. unisulcata* is (Fig. 64:15):

- doublure with bottle-like outgrowths located in pits (Pl. 10D; plesiomorphic state: doublure smooth, without outgrowths and pits, as in the ground pattern of Euphosphatocopina and Phosphatocopina).

Bottle-like outgrowths are located on the ventral part of the doublure in *H. unisulcata*. The outgrowths are located in shallow pits in some specimens, but this appearance is somewhat uncertain as the pits (Pl. 10D) may be a preservational artefact (see *H. unisulcata* description). Outgrowths on the doublure occur also, e.g. in *Waldoria rotundata*, *H. necopina* and *H. curvispina*, but the outgrowths of *H. unisulcata* differ from all other outgrowths on the doublure by their specific shape. Moreover, the distribution of such structures among phosphatocopine species implies convergent evolution.

Dorsospinata new taxon (NN3 + NN5)

The name Dorsospinata is chosen for the characteristic spines present as an autapomorphy in the ground pattern of this phosphatocopine taxon.

The autapomorphies of Dorsospinata new taxon are (Figs. 64:16, 67:17):

- interdorsum anteriorly drawn out into a spine of at least one sixth the length of the valves, directed antero-dorsally (plesiomorphic state: anterior area of the interdorsum with a small dome-like thickening, as in Hesslandonidae);
- interdorsum posteriorly drawn out into a spine of at least one sixth the length of the valves, directed postero-dorsally (plesiomorphic state: posterior area of the interdorsum with a small dome-like thickening, as in Hesslandonidae);

- maximum width of the doublure posteriorly (plesiomorphic state: postero-ventrally, as in Hesslandonidae).

An autapomorphy of the Dorsospinata is the anterior and posterior interdorsal spine-like outgrowths of at least one sixth the shield length (Figs. 64:16, 67:17). In all investigated species of *Hesslandona*, excluding *H. unisulcata*, *H. toreborgensis* n. sp., and *H. angustata* n. sp., the anterior and posterior ends of the interdorsum are drawn out into a spine of at least one ninth the shield length. *Hesslandona unisulcata* lacks both spines. *Hesslandona toreborgensis* n. sp. has significantly shorter spines, and *H. angustata* n. sp. lacks spines completely (see below). An anterior spine is not known from any other phosphatocopine species investigated, whereas a posterior spine also occurs in *Vestrogothia spinata*, originating from an axially elongated triangular plate. This species, however, has no interdorsum, but the spine stems from a triangular plate.

According to the phylogenetic analysis, it is assumed that the posterior spine of *Vestrogothia spinata* developed independently and does not indicate a principle precursor of spines prior to completion of the interdorsum. In both sister taxa within Dorsospinata, species occur with spines as long as one sixth the shield length, such as *H. kinnekullensis* in NN3 and *H. ventrospinata*, *H. curvispina* n. sp. and *H. necopina* in NN5 (cf. Table 40). If shorter spines belonged to the ground pattern of Dorsospinata, longer spines would have developed independently three times. Alternatively, if longer spines belonged to the ground pattern of Dorsospinata, shorter spines would have developed independently twice. In this conflicting decision, the more parsimonious explanation, according to the phylogenetic analysis, is preferred: i.e. the spines were at least one sixth the shield length in the ground pattern of Dorsospinata.

Another autapomorphy is the posteriorly located maximum width of the doublure (Figs. 64:16, 67:17). The doublures of *Falites fala*, *Waldoria rotundata*, *Cyclotron lapworthi*, *Veldotron bratteforsa*, and *H. unisulcata* have their maximum width postero-ventrally (Pls. 2A, 40C). In *Trapezilites minimus*, the maximum width of the doublure is ventral, probably an autapomorphy of this species. In all dorsospinate species, the maximum width of the doublure is posterior (Pls. 15A, 27A). Due to the distribution of this character state within the Euphosphatocopina, the plesiomorphic state is regarded as the location of the maximum width on the postero-ventral part of the doublure, while a posterior location of the maximum width is an apomorphy.

NN3 (*Hesslandona kinnekullensis* + NN4)

The autapomorphies of NN3 are (Fig. 67:18):

- free shield margin curving back dorsally equally anteriorly and posteriorly (Pl. 16A; plesiomorphic state: shield margin with back-curvature more posteriorly, as in the ground pattern of NN1);
- surface of the valves lacks lobes (Pl. 16A; plesiomorphic state: valves with prominent lobe L, as in the ground pattern of NN1).

The curving back of the free shield margin antero-dorsally to an equal degree as postero-dorsally (cf. Fig. 55) is an autapomorphy of NN3. As shown above, the outline of the shield of all species comprising the taxon NN3 is more or less antero-posteriorly symmetrical (cf. Pls. 22A, 26A). This is also true for *Waldoria rotundata* and *Trapezilites minimus*, whereas the free dorsal margin curves back more strongly posteriorly than anteriorly in all other investigated phosphatocopines, such as, e.g. *Falites fala* and *H. unisulcata* (cf. Pls. 1A, 39A). The plesiomorphic state of this character, as in the ground pattern of Euphosphatocopina, is no distinct back-curvature. The position of *Waldoria rotundata* and *Trapezilites minimus* within the resulting phylogeny of Phosphatocopina does not imply a close relationship between both species, and in the ground pattern of the Hesslandonina the back-curvature at the posterior free margin is stronger than that at the anterior (see above). Therefore, the equal curvature of the free anterior and posterior dorsal margins of the valves is regarded as having been independently achieved in NN3, *Waldoria rotundata* and *Trapezilites minimus* (see above).

Another autapomorphy of NN3 is the smooth shield surface (Fig. 67:18). A single lobe is present in the ground pattern of NN1 (Fig. 64). The valves of *H. kinnekullensis*, *H. toreborgensis* n. sp., *H. suecica* n. sp., and *H. angustata* n. sp. are entirely smooth (Pl. 16A, 22A, 25A, 31A), while the valves of all other hesslandonine species, except *H. necopina*, show lobes (cf. Pl. 3A, 17B). *Hesslandona necopina* belongs to a monophyletic unit also comprising lobe-bearing species such as *H. ventrospinata* and *H. trituberculata*. *Hesslandona necopina* plus the stem species of NN3 must have suppressed the development of lobes or lost the lobes convergently.

Hesslandona kinnekullensis

The autapomorphies of *H. kinnekullensis* are (Fig. 67:19):

- anterior interdorsal spine directed dorsally (Pl. 16A; plesiomorphic state: directed anteriorly, as in Dorsospinata);
- posterior interdorsal spine directed dorsally (Pl. 16A; plesiomorphic state: directed posteriorly, as in Dorsospinata).

The dorsally directed anterior and posterior interdorsal spines are autapomorphic to *H. kinnekullensis*. The appearance of such spines is treated separately in the

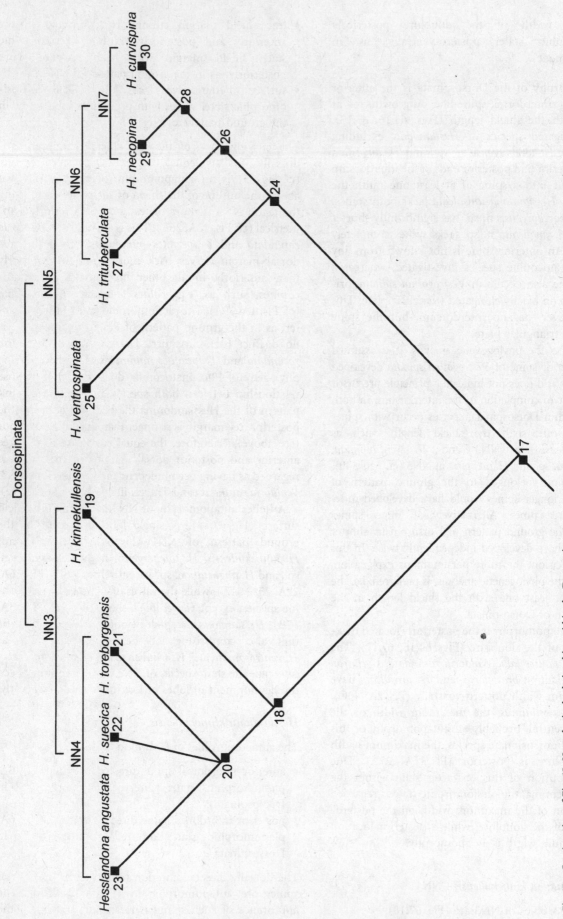

Fig. 67. Continuation of Fig. 65. Hypothesised relationships within the Dorsospinata new name.

anterior and posterior ends as both dorsal ends of the shield can occur in different conditions. For example, as can be observed in *H. toreborgensis* or in *Vestrogothia spinata*, only the posterior end is drawn out into a distinct spine. *Hesslandona kinnekullensis* is the only species in which the anterior and posterior interdorsal spines are directed more or less straight dorsally (Pl. 16A). In all other investigated species the anterior spine is more or less antero-dorsally directed, as may characterise the ground pattern of the Dorsospinata.

NN4 (*Hesslandona toreborgensis* n. sp., *H. angustata* n. sp. + *H. suecica* n. sp.)

The autapomorphies of NN4 are (Fig. 67:20):

- anterior interdorsal spine distinct but shorter than one ninth the length of the valves (Pl. 23A; Table 40; plesiomorphic state: length of the spine between one ninth and one sixth the length of the valves, as in the ground pattern of Dorsospinata);
- posterior interdorsal spine distinct but shorter than one ninth the length of the valves (Pl. 23A; Table 40; plesiomorphic state: length of the spine between one ninth and one sixth the length of the valves, as in the ground pattern of Dorsospinata).

The relationships of the species assigned to the unnamed taxon NN5 remain unclear.

An anterior spine of less than one ninth the shield length is an autapomorphy of NN4 (Fig. 67:20). From the analysis, a spine arising from the anterior end of the interdorsum being at least as long as one ninth the length of the shield represents an autapomorphy of the Dorsospinata (see above). Within the Dorsospinata, several taxa, such as *H. suecica* n. sp. and *H. trituberculata*, have shorter anterior spines, while *H. toreborgensis* n. sp. has only a dome-like anterior thickening of the interdorsum (Pl. 31A). *Hesslandona angustata* does not have spines at all (cf. Table 40). The distribution of the species within the taxon Dorsospinata indicates, first, the reduction of originally long spines, as in *H. kinnekullensis*, to a complete loss in *H. angustata* n. sp. through phylogenetic stages with small spines like *H. suecica* and dome-like thickenings as in *H. toreborgensis* n. sp., and, second, the independent reduction of the spines twice, in the ground pattern of NN4 within the taxon NN3 and in *H. trituberculata* as a member of NN6 within NN5.

Another autapomorphy of NN4 is the short posterior spine of less than one ninth the shield length (Fig. 67:20). From the phylogenetic analysis, a long spine arising from the posterior end of the interdorsum being at least as long as one ninth the length of the shield occurs in the stem species of the Dorsospinata (see above). However, within the taxon Dorsospinata, several taxa, such as *H. suecica* n.

sp., *H. trituberculata* and *H. toreborgensis* n. sp., have shorter posterior spines, and *H. angustata* does not have spines at all (cf. Table 40). As explained for the anterior spine the distribution of the species within the taxon Dorsospinata suggests an independent reduction of the spines twice, mainly in the stem species of NN4 and in *H. trituberculata*. The short posterior spine in the stem species of NN4 might, therefore, be an intermediate step to complete loss, as occurs in *H. angustata* n. sp.

Hesslandona toreborgensis n. sp.

The autapomorphies of *H. toreborgensis* n. sp. are (Fig. 67:21):

- anterior interdorsal spine much reduced to a dome-like thickening (Pl. 31A; plesiomorphic state: distinct spine, but shorter than one ninth the length of the valves, as in the ground pattern of NN4);
- posterior interdorsal spine much reduced to a short, pointed protrusion (Pl. 31A; plesiomorphic state: distinct spine, but shorter than one ninth the length of the valves, as in the ground pattern of NN4).

The dome-like thickening instead of a spine at the anterior end of the interdorsum is an autapomorphy of *H. toreborgensis* n. sp. (Fig. 67:21). The anterior end of the interdorsum of *H. toreborgensis* n. sp. is slightly domed without being drawn out into a distinct spine (Pl. 31A). In the ground pattern of NN3, the anterior end of the interdorsum is drawn out into a long spine, retained from the ground pattern of the Dorsospinata. The spine is somewhat reduced in length in the ground pattern of NN5 and retained in *H. suecica*. This implies that the phylogenetic reduction of the anterior spine present in the ground pattern of NN4 was continued in *H. toreborgensis* n. sp. to its almost complete reduction.

Another autapomorphy of *H. toreborgensis* n. sp. is the short protrusion instead of a spine at the posterior end of the interdorsum (Fig. 67:21). In fact, the posterior spine of the interdorsum of *H. toreborgensis* n. sp. does not exceed 1/15 the shield length (Pl. 31A), by contrast to the situation in *H. suecica* (Pl. 23C, cf. Table 41). In the ground pattern of NN3, the posterior end of the interdorsum is drawn out into a distinct spine, retained from the ground pattern of the Dorsospinata new taxon and retained in *H. kinnekullensis*. The spine is between one sixth and one ninth the shield length in the ground pattern of NN4, as present in *H. suecica*. This implies that the reduction of the posterior spine characterises the ground pattern of NN4 and was continued in *H. toreborgensis* n. sp., but not to the extreme as in the anterior interdorsal outgrowth.

Hesslandona suecica n. sp.

No autapomorphies were found for this species (Fig. 67:22). *Hesslandona suecica* n. sp. has retained the short anterior and posterior spines, as proposed for the ground pattern of NN4 – in contrast to *H. toreborgensis* n. sp. and *H. angustata* n. sp. which reduced both outgrowths. The investigation of soft part characters and ontogeny in the second paper on Phosphatocopina will also probably yield autapomorphies for *H. suecica* n. sp.

Hesslandona angustata n. sp.

The autapomorphies of *H. angustata* n. sp. are (Fig. 67:23):

- no anterior interdorsal spine (Pl. 26B; plesiomorphic state: distinct spine, shorter than one ninth the length of the shield, as in the ground pattern of NN3);
- no posterior interdorsal spine (Pl. 26D; plesiomorphic state: distinct spine, shorter than one ninth the length of the shield, as in the ground pattern of NN3);
- width of the interdorsum less than 1/25 the length of the valves (Pl. 26A; plesiomorphic state: between 1/10 and 1/20 the length of the valves, as in the ground pattern of Hesslandonidae).

An autapomorphy of *H. angustata* n. sp. is the lack of interdorsal spines. In fact, *H. angustata* is the only species within the Hesslandonidae (Fig. 64) with an interdorsum not drawn out into a spine or thickening at either end. *Hesslandona angustata* is, nevertheless, a member of the taxon Dorsospinata, belonging, together with *H. toreborgensis* n. sp. and *H. suecica* n. sp., to a monophyletic taxon NN4 within the taxon NN3 (Fig. 67). Outside the Dorsospinata, spines are only present in *Vestrogothia spinata*, in which the posterior plate – an interdorsum is not present – is drawn out into a slender spine (Pl. 42A). The most parsimonious explanation is that *H. angustata* has lost both the anterior and posterior interdorsal spines. In all other phosphatocopine species with no interdorsal spines, such as the species of Cyclotronidae, *Trapezilites minimus*, and *H. unisulcata* (Fig. 64), the lack of spines is, according to the positions of these species in the phylogenetic tree (Fig. 64), a plesiomorphy.

The narrow interdorsum of less than 1/25 the shield length is an autapomorphy of *H. angustata* (Fig. 67:23). *Hesslandona angustata* has the narrowest interdorsum among the investigated species. The interdorsum is 1/36 the length of the shield (Table 40). In the ground pattern of Hesslandonina, where the interdorsum represents an autapomorphy, the interdorsum has a width of less than 1/25 the shield length. *Waldoria rotundata* and *Veldotron bratteforsa* as well as *Cyclotron lapworthi* have such narrow interdorsa, measuring 1/30 and 1/26 the shield length, respectively [cf. Table 40; Williams *et al.* (1994a) for *Cyclotron lapworthi*]. *Waldoria rotundata* and *Veldotron bratteforsa* differ from *H. angustata* in the presence of valve lobes, the outline of the shield and the

shape of the doublure. Their position in the tree resulting from the phylogenetic analysis, being grouped together as Cyclotronidae, is clearly distant from *H. angustata* n. sp. (cf. Figs. 64, 67). Moreover, an interdorsum with a width of between 1/20 and 1/10 the shield length characterises the ground pattern of the Dorsospinata. The relationships of *H. angustata* suggest that the narrow interdorsum is reductive. The loss of interdorsal spines in *H. angustata* n. sp. probably correlates with the narrowing of the interdorsum.

NN5 (*Hesslandona ventrospinata* + NN6)

The autapomorphies of NN5 are (Fig. 67:24):

- shield valves with prominent lobes L_2 and L_3 in addition to lobe L_1 (Pl. 17B; plesiomorphic state: shield valves with only lobe L_1, as in the ground pattern of NN2);
- doublure with conical outgrowths on the surface (Pl. 15C; plesiomorphic state: doublure smooth, without outgrowths, as in the ground pattern of Phosphatocopina).

Lobes L_2 and L_3 in addition to lobe L_1 is an autapomorphy of NN5 (Fig. 67:24). Lobe L_1 is an autapomorphy in the ground pattern of NN1 (Fig. 64:5). Additional lobes occur in several phosphatocopine species, four more in *Waldoria rotundata*, five more in *Veldotron bratteforsa*, *H. ventrospinata* and *Cyclotron lapworthi*, and two more in *Falites fala*, *H. curvispina* n. sp., and *H. trituberculata*. The existence of four to five more lobes is reconstructed as an autapomorphy in the ground pattern of Cyclotronidae (Fig. 64:8). The presence of lobes L_2 and L_3 in *Falites fala* has, according to the phylogenetic analysis, evolved independently and has to be regarded as an autapomorphy of this species. All other mentioned species belong to the unnamed taxon NN4, which also contains the lobe-less species *H. necopina*. This species is the sister taxon of *H. curvispina*. The assumption of a lobe-less shield in the ground pattern of NN5 would imply, first, the autapomorphical suppression or loss of lobe L_1 and, second, the acquisition of all three lobes in at least three different lineages (cf. Fig. 67). The most parsimonious explanation is to regard the occurrence of two additional lobes as an autapomorphy in the ground pattern of NN5, while *H. necopina* has probably suppressed the development of all lobes or lost the lobes entirely.

A further autapomorphy of NN5 is the presence of conical outgrowths on the surface of the doublure (Fig. 67:24). Conical outgrowths on the doublure occur only in *H. necopina*, *H. trituberculata*, *H. ventrospinata*, and *H. curvispina* n. sp., comprising all species of NN5 (Pls. 18D, E, 20D, 30A, B), and in the clearly unrelated

Waldoria rotundata (Pl. 36A) among the Cyclotronidae (cf. Fig. 64). The outgrowths of *Waldoria rotundata* can thus be clearly identified as convergent developments. The outgrowths of the doublure of *H. unisulcata* (Pl. 3C) are different in shape and are probably an autapomorphy of this species (see above).

Hesslandona ventrospinata

A single, but very distinctive, autapomorphy could be identified for *H. ventrospinata* (Fig. 67:25):

- postero-ventral margin of both valves with a pair of interlocking triangular outgrowths (Pl. 19A; plesiomorphic state: margins smooth, without any outgrowths, as in Phosphatocopina).

The postero-ventral margins of both valves in *H. ventrospinata* are drawn out into a pair of triangular outgrowths, which interlock between the valves. Such outgrowths do not occur in any other investigated phosphatocopine species; the marginal single spiny outgrowth of the left valve of *Vestrogothia spinata* has, as stated above, to be regarded as a different structure and an autapomorphy of this species. The species of *Bidimorpha* Hinz-Schallreuter, 1993 do have similar outgrowths to that of *H. ventrospinata*, but they could not be included in the phylogenetic analysis because of a lack of data. Consequently, the triangular outgrowths have to be considered an autapomorphy of *H. ventrospinata* in this paper. Consideration of the species of *Bidimorpha* will be part of the second paper on the Phosphatocopina.

NN6 (*Hesslandona trituberculata* + NN7)

The autapomorphy of NN6 is (Fig. 67:26):

- anterior margin of the valves rather straight (Pl. 13A; plesiomorphic state: curved, as in the ground pattern of Phosphatocopina).

The anterior shield margin of *H. trituberculata*, *H. necopina*, *H. curvispina*, the species within NN6, and in *Vestrogothia spinata* is rather straight, whereas in all other investigated phosphatocopine species the anterior shield margin is distinctly curved. If one assumes a straight anterior shield margin in the ground pattern of Euphosphatocopina, the condition in *Vestrogothia spinata* would be plesiomorphic, which further implies an independent, change in this character state six times, i.e. in *Falites fala*, the Cyclotronidae, *Trapezilites minimus*, *H. unisulcata*, *H. ventrospinata*, and in the stem species of NN3 (cf. Figs. 64, 67). The more parsimonious explanation, which is followed here based on the phylogenetic analysis, is that a curved anterior shield margin occurred in the ground pattern of Phosphatocopina. Hence, the independent change into a straight anterior

shield margin developed only twice, i.e. in *Vestrogothia spinata* and in the stem species of NN6.

Hesslandona trituberculata

The autapomorphies of *H. trituberculata* are (Fig. 67:27):

- valves gaping posteriorly (Pl. 17F; plesiomorphic state: valves close tightly along their whole free length, as in the ground pattern of Euphosphatocopina);
- anterior interdorsal spine shorter than one nin the length of the valves (Pl. 17E; plesiomorphic state: anterior spine equal to or longer than one sixth the length of the valves, as in the ground pattern of Dorsospinata);
- posterior interdorsal spine shorter than one nin the length of the valves, directed posteriorly (Pl. 17F; plesiomorphic state: posterior spine equal to or larger than one sixth the length of the valves, spine directed postero-dorsally, as in the ground pattern of Dorsospinata).

A posteriorly gaping shield is an autapomorph of *H. trituberculata* (Fig. 67:27). As stated above, a shield with a closing free margin must be interpreted the plesiomorphic condition within the ground pattern of Euphosphatocopina, Hesslandonina, Hesslandonidae, *Hesslandona*, Dorsospinata, and NN4. The gaping shields of *Veldotron bratteforsa* (Pl. 38B) and *H. trituberculata* (Pl. 17D) have to be assumed to be independent acquisitions of both species.

Another autapomorphy of *H. trituberculata* the shortness of the anterior and posterior interdorsal spines (Fig. 67:27). Spines arising from the anterior and posterior ends of the interdorsum with at least one sixth the length of the shield have been identified as an autapomorphy in the ground pattern of Dorsospinata (see above). Within this group, *H. trituberculata*, *H. toreborgensis* n. sp., and *H. suecica* n. sp. do not have such long spines, and *H. angustata* does not have spines at all (cf. Table 40). This distribution of the species within the Dorsospinata indicates the independent reduction of the spines twice, in the ground pattern of NN4 (= *H. toreborgensis* n. sp. + *H. suecica* n. sp. + *H. angustata* n. sp.) and in *H. trituberculata*

NN7 (*Hesslandona necopina* + *Hesslandona curvispina*)

The autapomorphies of NN7 are (Fig. 67:28):

- maximum length of the shield along the dorsal rim (cf. Pl. 13A; plesiomorphic state: maximum length of the shield between the dorsal rim and the midline, as in the ground pattern of Phosphatocopina);
- posterior interdorsal spine with scale-like spinules (cf. Pl. 14C; plesiomorphic state: spine smooth, as in the ground pattern of Phosphatocopina).

The location of the maximum length along the dorsal rim is an autapomorphy of NN7. In *H. necopina* and *H. curvispina*, the maximum length of the shield is located on the dorsal rim (Pls. 13A, 28A). In all other investigated phosphatocopine species, the maximum length is between the dorsal rim and the dorso-ventral midline and more or less significantly longer than the dorsal rim (cf. Table 40). The location of the maximum length of the shield within Phosphatocopina has probably been subject to several changes. The distribution of this character within the studied Phosphatocopina and the species that were not considered in the analysis, points to an original location of the maximum length in the ground pattern of Phosphatocopina between the dorsal rim and the midline, as it occurs in all phosphatocopine lineages apart from that leading to NN7.

A further autapomorphy of NN7 is the presence of outgrowths on the posterior interdorsal spine (Fig. 67:28). The posterior spines of the interdorsum of *H. curvispina* n. sp. as well as that of *H. necopina* bear small, scale-like outgrowths on their surface (Pls. 14C, 29A, B). Such outgrowths are not known from any other investigated phosphatocopine species. In all other species, both the anterior and the posterior spines are smooth. The most parsimonious explanation is that smooth interdorsal spines represent the plesiomorphic character state in the ground pattern of the Dorsospinata.

Hesslandona necopina

The autapomorphies of *H. necopina* are (Fig. 67:29):

- shield valves smooth, without prominent lobes L_1, L_2 and L_3 (Pl. 13A; plesiomorphic state: shield valves with lobes L_1, L_2 and L_3, as in the ground pattern of NN5);
- groove on the outer rim of the shield (Pl. 13E; Fig. 32; plesiomorphic state: shield without such a rim, as in the ground pattern of Phosphatocopina).

A smooth shield is an autapomorphy of *H. necopina* (Fig. 67:29). The shield of *H. necopina* is entirely smooth (Pl. 13A), but lobes are present, for instance, in its sister taxon *H. curvispina* and many other hesslandonids. A single lobe L_1 was discussed as being an autapomorphy of NN1 (Fig. 64:5), and the occurrence of additional lobes L_2 and L_3 is regarded as an autapomorphy of NN5 (Fig. 67:24). Therefore, the absence of lobes in *H. necopina* is interpreted as apomorphic.

A further autapomorphy of this taxon is the presence of a groove on the outer rim of the shield (Pl. 13E; Figs. 28, 67:29). A similar structure is not present in any other investigated phosphatocopine species.

Hesslandona curvispina n. sp.

The autapomorphy of *H. curvispina* n. sp. is (Fig. 67:30):

- posterior interdorsal spine curving postero-ventrally (Pl. 31A; plesiomorphic state: spine directed postero-dorsally, as in the ground pattern of Dorsospinata).

The autapomorphy of *H. curvispina* is the postero-ventral curvature of the posterior interdorsal spine (Fig. 67:30). This posterior spine arising from the interdorsum is very long and strongly curved along its whole length, such that it is directed postero-ventrally (Pl. 28A; Fig. 41A, B). No other studied phosphatocopine species has a comparably long and curved posterior spine. In the ground pattern of Dorsospinata, the posterior interdorsal spine was probably directed straight in a postero-dorsal direction, as present in, e.g. *H. necopina* within NN5 and *H. suecica* in NN3 (cf. Fig. 67).

Consequences of the proposed phylogeny of Phosphatocopina

Since the taxon Phosphatocopina was erected by Müller (1964a), the unity of Phosphatocopina has been generally accepted in the literature – although not in the sense of a monophyletic taxon following the concept of phylogenetic systematics introduced as a method to obtain natural relationships (Hennig 1950). More recently, Walossek & Müller (1998a) and Walossek (1999) proposed the monophyly of Phosphatocopina in a phylogenetic sense; and Walossek (1999) also listed a set of autapomorphies. The current investigation confirms and emends the hypotheses of a monophyletic Phosphatocopina *sensu* Walossek & Müller (1998a) and Walossek (1999).

Hinz-Schallreuter (1993c) regarded, without giving detailed arguments for this decision, *Falites* as closely related to *Trapezilites* and *H. unisulcata* and used the name Falitidae Müller, 1964 to combine these three taxa. The phylogenetic analysis presented herein has, however, shown that these species are members of different lineages (Fig. 68). The taxon Falitidae is therefore paraphyletic (Fig. 68) and should be abandoned.

The name "Vestrogothiidae" was established by Kozur (1974) to comprise only species of *Vestrogothia*, such as *Vestrogothia spinata* and *Vestrogothia longispinosa* Kozur, 1974. Subsequently, the taxa *Cyclotron*, *Veldotron* and *Waldoria* were included in the Vestrogothiidae (Williams & Siveter 1998). This work shows that the members of the traditional Vestrogothiidae belong to different lineages within the Phosphatocopina (Fig. 68). Consequently, such a wide concept of "Vestrogothiidae" leads to a polyphyletic taxon, and the taxon name "Vestrogothiidae" has no place in a natural system.

The assumption of Adamczak (1965), who proposed that the hesslandonid species of Phosphatocopina should rather be ephippia of Cladocera "or similar findings", has already been rejected by Müller (1979a) with his discovery of soft part preserved hesslandonid species (cf. Pl. 7A).

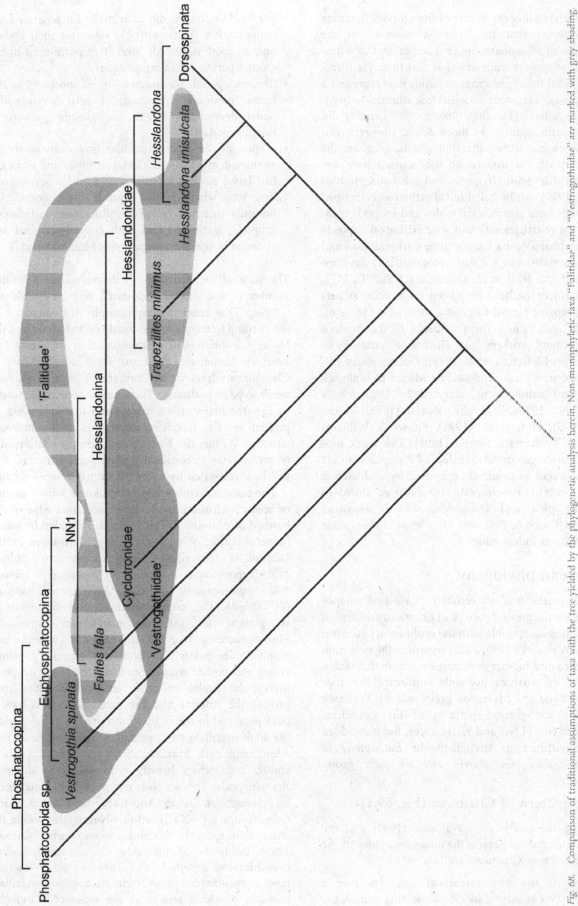

Fig. 68. Comparison of traditional assumptions of taxa with the tree yielded by the phylogenetic analysis herein. Non-monophyletic taxa "Falitidae" and "Vestrogothiidae" are marked with grey shading.

Misunderstanding the nature of the exopods, Landing (1980) claimed that the "cirri-like nature" of the appendages of Phosphatocopina is unlike that of ostracodes – as they were understood at that time. He therefore suggested that "phosphatocopinids may represent a vagrant group of crustaceans with close affinities to barnacles (Cirripedia)" (Landing 1980, p. 757). Landing did not explain his opinion in more detail. However, this present work has shown that Phosphatocopina are the sister group of Eucrustacea, so they cannot have any synapomorphies with cirripedes and the assumption of Landing (1980) can be ruled out. Furthermore, cirriped thoracopods have symmetrical endo- and exopods composed of many articles with stiff setae arising on opposite sides. Phosphatocopina have a three-part endopod and a multi-annulated exopod that ontogenetically develops into a paddle, at least in *H. unisulcata* (cf. Pls. 7C, 11F).

Phosphatocopina have long been understood as part of the Palaeozoic record for, and Cambrian evidence of, Ostracoda and, hence, the precursors of the modern forms by many workers (e.g. Hinz-Schallreuter 1998; Ziegler 1998; McKenzie *et al.* 1999). Doubts about this assumption were first proposed by Müller & Walossek (1991a) and subsequently accepted by Walossek & Müller (1992, 1998a, b), Walossek *et al.* (1993), Müller *et al.* (1995), Hou *et al.* (1996), Siveter & Williams (1997) and Williams & Siveter (1998). This work supports the sister-group relationship of Phosphatocopina and Eucrustacea as originally proposed by Walossek & Müller (1998a). Consequently, the bivalved shield of Phosphatocopina and Ostracoda, as well as other bivalved arthropods (see also Hou *et al.* 1996), must have developed independently.

Crustacean phylogeny

The phylogenetic analysis resulted in a set of autapomorphies for the ground pattern of Crustacea and several ingroups. It is compatible with the evolutionary scenario drawn by Walossek (1999, 2002) regarding the evolution of Crustacea and further development within this taxon. Moreover, the analysis not only confirmed the apomorphic characters presented previously by Walossek (1999) for the ground patterns of the Crustacea, Labrophora (his N.N.) and Eucrustacea, but also added new apomorphies to strengthen the monophyly of Phosphatocopina (see above) and its sister group Eucrustacea.

Ground pattern of Crustacea (Fig. 69:1)

Combining the analysis by Walossek (1999) and the phylogenetic analysis herein, the autapomorphies of the ground pattern of Crustacea are (Fig. 69:1):

- basipod of the post-antennular limbs bearing a movable (via musculature), setae-bearing endite, the proximal endite, medio-proximally for separate food manipulation (plesiomorphic state: no such endite, only basipod as a single limb stem portion, as in the ground pattern of Euarthropoda);

- five endopodal podomeres in all post-antennular limbs (plesiomorphic state: at least seven endopodal podomeres, as in the ground pattern of Euarthropoda);

- exopod of the post-antennular limbs can be multi-annulated, at least in the anterior limbs and from the first larval stages onwards, each annulus bearing one long seta, which points towards the endopod, i.e. inwardly oriented (plesiomorphic state: leaf-shaped exopods, undivided and with circum-marginal setation, as in the ground pattern of Euarthropoda).

The state of the character "antennula with a limited number of rod-shaped segments", noted by Walossek (1999, p. 7) as another autapomorphy of Crustacea may – according to recent studies made on material from the Lower Cambrian Chengjiang fauna in the Early Life Research Center of Professor Dr Chen Junyuan in Chengjiang, China, by us – rather be plesiomorphic and needs to be re-evaluated. The consequence of this would be that the filamentous, multi-annulated antennula, as present in, e.g. trilobites, represents an apomorphic structure. As this also had a bearing on the relationships of atelocerates (traditionally myriapods + insects), this will be investigated by us in the future in more detail.

The proximal endite is a plate-like or lobate, setated or spiny protrusion at the inner proximal edge of the basipod of post-antennular limbs, surrounded by membranous cuticle of all stem lineage derivatives of the Labrophora described, i.e. †*Cambrocaris baltica*, †*Cambropachycope clarksoni*, †*Goticaris spinosa*, †*Henningsmoenicaris scutula*, †*Martinssonia elongata* (cf. Table 2). The same structure, also with setation, is present in the post-mandibular limbs of Phosphatocopina (e.g. Pls. 10B, 36C; for antenna and mandible see below). In Entomostraca, the proximal endite and similar separate endites are present on post-mandibular limbs and different morphologies concerning the antenna and the mandible, which have a coxa proximal to the basipod, occur. In Cephalocarida, the adult maxillula has a proximal endite modified into a long outgrowth. Maxillae always retain their proximal endite, even when heavily reduced, as in different Branchiopoda. Thoracopods may retain proximal endites throughout, as in Anostraca, Conchostraca and Calmanostraca ($\hat{=}$ Kazcharthra + Notostraca), while the latter taxon splits the basipod into two parts (Walossek 1993). Inside the Maxillopoda, the pattern is diverse: branchiurans, copepods and myodocopid ostracodes have a two-divided limb stem in the post-maxillary limbs, a proximal endite in the sense of a movable

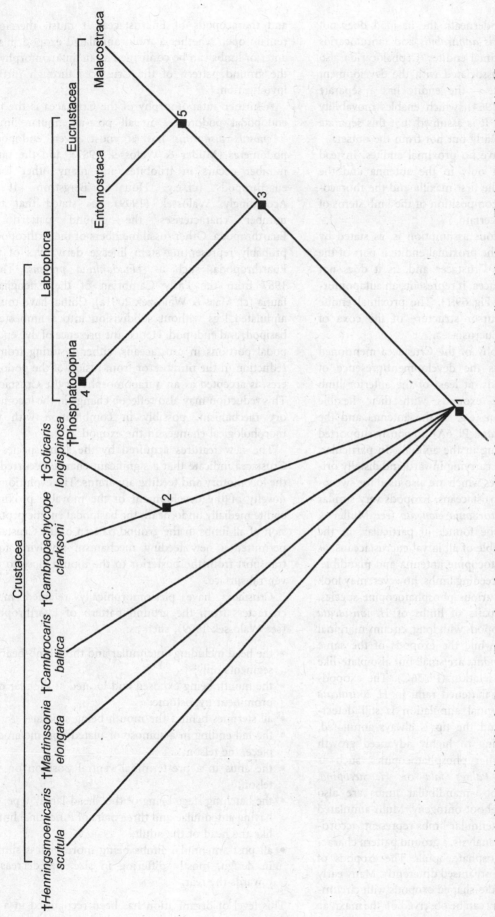

Fig. 69. Hypothetical relationships within the Crustacea. The numbers refer to the autapomorphies of the respective stem species, as discussed in the text. The sister-group relationship of Entomostraca and Malacostraca is not discussed herein, as this work concentrates on the phosphatocopines [see Walossek (1993, 1999) and Walossek & Müller (1998a, b) for a discussion and references].

protrusion medially underneath the basipod does not occur, while †*Bredocaris admirabilis* and tantulocarids have very distinct proximal endites. Cephalocarida also show another feature associated with the development of the proximal endite – the endite has a separate musculature (Hessler 1964), which enables movability (Sanders 1963, fig. 10). It is assumed that this separate action was established early but not from the outset.

The Malacostraca have no proximal endites. Instead they have a coxa, not only in the antenna and the mandible, but also in the first maxilla and the thoracopods throughout. The composition of the limb stems of the pleopods is still uncertain.

The most parsimonious assumption is, as stated by Walossek (1999), that the proximal endite is part of the ground pattern of the Crustacea and, as it does not occur outside the Crustacea, it represents an autapomorphy of the Crustacea (Fig. 69:1). The proximal endite seems to be the precursor structure of the coxa of different limbs of the Eucrustacea.

Another autapomorphy of the Crustacea mentioned by Walossek (1999) is the development/presence of multi-annulated exopods, at least of the anterior limb pairs (Fig. 69:1). The exopods with their flexible annulation and setation (e.g. in the antenna and the mandible of *H. unisulcata*; Pl. 6A) obviously supported and improved swimming in the animals. In particular, small-scale annulations carrying inward and distally oriented setae are structures which are also used for sweep-net feeding in Recent crustaceans. Exopods very similar to those of, e.g. †*Martinssonia elongata* (see Müller & Waloszek 1986a) can be found, in particular, in the antenna and the mandible of all larval eucrustaceans, as well as in the phosphatocopine antenna and mandible. The exopods of the succeeding limbs, however, may look quite different in the various phosphatocopine species. The post-mandibular pairs of limbs of *H. unisulcata* have huge plate-like exopods with long, circum-marginal setae (Pls. 11F, 12G), while the exopods of the same limbs in *Waldoria rotundata* are small but also plate-like with circum-marginal setation (Pl. 36C). The exopods of these limbs start as flattened rami in *H. unisulcata* (Pl. 6A), but their original annulation is still detectable in early larvae, and the tip is always annulated, even in limbs belonging to highly advanced growth stages (Pl. 11F). In other phosphatocopines, such as *Vestrogothia spinata*, *Falites fala* or *H. necopina*, the exopods of the post-mandibular limbs are also multi-annulated throughout ontogeny. Multi-annulated exopods on all post-antennular limbs represent, according to the phylogenetic analysis, a ground pattern character of at least the Phosphatocopina. The exopods of other euarthropods are organised differently. Many early euarthropods have paddle-shaped exopods with circum-marginal setation as they can be observed on the maxillae

and thoracopods of Eucrustacea. It must, therefore, remain open whether a multi-annulated exopod in the anterior limbs can be confirmed as an autapomorphy in the ground pattern of the Crustacea through further investigation.

A further autapomorphy of the Crustacea is the five endopodal podomeres in all post-antennular limbs. †*Agnostus pisiformis* has at most seven endopodal podomeres (Müller & Walossek 1987), and the same number occurs in trilobites and many other fossil euarthropods (cf. e.g. Hou & Bergström 1997). Accordingly, Walossek (1999) has stated that this number characterises the ground pattern of Euarthropoda. Other fossil members of the Arthropoda, probably representing stem lineage derivatives of the Euarthropoda, such as †*Fuxianhuia protensa* Hou, 1987 from the Early Cambrian of the Chengjiang fauna (cf. Maas & Waloszek 2001a), China, have multi-annulated legs without subdivision into a limb stem, basipod, and endopod. Hence, the presence of five endopodal portions in crustaceans, either resulting from a reduction in the number or from fusion of the podomeres, is accepted as an autapomorphy of the Crustacea. This reduction may also reflect a change in the locomotory mechanism, possibly in combination with the morphological changes in the exopod.

The new features acquired by the stem species of Crustacea indicate that a significant change occurred in the locomotory and feeding apparatus. The evolutionary novelty of the development of the movable proximal endite medially underneath the basipodal enditic protrusion of all limbs in the ground pattern of the Crustacea permitted a new feeding mechanism involving food transport from the posterior to the mouth close to the ventral surface.

Crustacea have plesiomorphically retained many characters from the ground pattern of Euarthropoda (see Walossek 1999), such as:

- the head including antennular and three limb-bearing segments only;
- the mouth being exposed and located at the rear of a prominent hypostome;
- all sternites behind the mouth being separate;
- the tail ending in a spinose or plated non-metameric piece, the telson;
- the anus in a pre-terminal ventral position on the telson;
- the hatching stage being of the "head-larva" type, i.e. having antennulae and three pairs of functional limbs, like the head of the adult;
- all post-antennular limbs being more or less similar in design, mostly differing in size, i.e. decreasing towards the rear.

This level of organisation has been recognised in a set

of Upper Cambrian "Orsten" fossils: †*Martinssonia elongata*, †*Henningsmoenicaris scutula*, †*Goticaris longispinosa*, †*Cambropachycope clarksoni*, and †*Cambrocaris baltica* (cf. Walossek & Müller 1990, 1998a, b; Walossek 1999). Two of these forms, †*Cambropachycope clarksoni* and †*Goticaris longispinosa*, have their mouth lying directly on the ventral surface. Because a hypostome, as a prominent feature in the anterior head region, is present in three other species and also in the Phosphatocopina and various fossil euarthropods (cf. Hou & Bergström 1997), this structure is probably part of the ground pattern of the Euarthropoda. Accordingly, †*Cambropachycope clarksoni* and †*Goticaris longispinosa* have lost the hypostome. Furthermore, both share a huge single facetted eye in front of the head which is drawn out postero-distally with an anteriorly turned small hook ventrally. These two features and uniramous post-cephalic limbs (four in all) are characters that strongly support their sister-group relationship (Fig. 69:2; Walossek 1999). Further investigations on all derivative species are under way. The hypostome of Eucrustacea is, due to the development of the labrum and the different mouth position, less prominent in phosphatocopines than in Euarthropoda, and even less so in the Eucrustacea.

The ground pattern of Labrophora Siveter, Waloszek & Williams, 2003 (Fig. 69:3)

The next step on the evolutionary line towards the Eucrustacea is marked by the stem species of the taxon embracing the Phosphatocopina and Eucrustacea, for which Siveter *et al.* (in press) have proposed the name Labrophora. Walossek & Müller (1990) still kept Phosphatocopina as stem lineage crustaceans. They hence considered some characters now proposed for Labrophora as part of the eucrustacean ground pattern. This is now modified in light of the new discoveries. Accordingly, autapomorphies in the ground pattern of Labrophora (Fig. 69:3) are seen in:

- bulged structure, the labrum, present as an outgrowth at the rear of the hypostome extending above the mouth, possibly originating from the circum-oral membrane, labrum with glandular openings and sensilla (plesiomorphic state: no labrum present, only a hypostome with a mouth opening at its rear, as in the ground patterns of Crustacea and Euarthropoda);
- mouth opening at the posterior end of the hypostome/labrum complex, close to the ventral surface and underneath the slightly overhanging distal part of the labrum (plesiomorphic state: mouth exposed at the rear of the hypostome, as in the ground patterns of Crustacea and Euarthropoda);
- sternum as a sclerotic plate, the product of fusion of the sternites of the first three post-antennular seg-

ments (plesiomorphic state: isolated sternal plates, sternites, as in the ground patterns of Crustacea and Euarthropoda);
- mandibular sternal portion with a pair of humps, paragnaths (plesiomorphic state: lacking in the ground patterns of Crustacea and Euarthropoda);
- fine hairs on the labrum, sternum, paragnaths and enditic surfaces of all head limbs and also on the setae and spines (plesiomorphic state: no such hairs on the isolated sternitic plates, as in the ground patterns of Crustacea and Euarthropoda);
- proximal endite of the first post-antennular limb (antenna) enlarged and more strongly sclerotised to form a separate stem portion proximal to the basipod, named the coxa; coxa antero-posteriorly compressed with a medially extended, spine-bearing edge (plesiomorphic state: proximal endite as a movable, setae-bearing, bluntly rounded median outgrowth medially under the basipod, as in the ground pattern of Crustacea);
- proximal endite of the second post-antennular limb (mandible) enlarged and more strongly sclerotised to form a separate stem portion proximal to the basipod, named the coxa; coxa antero-posteriorly compressed with an obliquely oriented, medially extended, spine-bearing edge (plesiomorphic state: proximal endite as a movable, setae-bearing, bluntly rounded median outgrowth medially under the basipod, as in the ground pattern of Crustacea);
- five limb-bearing head segments, including the segment of the trunk-limb-shaped third and fourth limbs ("maxillula" and "maxilla") (plesiomorphic state: four limb-bearing head segments, as in the ground patterns of Crustacea and Euarthropoda).

The name-giving autapomorphy of the Labrophora is the labrum (Fig. 69:3). A labrum as a bulged outgrowth at the rear of the hypostome extending above the mouth is present in all investigated representatives of the Phosphatocopina and all eucrustacean species (cf. Pl. 3A). In Phosphatocopina, its posterior surface shows pores, possibly glandular openings, sensilla and papilliform structures (cf. Pls. 3G, 24D). It is well known that Eucrustacea have slime glands in their labrum and also chemoreceptors, but until now no SEM picture similar to those of the Cambrian fossil Phosphatocopina has ever been published. In the derivatives of the eucrustacean stem lineage a labrum is absent – as all structures associated with the new feeding system of Labrophora. The presence of only the hypostome, either as a bulged or plate-like structure, is also true for other euarthropod taxa such as trilobites, naraoiids, agnostids, eodiscids (Müller & Walossek 1987). The hypostome is never lost in Labrophora, which is the reason why the hypostome and the labrum cannot be equivalent as several authors erroneously believe (e.g. Ritterbush 1983; Scholtz 1995, 1998). Both structures have a different phylogenetic origin, and they do not substitute each other. The hypostome is the attachment area of the antennula and the antenna; hence, it is either structurally retained, as in ostracodes, or shortened, as in all other eucrustaceans.

Again, †*Agnostus pisiformis* shows a structure antero-ventrally on the hypostome, consisting of two bulbs (Fig. 70A, B). This is very similar to a structure present in phosphatocopines in the same position, interpreted herein as the median eye (cf. Pls. 3A, C, 6B). A similar organ is present at least in †*Henningsmoenicaris scutula* as a representative of the stem lineage of the Labrophora (Walossek & Müller 1990). It is still unclear whether this structure is homologous to the naupliar eye of Eucrustacea, and it is not the scope of this paper to go further in detail.

A hypostome was probably present in the ground pattern of Euarthropoda (cf. Walossek 1999). The origin of the hypostome and the labrum has long been the subject of discussion (see, e.g. Lauterbach 1973; Kukalova-Peck 1992, 1998; Bitsch 1994; Dewel *et al.* 1999), and the hypostome and the labrum are frequently confused in the literature (see above). Scholtz (1995, 1998) claimed that the hypostome, which he misunderstood as the labrum, was the tip of the acron that moved postero-ventrally during the evolution of the Euarthropoda. This theory has some major problems. An acron, although never explicitly observed, is assumed to be present in arthropods only because of the assumption of the close relationship of Arthropoda and Annelida, forming a taxon Articulata. Doubts about this

appeared very recently, and there is good evidence for a close relationship of Arthropoda and Nemathelminthes (Aguinaldo *et al.* 1997; Schmidt-Rhaesa *et al.* 1998; Manuel *et al.* 2000; Valentine & Collins 2000), the so-called Ecdysozoa theory. This consequently would imply that an acron cannot exist in arthropods. A similar suggestion to that of Scholtz, but without using the term acron, was proposed by Dewel *et al.* (1999). However, the origin of the hypostome remains subject to future investigations.

Walossek & Müller (1990) and Walossek (1999) proposed that the labrum of crustaceans originated from the circum-oral membrane, as can be seen in, e.g. the Cambrian †*Agnostus pisiformis* (Müller & Walossek 1987, pl. 12, fig. 5), which became drawn out distally. The position of the mouth may help in this aspect. In trilobites, †*Agnostus pisiformis* and the derivatives of the eucrustacean stem lineage, the mouth is located in an exposed position at the rear of the hypostome (Fig. 70B). In Phosphatocopina and Eucrustacea, the mouth area is located at the posterior end of the hypostome/labrum complex close to the ventral surface and underneath the slightly overhanging distal part of the labrum. The labrum in phosphatocopines is fully developed from the start of ontogeny onwards, and the position of the mouth in phosphatocopines and eucrustaceans implies that the labrum represents an outgrowth of the area above the mouth opening, representing an autapomorphy of the Labrophora (Fig. 69:3).

It is noteworthy to mention that the term "labrum" has also been applied for a structure in the head of atelocerates. It is only for historical reasons that both structures in Atelocerata and Labrophora are termed

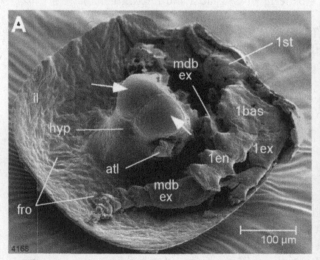
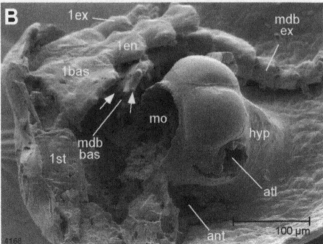

Fig. 70. UB 840 (Müller & Walossek 1987, pl. 11, figs. 3, 4). †*Agnostus pisiformis* [Linné, 1757] Wahlenberg, 1822, the leading fossil for the Cambrian *Agnostus pisiformis* Zone (Zone 1), from Gum, Kinnekulle of the Upper Cambrian "Orsten" of Sweden. A: Antero-ventral view of the cephalic shield, displaying the globular hypostome (hyp) and partly preserved right cephalic appendages. The arrows point to the bulbous areas on the ventral surface of the hypostome (cf. Pls. 2E, 3A, 9A). The antennula (atl) is preserved with only its proximal articles. B: Latero-ventral view of the same specimen, the membrane around the mouth (mo) is not preserved, the antennula (atl) and antenna (ant) of the left side are only represented by their insertion areas. The mandibular basipod shows a guiding set of setae (arrows) as the limb stem in phosphatocopines (cf. Fig. 67B), which will be discussed in detail in the second paper on Phosphatocopina.

equally, it has never been proved whether they really originated from the same phylogenetic level. Arachnologists traditionally call the hypostome of spiders and allies the "labrum" (Dunlop, pers. comm. 2002).

Another autapomorphy of the Labrophora is the sternum. This is a sclerotic plate resulting from fusion of the sternites of the first three post-antennular segments (Fig. 69:3). The ventral body surface between the hypostome/labrum complex and the second pair of post-mandibular limbs in Phosphatocopina and Eucrustacea is a single sclerotic plate, the sternum (cf. Pls. 15D, 24A). This plate develops ontogenetically by fusion of the single sternites of the segments of the antenna, mandible and maxillula in the branchiopod crustacean †*Rehbachiella kinnekullensis* (Walossek 1993). In all investigated phosphatocopines, it is a single plate throughout ontogeny (Pls. 3F, 12B, 25B). The derivatives of the labrophoran stem lineage as well as other arthropods, such as †*Agnostus pisiformis*, have isolated sternal plates in their head region throughout ontogeny, the sternites, retained from the ground patterns of Crustacea and Euarthropoda (Walossek 1999).

Also exclusive to Labrophora are the paragnaths (Fig. 69:3), present in all Phosphatocopina and Eucrustacea (cf. Pls. 3F, 36B). These structures may have formed in association with the oblique mandibular coxal gnathobase. The cuticular surface of the paragnaths is adorned with fine hairs, possibly to guide food from the posterior or to clean the setae of the antenna and the mandible (Pl. 3F).

Furthermore fine hairs occur on the anterior surface of the sternum and the sides of the labrum (cf. Pls. 3F, 4C, 34D, E), on the enditic surfaces of all post-antennular limbs (cf. Pl. 8D, E) and on the setae and spines of these limbs (cf. Pl. 12C). The labrum and the sternum of Phosphatocopina and Eucrustacea show groups or rows of fine hairs. Such hairs have not yet been observed in any of the stem lineage derivatives, as well as in earlier evolutionary levels. Also, these structures are obviously associated with a significant change in feeding mechanisms at this level.

In the stem species of the Labrophora, the antenna and the mandible have limb stems comprising two elements, the coxa and the basipod (Fig. 69:3), while only a proximal endite is present in the ground pattern of the Crustacea.

The proximal part of the crustacean – actually the arthropod – limb has long been subject to dispute and needs special consideration. Different hypotheses have been proposed in the literature to explain how the limb stem evolved. Hansen (1893) was the first to discuss the evolution of arthropod limbs in detail. He believed that the limb stem of crustaceans was originally a three-part structure. Later on, Hansen (1925, 1930) regarded the

most proximal part of the limb stem as a reference point and introduced the term "praecoxa" for it. Størmer (1939), based on Hansen (1925, 1930), regarded the precoxa and coxa as reference points of the limb. He illustrated a "crustacean limb", idealised from a syncarid, with a stem consisting of three portions, precoxa, coxopodite, and basipodite, a five-segmented endopodite, and an exopodite with a uniform proximal portion and an annulated distal portion (Fig. 71). Laterally, a pre-epipodite and an epipodite inserted on the precoxa and the coxopodite (Fig. 71). For comparison he illustrated a trilobite limb that consisted of a main shaft with a ramus arising on the outer edge of the proximal portion, his pre-coxa. Therefore, he labelled the ramus the pre-epipodite. Størmer used the term telopodite for the inner branch distal to the coxa, while not applying the word basipod(ite) for any limb portion (Fig. 71). He concluded that the outer ramus of crustaceans and trilobites cannot be homologous as they arise from different portions, from his basipodite in crustaceans from his pre-coxa in trilobites (Fig. 71). According to these assumed non-homologies between the legs, the podomeres of the inner branch (his endopodite and/or telopodite) also received distinctive names.

Snodgrass (1958) rejected the assumption of a "precoxal" element of Størmer and generally labelled the most proximal portion of the arthropod limb stem as the coxa (his fig. 19), which he used as a reference point.

Hessler & Newman (1975) created a different scenario. Based on the morphology of the post-maxillulary limbs of the Cephalocarida, which they considered to be "the most primitive crustaceans" following Sanders (1957, 1963), they rejected a three-part limb stem and proposed an undivided limb stem in their "urcrustacean" limb (their fig. 11). Their limb consists of a proximal protopod with five median enditic protrusions, a medially inserting endopod consisting of six portions plus a set of three distal setal claws, a latero-distally inserting exopod and a latero-proximally inserting epipodite. Although Sanders (1963) drew the proximal endite precisely, Hessler & Newman (1975) did not include it in their scheme (this would mean six rather than five median endites).

Itô (1989) investigated muscle signature patterns in copepods. He concluded from his studies that the basipod in copepod thoracopods formed by the fusion of the proximal part of the original exopod and the first segment of the originally four-segmented endopod.

Walossek & Müller (1990) regarded the portion that carries the two rami as a reference point and used the term basipod. Furthermore, they discovered a setose endite medio-proximally to the basipod, which they called the "proximal endite", present in several of the Upper Cambrian "Orsten" crustaceans. Because this endite has been explicitly mentioned by Calman (1909)

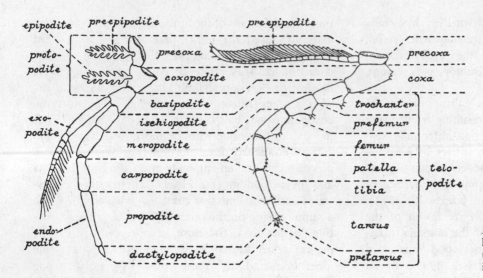

Fig. 71. Interpretation of the arthropod limb morphology in the sense of Størmer (1939). Left crustacean limb; right trilobite limb [after Størmer (1939, fig. 1)].

for Branchiopoda and is developed in Cephalocarida as well (Sanders 1963, figs. 6, 10), Walossek & Müller (1990) developed the hypothesis that this proximal endite is the structure from which the coxa of several limbs developed. In a first evolutionary step, a coxa should have developed in the antenna and the mandible in the ground pattern of Phosphatocopina plus Eucrustacea (Walossek 1999). All post-mandibular limbs retain the proximal endite. Walossek (1999) argued that the limb stem in Euarthropoda consists of only the basipod and that the proximal endite, and hence the coxa, is newly acquired by the Crustacea and/or the Labrophora.

All of these suggestions, except that of Walossek & Müller (1990), are unsatisfying because they are not based on real, existing evidence. Hansen (1925), in assuming a three-divided limb stem as original, may either have misunderstood the basal joint membrane as a limb portion (e.g. his pls. IV, fig. 1c, d; pl. 5, fig. 1c; pl. 7, fig. 2c; pl. 8. fig. 11c) or misunderstood the whole limb morphology completely (e.g. his pl. III, fig. 2g). Taxa with three-divided limb stems like the stomatopods in their thoracopods VI–VIII indeed occur, but their systematic relationship demands the assumption of an ingroup autapomorphy rather than a ground pattern character. Nevertheless, several subsequent authors continued to use the term "praecoxa" – or "pre-coxa" – as the most proximal portion of the limb stem of different arthropods and this has been upheld by the most recent literature (e.g. Carpentier & Barlet 1959; Kaestner 1967; Moritz 1993; Boxshall 1998; Cohen *et al.* 1998; Schram & Koenemann 2001). Others denied the presence of a "praecoxa" in the Crustacea (e.g. Heegaard 1945; Snodgrass 1956, 1958; Hessler & Newman 1975; Itô 1989; Walossek 1993), but for different reasons.

Cohen *et al.* (1998, p. 251) wrote that "the coxa (or

pre-coxa) is by definition the most basal part of the limb", which does not help any further.

If one regards the coxa as the most proximal portion and reference point of a limb, there is one striking problem – which only Walossek & Müller (1990) solved: if a hypothetical euarthropod limb stem had one portion, then the rami would insert on this portion; if there were two stem portions, the rami would insert on the distal of these two portions. Subsequent investigations of trilobite legs proved Størmer wrong in that these have a uniform stem in all post-antennular limbs (e.g. Cisne 1975; Whittington 1975; Whittington & Almond 1987; Ramsköld & Edgecombe 1996). This also holds true for †*Agnostus pisiformis* (Fig. 70A, B; Müller & Walossek 1987) and for various other Cambrian euarthropods (Hou & Bergström 1997). On the other hand, phosphatocopines and eucrustaceans have two portions, at least in the antenna and the mandible, the coxa and the basipod. Adopting the traditional terminology, this would imply that the coxa carries the rami in all post-antennular trilobite limbs and in post-mandibular limbs of phosphatocopines and eucrustaceans. But in the antenna and the mandible of phosphatocopines and eucrustaceans as well as in the thoracopods of malacostracans, the basipod carries the rami. In this case, one had to assume a shift of the rami from one portion to another, and the appearance of an additional element in the limb stem still has to be explained.

The assumption of Walossek & Müller (1990) that the coxa developed from a proximal endite within the Crustacea rested on two observations, the stem lineage derivatives and the ontogeny of Recent entomostracans. Indeed their hypothesis does not require a shift of the rami, because the rami always insert on the basipod. The basipod is, hence, a limb stem portion that can be found in every euarthropod that has two rami.

Accordingly, there is no coxal portion in the ground pattern of Euarthropoda. Neither trilobites nor chelicerates have a coxal portion on any limb, but they retained the single limb stem portion, the basipod, and the two rami [see also Walossek & Müller (1998a, fig. 12.9) for a limb of the chelicerate *Limulus polyphemus* (Linné, 1758)]. The reference point is the oblique insertion of the exopod on the basipod [see e.g. Hou & Bergström (1997, fig. 43, basipod labelled as "en1") for a post-antennular appendage of *Naraoia longicaudata* Zhang & Hou, 1985], also observable in phosphatocopines (cf. Pl. 11F; Figs. 24, 25).

Only crustaceans have limbs in which more than one stem portion is present. Atelocerates (myriapods and hexapods) do not have two rami, so the reference point is not detectable, and it is still unclear which limb design must be reconstructed for the ground pattern of the Atelocerata or the ground pattern of the stem species of the Atelocerata plus their as yet unknown sister group. Consequently, for the moment, the explanation of how additional limb stem portions could have developed, can be restricted to crustaceans.

Indeed, various crustaceans have a different design of their limb stems compared with that in the ground pattern of Euarthropoda. And this design is not the same in the different limbs. The antenna and the mandible have a two-part limb stem in all eucrustaceans and phosphatocopines, although both portions may be fused in some lineages, such as the Euphosphatocopina and Euphausiacea, representing autapomorphies of both taxa (cf. Maas & Walossek 2001b). Post-mandibular limbs may have either uniform limb stems, such as in the Cephalocarida, Branchiopoda, podocope Ostracoda, or the limb stems are two-part structures, such as in the maxillula, maxilla and first eight thoracopods of Malacostraca. Three-part limb stems as described for some Copepoda and myodocopine Ostracoda are probably based on misinterpretations of arthrodial membranes as limb portions (e.g. Boxshall (1998) uses the term "pre-coxa" for the arthrodial membrane of copepod limbs).

The post-antennular limbs of the derivatives of the Labrophora, †*Martinssonia elongata*, †*Henningsmoenicaris scutula*, †*Goticaris longispinosa*, †*Cambropachycope clarksoni*, and †*Cambrocaris baltica*, have a uniform limb stem with a proximal endite underneath (Walossek & Müller 1990; Walossek & Szaniawski 1991). A proximal endite is also present in the post-mandibular limbs of, e.g. the "Orsten" fossils *Bredocaris admirabilis* Müller, 1983 (Maxillopoda) (Müller & Walossek 1988) and *Rehbachiella kinnekullensis* (Branchiopoda) (Walossek 1993), euanostracans (Calman 1909; Walossek 1993) and tantulocarids (Huys 1991). In the ontogeny of †*Rehbachiella kinnekullensis* and the fresh water copepod *Eudiaptomus gracilis*

(G.O. Sars, 1863), the coxa of the mandible develops from the proximal endite of the naupliar stages (Walossek 1993; Mayer 2002, unpublished diploma work).

Consequently, the ontogeny and morphology of crustaceans point to the explanation for the presence of a coxal element at least in the mandible of Phosphatocopina and Eucrustacea by the development from a proximal endite in the sense of Walossek & Müller (1990). Because the limb stem of both the antenna and the mandible are rather similar in Phosphatocopina (see Fig. 58A, B), the proximal endite can be regarded as a precursor of the coxa, not only in the mandible but also in the antenna of the Labrophora.

Schram & Koenemann (2001, fig. 1) illustrated a hypothetical crustacean appendage with a three-part limb stem, epipodites and symmetrical rami. Drawn previously by Schram (1986), such a hypothetical limb has no similarity to any fossil or Recent crustacean, or even any arthropod limb. These authors neither considered the derivatives of the eucrustacean stem lineage nor did they mention the proximal endite, so omitted important existing evidence. Epipodites are absent in all derivatives of the eucrustacean stem lineage as well as in the phosphatocopines, and there is no clear evidence for the existence of epipodites in the ground pattern of Entomostraca or Malacostraca as sister taxa within the Eucrustacea. A five-segmented endopod has to be reconstructed for the ground pattern of Crustacea, as is retained in Entomostraca and Malacostraca, at least in all post-maxillulary limbs. A three-part endopod as an "urform", as drawn by Schram & Koeneman (2001), is therefore simply at odds with known data. The exopod in the ground pattern of Crustacea was either annulated or paddle-shaped at least in the anterior limbs, as discussed above. An "ur"-exopod being symmetrical to the endopod and also with two to three portions as shown by Schram & Koenemann (2001, fig. 1) cannot have characterised the ground pattern of Euarthropoda, Crustacea or Eucrustacea.

In the stem species of Labrophora the head includes five limb-bearing segments (Fig. 69.3). The original interpretation of five limb-bearing head segments in the ground pattern of Euarthropoda (cf. Størmer 1939) was rejected by Cisne (1975). He showed that the head of the trilobite †*Triarthrus eatoni* (Hall, 1863), consists of only four limb-bearing segments, the segment of the antennula (traditionally named "antenna" in trilobite terminology) plus three more. The same number of segments is recorded from various other trilobites and other arthropods [cf. Whittington (1975) for †*Olenoides serratus* (Rominger, 1887); Briggs et al. (1979) for †*Aglaspis spinifer* Raasch, 1939; Müller & Walossek (1987) for †*Agnostus pisiformis*]. The representatives of the labrophoran stem lineage also have a head with no more than four segments, except for †*Henningsmoenicaris scutula*, which has a head consisting of five segments (cf. Walossek & Müller 1990), implying that a five-segmented head was already developed in the stem lineage of the Labrophora. Although much investigated (e.g. Abzhanov *et al.* 1999; Eriksson & Budd 2001), this feature has to be investigated in more detail in the future.

Walossek (1999) considered the tail end, including the telson and furcal rami, as a character of Eucrustacea, but it can also now be demonstrated for Phosphatocopina. A furca is observed in two phosphatocopine species, i.e. *H. unisulcata* (Pls. 8F, 12D) and *Waldoria rotundata* (Pl. 36D). It is clearly present in various eucrustaceans, such as Cephalocarida, Copepoda, Branchiopoda, Leptostraca and Euphausiacea, although in the latter taxon the furca is fused to the telson (Maas & Waloszek 2001b). Several eucrustaceans do not show a furca, but the distribution of this character among eucrustaceans implies that it is a ground pattern character of this taxon. In accordance with the new phylogenetic analysis, the furca has to be considered as part of the ground pattern of Labrophora instead and is plesiomorphically retained in eucrustacean ingroups. A furca is not present in the stem lineage derivatives of the Labrophora, and it is absent in earlier lineages, there the tail ends in a plate or single spine, with the anus opening ventrally and pre-terminally, as in the ground patterns of Crustacea and Euarthropoda. The furca is thus reconstructed as a synapomorphy of Phosphatocopina and Eucrustacea (Fig. 69:3).

The position of the anus could not be observed in the studied phosphatocopines. It therefore remains uncertain whether the terminal position of the anus, which is given as an autapomorphy of the Eucrustacea by Walossek (1999), is part of the ground pattern of the Labrophora.

In light of the morphology of Phosphatocopina, Labrophora have retained at least two features from the ground pattern of Euarthropoda:

- a "head-larva" *sensu* Walossek & Müller (1990) having antennulae and three pairs of functional limbs;
- the similarity in shape of all post-mandibular limbs (first two post-antennular limbs different from the succeeding ones: antenna and mandible).

Walossek & Müller (1990) postulated that the post-embryonic development in the ground pattern of the Euarthropoda should have started with a larva having the same segmental number as the adult head (Walossek 1999; Maas & Waloszek 2001a). This kind of larva also represents the first growth stage of phosphatocopines, and thus represents, in accordance with the phylogenetic analysis, a plesiomorphy in the ground pattern of the Labrophora.

In the ground pattern of Labrophora, the first two post-antennular pairs of limbs, i.e. the antennae and the mandibles, are of quite different design compared with the succeeding pairs of limbs – because of the development of a coxal portion proximal to the basipod. This character is already present in the earliest larva, as has been demonstrated for the Phosphatocopina, together with other changes in the feeding apparatus, such as the development of the labrum and the sternum with paragnaths. All post-mandibular pairs of limbs remain of the same design and comparable with that present in the ground pattern of the Crustacea, consisting of a basipod with a proximal endite underneath and two rami. Another two characters from the ground pattern of the Crustacea are retained in the Labrophora, namely:

- exopod of post-antennular limbs multi-annulated, each annulus with one seta pointing towards the endopod (plesiomorphic state: exopods leaf-shaped with marginal setation, as in the ground pattern of Euarthropoda);
- number of endopodal articles of post-antennular limbs maximally five (plesiomorphic state: seven endopodal articles, as in the ground pattern of Euarthropoda).

As stated above, the exopods of the post-antennular limbs in Phosphatocopina resemble the exopods of at least the anterior limbs of †*Martinssonia elongata*, representing a plesiomorphic condition. The endopod of all phosphatocopine limbs consists of no more than three portions. That of †*Martinssonia elongata*, as a representative of the stem lineage of the Labrophora has five articles (Müller & Walossek 1986a). More than five portions do not occur in the endopods of the post-mandibular limbs of eucrustaceans. Accordingly the endopods of at least the post-mandibular limbs of the Labrophora retained five articles from the ground pattern of the Crustacea.

Ground pattern of Phosphatocopina

The monophyly of the Phosphatocopina is founded on several autapomorphies (Figs. 64:1, 69:4), as presented above. Phosphatocopina retained the hatching "head-larva" having three pairs of post-antennular limbs. The term "head-larva" of Walossek & Müller (1990) applied to euarthropod larvae that consist of the same number of segments as the head of their adults, is a little confusing when applied to phosphatocopines as their head "already" consists of five segments as is characteristic for Labrophora. Therefore, within the Crustacea it would also be a "short-head-larva", but this term is already preoccupied by the eucrustacean orthonauplius (which, however, is an even shorter "short-head-larva" having only three pairs of appendages). This will require some terminological adjustments in the future.

Phosphatocopina also retain several other plesiomorphies from the ground pattern of the Crustacea, such as the multi-annulated exopod of post-antennular limbs, but this is a doubtful character as discussed above.

Ground pattern of Eucrustacea

The name Eucrustacea has been used in the literature for three taxa of different composition:

- Kingsley (1894) introduced it in arthropod taxonomy. He distinguished within the well-established taxon Crustacea Brünnich, 1772 (see footnote on p. 18) the proper crustaceans, which he called Eucrustacea, and the well-established †Trilobita Walch, 1771. His work was mostly ignored by subsequent workers (see, e.g. Eastman 1913).
- Walossek & Müller (1998b) and Walossek (1999, 2002) used the name Eucrustacea for the crown-group of Crustacea, the monophylum embracing all crustacean taxa with Recent derivatives, and established its monophyletic status by proposing a set of autapomorphies (see below).
- Ax (1999) used the name Eucrustacea for a taxon combining all living crustaceans except the Remipedia, based on a set of autapomorphies (but see below).

A close relationship between Crustacea and †Trilobita, as proposed by, e.g. Kingsley (1894) and many other authors of that age, is not assumed anymore. †Trilobita are now thought to be more closely related to the Chelicerata (Lauterbach 1973; Hou & Bergström 1997; cf. Hessler & Newman 1975) or at best members of an arachnatan clade (Lauterbach 1980, 1983; Ax 1985). Nevertheless, this relationship is not really clear, and the arguments given (Lauterbach 1983) are either weak or have been rejected (Fortey & Whittington 1989). However, this question was not touched by the phylogenetic analysis of the Phosphatocopina, and it is not discussed here.

Walossek & Müller (1998b) and Walossek (1999, 2002) proposed the monophyly of all crustaceans with living derivatives, the so-called crown-group Crustacea, which they named Eucrustacea. This proposal was based on of a set of autapomorphies, such as the nauplius as the first larval stage and the maxillula as a modified limb different from all succeeding limbs. The concept of their "Eucrustacea" is, therefore, rather close to that of Kingsley (1894), although Kingsley did not argue in a phylogenetic sense of course. The monophyly of the "Eucrustacea" can be supported by the phylogenetic analysis performed in this work, and the autapomorphies of Walossek & Müller (1998b) and Walossek (1999) are confirmed.

Ax (1999) presented only one autapomorphy for his "Eucrustacea". He reconstructed a division of the body into a limb-bearing thorax and a limb-less abdomen as a newly acquired character in the ground pattern his Eucrustacea. This sets off Remipedia, which have no limb-less body region. Such a division of the body is, however, also developed in various euarthropods and even derivatives of the stem lineage of Euarthropoda, such as †*Fuxianhuia protensa* (Ax did not consider fossil forms). The Remipedia are cave-living crustaceans that are highly adapted and specialised (e.g. Yager & Humphreys 1996). Although some upheld a very basal position of this group (e.g. Schram & Emerson 1990; Nielsen 1995; Ax 1999), molecular data (Spears pers. comm. 2001) point to a close relationship with copepods within the Maxillopoda. A subdivision of the body into a limb-bearing thorax and a limb-less abdomen might therefore be much more influenced by functional requirements and is frequently developed and is, at least at present, a doubtful character for a phylogenetic analysis. Therefore, the subdivision of the body was not coded herein. Ax's (1999) usage is also problematical because all other characters of the Remipedia in common with entomostracans and copepods should therefore be convergently derived, which causes many conflicting arguments and a much less parsimonious decision of a sister-group relationship.

Because the investigations on the Phosphatocopina and the phylogenetic analysis could confirm the autapomorphies given by Walossek (1999) for his "Eucrustacea", the name Eucrustacea in the sense of Walossek (1999) is used, rejecting the meaning *se*... Ax (1999). Characters newly characterising the Eucrustacea could be found in the caudal region.

The autapomorphies of Eucrustacea, according to Walossek (1999) and the phylogenetic analysis are (Fig. 69:5):

- first post-mandibular pair of limbs modified into a "mouthpart", the maxillulae (but see below) (plesiomorphic state: trunk-limb-shaped first post-mandibular limb, with a basipod and a proximal endite medio-proximal to it, as present in the ground patterns of Euarthropoda, Crustacea and Labrophora);
- hatching stage is a nauplius larva (orthonauplius), termed "short-head-larva" by Walossek & Müller (1990) and having three pairs of limbs, the antennulae and two subsequent ones (plesiomorphic state: "head-larva" with four pairs of functional limbs, i.e. the antennulae and three subsequent pairs, as in the ground patterns of Euarthropoda, Crustacea and Labrophora);
- nauplius with a supra-anal flap carrying a dorso-caudal spine, the telson with a pair of latero-caudal and ventro-caudal spines (plesiomorphic state: not present).

As stated above, the ground pattern of the Labrophora

includes the modified state of the first two post-antennular pairs of limbs, i.e. the antennae and the mandibles. Their distinctive difference from the succeeding limbs indicates a significant change in the feeding and locomotory apparatus, particularly of the head. As a further step, and regarded as an autapomorphy in the ground pattern of Eucrustacea, the first post-mandibular pair of limbs (maxillulae) is also incorporated into the feeding apparatus and removed from the posterior system of homogenous limbs (Fig. 69:5). The first post-mandibular pair of limbs of eucrustacean species, the maxillula or "first maxilla", is different from the succeeding pairs of limbs, unlike that in Phosphatocopina and the stem lineage derivatives of the Labrophora. The maxillula of all entomostracans is, again, different from that of all malacostracan species (Walossek & Müller 1998b, fig. 5.7). Therefore, the morphological state of the maxillula in the ground pattern of the Eucrustacea cannot be reconstructed.

As frequently assumed in textbooks, the second pair of post-mandibular limbs, the maxillae or "second maxillae", should also already be modified and different from the succeeding pairs in the ground pattern of Crustacea (Barnes *et al.* 1993; Gruner 1993; Schminke 1996; Storch & Welsch 1997; Ax 1999). This limb is almost identical to the succeeding limbs not only in the living Cephalocarida (Sanders 1963) but also in †*Rehbachiella kinnekullensis* among Branchiopoda (Walossek 1993), †*Bredocaris admirabilis* and †*Dala peilertae* Müller, 1983 among Maxillopoda (cf. Müller & Walossek 1985a) and *Paranebalia longipes* (Wilemoes Suhm, 1878) among Malacostraca (Brattegard 1970). Again, the status of the ostracode maxilla is at least difficult, because of its trunk-limb shape in Myodocopida. Consequently, the maxilla must have been an ordinary "thoracopod", although incorporated in the head, still in the ground pattern of the Eucrustacea, as stated by Walossek & Müller (1990, 1998a) and Walossek (1993, 1999).

Another autapomorphy of the Eucrustacea is seen in the orthonauplius as the hatching stage (Fig. 69:5). Within the major eucrustacean taxa Entomostraca and Malacostraca, several species of various groups hatch as orthonauplii, i.e. a larva having three pairs of functional appendages. A labrum is developed and the antenna has a limb stem consisting of a coxa and a basipod. Such a larva characterises: the Anostraca including †*Rehbachiella kinnekullensis* among the Branchiopoda (Walossek 1993), the Copepoda and Cirripedia among the Maxillopoda (cf. e.g. Walossek *et al.* 1996), and the Euphausiidae and Dendrobranchiata among the Malacostraca (cf. e.g. Fraser 1936; Cockcroft 1995). All other eucrustacean species hatch at a stage with a different number of segments, e.g. the cephalocarid *Hutchinsoniella macracantha* Sanders, 1955 hatches as a metanauplius with six segments (Sanders 1963). At one extreme, astacids hatch as completely segmented small copies of the adults. The feeding status and locomotory status further complicate the pattern of morphologies. It is, however, important to refer to the ground pattern of a group. A hatching larva with more than three limb-bearing segments, representing the ground pattern character of the Eucrustacea, is assumed by Scholtz (2000) on the basis of gene expression investigations, which at first sounds similar to the hypothesis of Walossek & Müller (1990). However, (a) it should hold only for the Recent taxa (Scholtz ignores fossil evidence), (b) in this case the true nauplius must then have developed at least three times independently, and (c) alterations of this supposed original pattern must have occurred anyway (requires the assumption of both shortening and segment addition of the hatching stage). The most parsimonious assumption is, in accordance with Walossek (1999, 2002), that the true nauplius occurs first in the ground pattern of the Eucrustacea and is retained plesiomorphically in all major lineages, the Malacostraca, the Branchiopoda, and the thecostracan plus the copepod lineage of the Maxillopoda. Again, hatching with different segment numbers occurred in various lineages, from the pattern in cephalocarids up to the direct development in astacids or cladocerans, but it probably developed independently from the nauplius level onwards. A more detailed discussion of different strategies can be found in Walossek (1993).

Cohen *et al.* (1998) stated that the phosphatocopine larva should be an advanced nauplius and that the earliest larva (the orthonauplius) of phosphatocopines simply had not yet been discovered. A characteristic of the phosphatocopine collection from the investigated "Orsten" material of Sweden is that the smaller the stages are the more individuals of them have been found. We are convinced that even smaller stages than the presented head-larvae (Fig. 62) would have been found in the material if present. Indeed, true nauplii are known from the "Orsten" in †*Rehbachiella kinnekullensis* (Walossek 1993), and the type-A larvae (Müller & Walossek 1986b) from Zones 1 and 5 of the Upper Cambrian (cf. Table 2) are orthonauplius-like. But the type-A larvae have no characters by which they could be the first growth stage of phosphatocopines. Other larvae of comparable size are also present and give a rather detailed picture of morphologies truly developed at this stage (cf. Müller & Walossek 1985a, 1987, 1988; Walossek & Müller 1990, 1994).

In the above-mentioned crustacean taxa with an orthonauplius as their hatching stage, it is traditionally distinguished between several "naupliar" stages. For example, the first two growth stages of the Euphausiidae are traditionally called "nauplius I" and "nauplius II" followed by one "metanauplius" stage. Yet, only the "nauplius I" is a true nauplius. The first six larval stages of the Copepoda are traditionally called "nauplius I" to "nauplius VI". In both cases the "naupliar" stages subsequent to the orthonauplius may have more than three pairs of limbs, as has the so-called "nauplius" of the Ostracoda and the so-called "nauplius" of the Cephalocarida, both not being true nauplii. In a phylogenetic sense, the orthonauplius, e.g. the "nauplius I" of Euphausiidae and the "nauplius I" of Copepoda, equals the first growth stage in the ground pattern of Eucrustacea. The second stage, e.g. the "nauplius II" of Euphausiidae and the "nauplius II" of Copepoda, therefore cannot be combined with the first one in naming them the "second nauplius" or "second orthonauplius". In fact, the second larval stage already has a fourth pair of limbs in the form of a bud or even only as a seta. Dietrich (1915) therefore applied the terms "orthonauplius I" and "orthonauplius II" to the first stages of fresh water copepods. Ewers (1929) used the terms "first nauplius" or "orthonauplius" for the first larval stage and called the succeeding larval "naupliar" stages of *Cyclops* species "first metanauplius" to "fifth metanauplius". Consequently, he distinguished between larval stages that possess only three pairs of limbs and those with at least an anlage in the form of a seta of a second pair, but he was not consistent. This imprecise definition of an orthonauplius causes problems. For example, larval stages can be "skipped" during ontogeny (Williamson 1982; Walossek 1993). Phylogenetic implications of a crustacean nauplius were discerned by Dahms (2000), but he failed to recognise the difference between the true nauplius and the succeeding larval stages, which are not nauplii, even in the traditional sense. The terms "nauplius I", "nauplius II" and so on are, hence, misleading, for they imply a morphological or mere segmental unity of all these larvae. Only the first "naupliar"

stage in Copepoda, Anostraca, Cirripedia, Euphausiidae and Dendrobranchiata is the orthonauplius that characterises the ground pattern of Eucrustacea and is truly equal to the orthonauplius having no more than three functional pairs of appendages. Any other succeeding larval stages of any groups are not nauplii, but developmentally advanced and should receive other terms. In order to reach a more conclusive terminology, it is suggested to restrict the term orthonauplius to those larval stages that have three pairs of limbs and no even an anlage of a fourth pair on their undeveloped hind body, i.e. to the first larval stage in the ground pattern of Eucrustacea. In this context it is necessary to briefly note that the "head-larva" is not bypassed during the development of Eucrustacea. This larva has a functional fourth limb, but this stage is not reached in Eucrustacea before at least stage III or IV. Consequently, the second growth stage of Eucrustacea is also an autapomorphy of this taxon.

The morphological acquirements in the ground pattern of the Labrophora have important implications for the evolution of crustacean larvae. The whole locomotory and feeding apparatus of the Labrophora is basically retained in the four-legged larva of Phosphatocopina and in the orthonauplius of Eucrustacea. In hindsight it cannot be argued that the eucrustacean larva should be more ancestral than the "head-larva" because it has a segmental composition smaller than these, but uses the feeding apparatus morphology developed first by the stem species of the Labrophora. By no means can the orthonauplius be the most plesiomorphic larval type of Crustacea or even of the Euarthropoda, as has been occasionally proposed (e.g. Lauterbach 1988; Stone & Welsch 1997).

Another autapomorphy of the Eucrustacea, according to Walossek (1999), has not been coded in the phylogen-

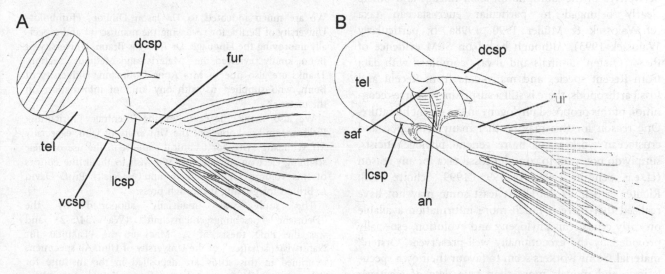

Fig. 72. Reconstruction of the eucrustacean telson (tel) with a furca (fur), latero-caudal spines (lcsp), ventro-caudal spines (vcsp) and supra-anal flap (saf) with dorso-caudal spine (dcsp). A: View from antero-lateral. B: View from postero-lateral.

etic analysis. This is the possession of a supra-anal flap carrying a dorso-caudal spine in the eucrustacean orthonauplii and a pair of ventro-caudal and latero-caudal spines 'on the telson (Figs. 69:5, 72). These features are well documented for the orthonauplius of †*Rehbachiella kinnekullensis* by Walossek (1993) and occur in various other fossil taxa, such as †*Bredocaris admirabilis* and †*Dala peilertae* among the Maxillopoda (Müller & Walossek 1985a, 1986b), and living taxa, such as copepods, cirrepedes, cephalocarids, leptostracans, and euphausiids.

In summary, the autapomorphies suggested by Walossek (1999) for the ground pattern of Labrophora (Eucrustacea + Phosphatocopina) and the autapomorphies for Eucrustacea could be confirmed and emended. There are some minor differences in the arrangement and assignment of particular characters in this work as compared with Walossek (1999) because of the improved knowledge on the Phosphatocopina. But these do not affect the proposed sister-group relationship of Phosphatocopina and Eucrustacea, supported by several significant synapomorphies, as listed and discussed above.

Conclusions

The recognition of the Phosphatocopina as a monophyletic taxon within the Crustacea and a sister group to the Eucrustacea requires a new consideration of morphological acquirements during the evolution of Crustacea up to the common stem species of the taxon Eucrustacea, embracing all crustaceans with living derivatives. The discovery of the autapomorphies in the ground pattern of Crustacea presented by Walossek (1999, 2002) was made possible mainly by studies of Upper Cambrian "Orsten" fossils. Some of them could be recognised as derivatives of the labrophoran stem lineage and others clearly belonged to particular eucrustacean taxa (cf. Walossek & Müller 1990, 1998a, b; particularly Walossek 1993). Although based on SEM evidence of these "Orsten" animals and always combined with data from Recent species and many data from Recent and fossil arthropods, there is still a surprisingly sparse recognition of this proposed phylogeny in the recent literature. One reason for this may be that many arthropod and crustacean workers, or more general phylogeneticists, simply do not want to consider fossil taxa for any reason (cf. e.g. Schmitt 1994; Ax 1995, 1999; Scholtz 1998; Richter & Scholtz 2001). At least some may not have realised that there is much more information available on early crustacean phylogeny and evolution, especially provided by the exceptionally well-preserved "Orsten" material. Some workers seem to favour their own speculations and models more than true data of whatever source. But the impact of the "Orsten" material is again exemplified by the material of phosphatocopines. In fact, phosphatocopine morphology adds significantly more to our understanding of crustacean evolution. The development of characters could be determined with more detail, e.g. the anterior head region including the hypostome and the labrum, the transformation of the proximal endite to a coxa in the antenna and the mandible, the sternum and the situation in the hind body region.

Future investigations and prospects

From the entire material of approximately 50,000 phosphatocopine specimens isolated between 1975 and the present, about 2,500 specimens were selected for SEM studies for this investigation. The bulk of these specimens are preserved with soft parts. Through this extensive SEM study, our knowledge of the morphology and ontogeny of the Phosphatocopina could be significantly extended and the study provides the basis for any further studies on Phosphatocopina world-wide. A second paper on the Phosphatocopina will include:

- a detailed description of the ventral morphology and ontogeny of other species among the material having preserved soft parts, such as *H. ventrospinata*, *H. muelleri* n. sp., *H. angustata* n. sp., *Falites fala*, and *Vestrogothia spinata*;
- an extended phylogenetic analysis that includes not only the new evidence from morphological and ontogenetic investigations of other phosphatocopine species, but also all phosphatocopine species from all over the world;
- an evaluation of the evolution, ecology and life habits of the different phosphatocopine species.

Acknowledgements

We are much indebted to Dr Jason Dunlop, Humboldt-University of Berlin, for reviewing the manuscript and especially improving the language. Dr Andreas Braun, University of Bonn, kindly reviewed the "Material and Methods" section. Thanks are also due to Mrs Anne Goßmann, University of Bonn, who supplied us with any kind of information on the material.

We also thank the team of the "Zentrale Einrichtung Elektronenmikroskopie" of the University of Ulm, especially Mr Wolfgang Fritz, for technical support in the use of their scanning electron microscope. We wish to thank the editors of this volume for their help and especially Prof. David L. Bruton for guiding it through press.

The study was financially supported by the "Deutsche Forschungsgemeinschaft" (Wa 754/5-2) and was the PhD thesis of A. Maas at the "Fakultät für Naturwissenschaften" of the University of Ulm. All specimens examined in this study are deposited in the Institute for Palaeontology, Bonn under the numbers indicated on the plates and figures.

References

Abushik, A.F., Guseva, E.A., Ivanova, V.A., Kanygin, A.V., Kashevarova, N.P., Melnikova, L.M., Molostovskaya, I.I., Neustrueva, I.Y., Sidaravichene, N.V., Stepanaytys, N.E. & Chizhova, V.A. 1990: Ostrakody paleozoja. [Palaeozoic Ostracoda]. *Prakticeskoe rukovodstvo po mikrofaune SSSR* [Practical Manual on Microfauna of USSR] *4*, 1–356 (in Russian).

Abzhanov, A., Popadic, A. & Kaufman, T.C. 1999: Chelicerate *Hox* genes and the homology of arthropod segments. *Evolution and Development 1*(2), 77–89.

Adamczak, F. 1965: On some Cambrian bivalved Crustacea and egg cases of the Cladocera. *Stockholm Contributions in Geology 13*(3), 27–34.

Aguinaldo, A.M.A., Turbeville, J.M., Linford, L.S., Rivera, M.C., Garey, J.R., Raff, R.A. & Lake, J.A. 1997: Evidence for a clade of nematodes, arthropods and other moulting animals. *Nature 387*, 489–493.

Allen, J.R.L. & Rushton, A.W.A. 1968: Local exposures. *In* Sylvester-Bradley, P. C & Ford, T. D. (eds): *The Geology of the East Midlands*, 38–40. Leicester University Press, Leicester.

Andersson, A., Dahlman, B., Gee, D.G. & Snäll, S. 1985: The Scandinavian alum shales. *Sveriges Geologiska Undersökning Ser. C 56*, 1–50.

Andres, D. 1969: Ostracoden aus dem mittleren Kambrium von Öland. *Lethaia 2*, 165–180.

Ax, P. 1985: Stem species and the stem lineage concept. *Cladistics 1*(3), 279–287.

Ax, P. 1995: *Das System der Metazoa I*, 1–226. Akademie der Wissenschaften und der Literatur, Mainz. Gustav Fischer, Stuttgart.

Ax, P. 1999: *Das System der Metazoa II*, 1–384. Akademie der Wissenschaften und der Literatur, Mainz. Gustav Fischer, Stuttgart.

Barrande, J. 1872: *Système Silurien du centre de la Bohême. Supplement to Vol. I*. Prague and Paris, 1–647.

Barnes, R.S.K., Calow, P. & Olive, P.J.W. 1993: *The Invertebrates. A New Synthesis*, 2nd edn, I–VIII, 1–488. Blackwell Science, Oxford.

Bassler, R.S. & Kellett, B. 1934: Bibliographic index of Paleozoic Ostracoda. *Geological Society of America Special Papers 1*, I–XIII, 1–500.

Bate, R.H. 1972: Phosphatized Ostracodes with appendages from the Lower Cretaceous of Brazil. *Palaeontology 15*(3), 379A–393A.

Bate, R.H., Collins, J.S.H., Robinson, J.E. & Rolfe, W.D.I. 1967: Arthropoda: Crustacea. *In The Fossil Record*, 535–563. Geological Society of London, London.

Becker, G. & Adamczak, F.J. 2001: Schalenmerkmale verkieselter Ostracoden: Sind die ordovizischen Duplikaturen nur Artefakte oder eher evolutionäre Vorversuche? *Paläontologische Zeitschrift 75*(2), 151–162.

Bednarczyk, W. 1979: On the occurrence of Bradorina Raymond 1935 Ostracods in the Upper Cambrian of NW Poland. *Bulletin de l'Académie Polonaise des Sciences, Série des Sciences de la Terre 26*(3–4), 215–219.

Bengtson, S., Conway Morris, S., Cooper, B.J., Jell, P.A. & Runnegar, B.N. 1990: Early Cambrian fossils from South Australia. *Memoirs of the Association of Australasian Palaeontologists 9*, IV + 364 pp.

Berg-Madsen, V. 1985a: Middle Cambrian biostratigraphy, fauna and facies in southern Baltoscandia. *Acta Universitatis Upsaliensis – Abstracts of Uppsala Dissertations from the Faculty of Science 781*, I–II, 1–37.

Berg-Madsen, V. 1985b: A review of the Andrarum Limestone and the upper alum shale (Middle Cambrian) of Bornholm, Denmark. *Bulletin of the Geological Society of Denmark 34*(3/4), 133–143.

Bergström, J. 1992: The oldest arthropods and the origin of the Crustacea. *Acta Zoologica 73*(5), 287–291.

Bergström, J. & Gee, D.G. 1985: The Cambrian in Scandinavia. *In* Gee, D.G. & Sturt, B.A. (eds.): *The Caledonide Orogen – Scandinavia and Related Areas*, 247–271. John Wiley & Sons, Chichester.

Bitsch, J. 1994: The morphological ground plan of Hexapoda: a critical review of recent concepts. *Annales de la Société Entomologique de France (N.S.) 30*(1), 103–129.

Bowring, S.A. & Erwin, D.H. 1998: A new look at evolutionary rates in deep time: uniting paleontology and high-precision geochronology. *Geological Society of America Today 8*(9), 1–40.

Boxshall, G.A. 1992: Synopsis of group discussion on the Maxillopoda. *Acta Zoologica 73*(5), 335–337.

Boxshall, G.A. 1998: Comparative limb morphology in major crustacean groups: the coxa-basis joint in postmandibular limbs. *In* Fortey, R.A. & Thomas, R.H. (eds.): *Arthropod Relationships*. Systematics Association Special Volume Series 55, 156–167. Chapman & Hall, London.

Brattegard, T. 1970: Marine biological investigations in the Bahamas. 13. Leptostraca from shallow water in the Bahamas and southern Florida. *Sarsia 44*, 1–7.

Briggs, D.E.G. 1977: Bivalved arthropods from the Cambrian Burgess Shale of British Columbia. *Palaeontology 20*(3), 595–621.

Briggs, D.E.G. 1983: Affinities and early evolution of the Crustacea: the evidence of the Cambrian fossils. *In* Schram, F.R. (ed.): *Crustacean Phylogeny*, 1–22. A.A. Balkema, Rotterdam.

Briggs, D.E.G., Bruton, D.L. & Whittington, H.B. 1979: Appendages of the arthropod *Aglaspis spinifer* (Upper Cambrian, Wisconsin) and their significance. *Palaeontology 22*(1), 167–180.

Broili, F. 1924: I. Abteilung: Invertebrata. *In* Zittel, K.A. von (founder): *Grundzüge der Paläontologie (Palaeozoologie)*, 6th edn. R. Oldenbourg, München.

Brünnich, M.T. 1772: *Zoologiae fundamenta praelectionibus academicis accomodata*. Apud Friderich Pelt, Hafniae et Lipsiae (= Copenhagen and Leipzig), 1–254.

Calman, W.T. 1909: Crustacea. *In* Lankester, E. (ed.): *A Treatise on Zoology, Part VII*(3), 1–346. Adam and Charles Black, London.

Carpentier, F. & Barlet, J. 1959: The first leg segments in the Crustacea Malacostraca and the insects. *Smithsonian Miscellaneous Collections 137*, 99–115.

Chen Jun-yuan, Vannier, J. & Huang Di-ying 2001: The origin of crustaceans: new evidence from the Early Cambrian of China. *Proceedings of the Royal Society of London B 268*, 2181–2187.

Cisne, J.L. 1975: Anatomy of *Triarthrus* and the relationship of the Trilobita. *Fossils and Strata 4*, 45–63.

Clarkson, E.N.K., Ahlberg, P. & Taylor, C.M. 1998a: Faunal dynamics and microevolutionary investigations in the Upper Cambrian *Olenus* Zone at Andrarum, Skåne, Sweden. *Geologiska Föreningens i Stockholm Förhandlingar 120*, 257–267.

Clarkson, E.N.K., Taylor, C.M. & Ahlberg, P. 1998b: Stop 5. Andrarum. *Lund Publications in Geology 141*, 26–28.

Cockcroft, A.C. 1985: The larval development of *Macropetasma africanum* (BALSS, 1913) Decapoda, Penaeoidea reared in the laboratory. *Crustaceana 49*, 52–74.

Cohen, A.C., Martin, J.W. & Kornicker, L.S. 1998: Homology of Holocene ostracode biramous appendages with those of other crustaceans: the protopod, epipod, exopod and endopod. *Lethaia 31*(3), 251–265.

Cowie, J.W., Rushton, A.W.A. & Stubblefield, C.J. 1972: A correlation of Cambrian rocks in the British Isles. *Special Report of the Geological Society 2*, 1–42.

Dahms, H.-U. 2000: Phylogenetic implications of the Crustacean nauplius. *Hydrobiologia 417*, 91–99.

Dewel, R.A., Budd, G.E., Castano, D.F. & Dewel, W.C. 1999: The organization of the subesophageal nervous system in Tardigrades: insights into the evolution of the arthropod hypostome and tritocerebrum. *Zoologischer Anzeiger 238*, 191–203.

Dietrich, W. 1915: Die Metamorphose der freilebenden Süßwasser-Copepoden. I. Die Nauplien und das erste Copepodidstadium. *Zeitschrift für Wissenschaftliche Zoologie 113*, 252–324.

Earp, J.R. & Hains, B.A. 1971: *British Regional Geology: The Welsh Borderland.* Her Majesty's Stationary Office, London.

Eastman, C.R. (ed.) 1913: *Text-book of Paleontology*, Vol. 1, edn 2, English edition adapted from Zittel, K.A. von: *Grundzüge der Paläontologie* (in German). Macmillan & Co., London.

Eichbaum, K.W. 1979: Öland. *Der Geschiebe-Sammler 13*(2), 65–90.

Eriksson, B.J. & Budd, G.E. 2001: Onychophoran cephalic nerves and their bearing on our understanding of head segmentation and stem-group evolution in Arthropoda. *Arthropod Structure & Development 29*(3), 197–209.

Ewers, L.A. 1929: The larval development of freshwater Copepoda. *Abstracts of Doctor's Dissertations, Ohio State University Press 1*, 60–70.

Fleming, P.J.G. 1973: Bradoriids from *Xystridura* zone of the Georgina Basin, Queensland. *Publications of the Geological Survey of Queensland 356, Palaeontological Papers 31*, 1–9.

Fortey, R.A. & Whittington, H.B. 1989: The Trilobita as a natural group. *Historical Biology 2*, 125–138.

Fraser, F.C. 1936: On the development and distribution of the young stages of krill (*Euphausia superba*). *Discovery Reports 14*, 1–192.

Frenzel, P. 2000: Die benthischen Foraminiferen der Rügener Schreibkreide (Unter-Maastrichtium/NE-Deutschland) – Taxonomie, Stratigraphie und Paläoökologie. *Neue Paläontologische Abhandlungen 3*, 1–361.

Fryer, G. 1996: Reflections on arthropod evolution. *Biological Journal of the Linnean Society 58*, 1–55.

Geyer, G. 1998: Die kambrische Explosion [The Cambrian explosion]. *Paläontologische Zeitschrift 72*(1/2), 7–30.

Grönwall, K.A. 1902: Bornholms Paradoxideslag og deres Fauna. *Danmarks geologiske Undersøgelse II. Raekke 13*, 1–231.

Groom, T. 1902: On *Polyphyma*, a new genus belonging to the Leperditiadae, from the Cambrian Shales of Malvern. *Quarterly Journal of the Geological Society of London 58*, 83–88.

Gründel, J. 1981: Die Bradoriida. *Freiberger Forschungsheft C363*, 59–73.

Gründel, J. & Buchholz, A. 1981: Bradoriida aus kambrischen Geschieben vom Gebiet der nördlichen DDR. *Freiberger Forschungsheft C363*, 57–73.

Gruner, H.-E. 1993: Klasse Crustacea. *In* Gruner, H.-E. (ed.): *Band I: Wirbellose Tiere, 4. Teil: Arthropoda (ohne Insecta)*, 448–1030. Gustav Fischer, Jena.

Grygier, M.J. 1983. Ascothoracida and the unity of Maxillopoda. *In* Schram, F.R. (ed.): *Crustacean Issues, 1. Crustacean Phylogeny*, 73–104. Balkema, Rotterdam.

Gürich, G. 1929: *Silesicaris* von Leipe und die Phyllocariden überhaupt. *Mitteilungen des Mineralogisch-Geologischen Staatsinstitutes Hamburg 11*, 21–90.

Hansen, H.J. 1893: Zur Morphologie der Gliedmaßen und Mundteile bei Crustaceen und Insecten. *Zoologischer Anzeiger 16*, 193–198.

Hansen, H.J. 1925: On the comparative morphology of the appendages in the Arthropoda. A. Crustacea. *In* Hansen, H.J. (ed.): *Studies on Arthropoda II*, 1–175. Gyldendalske Boghandel, Kopenhagen.

Hansen, H.J. 1930: *Studies on Arthropoda III*, 1–163. Gyldendalske Boghandel, Kopenhagen.

Hartmann, G. 1963: Zur Phylogenie und Systematik der Ostracoden. *Zeitschrift für Zoologische Systematik und Evolutionsforschung 1*(1–2), 1–154.

Hartmann, G. 1966–1989: Ostracoda. *In* Gruner, H.-E. (ed.): *Dr. H. G. Bronns Klassen und Ordnungen des Tierreichs. Fünfter Band: Arthropoda. I. Abteilung: Crustacea. 2. Buch, IV. Teil.* Gustav Fischer, Jena [1. Lieferung 1966, pp. 1–216; 2. Lieferung 1967, pp. 217–408; 3. Lieferung 1968, pp. 409–568; 4. Lieferung 1975, pp. 569–786; 5. Lieferung 1989, pp. 787–1067].

Heegaard, P. 1945: Remarks on the phylogeny of the arthropods. *Arkiv för Zoologi 37A*(3), 1–15.

Hennig, W. 1950: *Grundzüge einer Theorie der phylogenetischen Systematik.* Deutscher Zentralverlag, Berlin.

Henningsmoen, G. 1953: Classification of Paleozoic straight–hinged Ostracods. *Norsk Geologisk Tidsskrift 31*, 185–291.

Henningsmoen, G. 1957: The trilobite family Olenidae. *Skrifter utgitt av det Norske Videnskaps-Akademi i Oslo 1. Mat.-Naturv. Klasse 1*, 1–362.

Hessler, R. R. 1964: The Cephalocarida. Comparative Skeletomusculature. *Memoirs of the Connecticut Academy of Arts & Sciences 16*, 1–97.

Hessler, R.R. & Newman, W.A. 1975: A trilobitomorph origin for the Crustacea. *Fossils and Strata 4*, 437–459.

Hicks, H. 1871: Description of new species of fossils from the Longmynd Rocks of St. David's. *Quarterly Journal of the Geological Society of London (I) 27*(4), 399–402.

Hill, D., Playford, G. & Woods, J. T. 1971. *Cambrian fossils of Queensland.* Queensland Paleontographical Society, Brisbane, 1–32.

Hinz, I. 1987: The Lower Cambrian microfauna of Comley and Rushton, Shropshire/England. *Palaeontographica A 198*(1–3), 41–100.

Hinz, I. 1991a: Ostrakoden aus kambrischen Geschieben. *Archiv für Geschiebekunde 1*(3/4), 231–234.

Hinz, I. 1991b: On *Ulopsis ulula* Hinz gen. et sp. nov. *Stereo-Atlas of Ostracod Shells 18*(17), 69–72.

Hinz, I. 1992a: On *Semillia pauper* Hinz. *Stereo-Atlas of Ostracod Shells 19*, 13–16.

Hinz, I. 1992b: On *Monasterium oepiki* Fleming. *Stereo-Atlas of Ostracod Shells 19*(29), 123–130.

Hinz, I. 1992c: On *Pejonesia sestina* (Fleming). *Stereo-Atlas of Ostracod Shells 19*(2), 5–8.

Hinz, I. 1993: Evolutionary trends in archaeocopid ostracods. *In* McKenzie, K.G. & Jones, P.J. (eds.): *Ostracoda in the Earth and Life Sciences*, 3–12. Proceedings of the 11th International Symposium on Ostracoda, Warrnambool, Victoria, Australia, 8–12 July 1991. Balkema, Rotterdam.

Hinz, I. & Jones, P.J. 1992: On *Tubupestis tuber* Hinz & Jones gen. et sp. nov. *Stereo-Atlas of Ostracod Shells 19*(3), 9–12.

Hinz, I. & Jones, P.J. 1994: *Gladioscutum lauriei* n. gen. n. sp. (Archaeocopida) from the Middle Cambrian of the Georgina Basin, central Australia. *Paläontologische Zeitschrift 68*(3/4), 361–375.

Hinz-Schallreuter, I. 1993a: Ostracodes from the Middle Cambrian of

Australia. *Neues Jahrbuch für Geologie und Paläontologie, Abhandlungen 188*(3), 305 326.

Hinz-Schallreuter, I. 1993b: Ein mittelkambrischer hesslandonider Ostrakod sowie zur Morphologie und systematischen Stellung der Archaeocopa. *Archiv für Geschiebekunde 1*(6), 329–350.

Hinz-Schallreuter, I. 1993c: Cambrian Ostracodes mainly from Baltoscandia and Morocco. *Archiv für Geschiebekunde 1*(7), 385–448.

Hinz-Schallreuter, I. 1994: "Dreiklappige" kambrische Ostracoden. Paläontologische Gesellschaft, 63. Jahrestagung 21–26 September 1994 in Prag, p. 13, Abstracts.

Hinz-Schallreuter, I. 1995: The early evolution of ostracods. *In* Riha, J. (ed.): *Ostracoda and Biostratigraphy*, 415. A. A. Balkema, Rotterdam.

Hinz-Schallreuter, I. 1996a: On *Trapezilites minimus* (Kummerow). *Stereo-Atlas of Ostracod Shells 23*, 85–88.

Hinz-Schallreuter, I. 1996b: On *Falites fala* Müller. *Stereo-Atlas of Ostracod Shells 23*, 89–94.

Hinz-Schallreuter, I. 1996c: Ein neuer ostrakodenähnlicher Arthropode aus einem vermutlich unterkambrischen Geschiebe. *Geschiebekunde aktuell 12*(1), 25–32.

Hinz-Schallreuter, I. 1997: Leben im Kambrium – die Welt der Mikrofossilien. *Berliner Beiträge zur Geschiebeforschung 1997*, 5–23.

Hinz-Schallreuter, I. 1998: Population structure, life strategies and systematics of phosphatocope ostracods from the Middle Cambrian of Bornholm. *Mitteilungen aus dem Museum für Naturkunde Berlin, Geowissenschaftliche Reihe 1*, 103–134.

Hinz-Schallreuter, I. & Jones, P.J. 1994: *Gladioscutum lauriei* n. gen. n. sp. (Archaeocopida) from the Middle Cambrian of the Georgina Basin, central Australia. *Paläontologische Zeitschrift 68*(3/4), 361–375.

Hinz-Schallreuter, I. & Koppka, J. 1996: Die Ostrakodenfauna eines mittelkambrischen Geschiebes von Nienhagen (Mecklenburg). *Archiv für Geschiebekunde 2*(1), 27–42.

Holm, G. 1893: Sveriges kambrisk-siluriska Hyolithidae och Conulariidae. [The Cambro-Silurian Hyolithidae and Conulariidae of Sweden]. *Sveriges Geologiska Undersökning Ser. C 112*, 1–129.

Holthuis, L.B. 1991: *Marine Lobsters of the World. An Annotated and Illustrated Catalogue of Species of Interest to Fisheries Known to Date.* FAO Species Catalogue, Vol. 13, FAO Fisheries Synopsis no. 125, Rome.

Hou Xian-guang 1987a: Early Cambrian large bivalved arthropods from Chengjiang, eastern Yunnan. *Acta Palaeontologica Sinica 26*(3), 286–298.

Hou Xian-guang 1987b: Oldest Cambrian bradoriids from eastern Yunnan. *In Stratigraphy and Palaeontology of Systemic Boundaries in China: Precambrian–Cambrian Boundary*, 537–545. Nanjing Institute of Geology and Palaeontology, Nanjing.

Hou Xian-guang & Bergström, J. 1997: Arthropods of the Lower Cambrian Chengjiang fauna, southwest China. *Fossils and Strata 45*, 1–116.

Hou Xian-guang, Siveter, Da.J., Williams, M., Walossek, D. & Bergström, J. 1996: Appendages of the arthropod *Kunmingella* from the early Cambrian of China: its bearing on the systematic position of the Bradoriida and the fossil record of the Ostracoda. *Philosophical Transactions of the Royal Society of London B 351*, 1131–1145.

Huo Shi-cheng 1956: Brief notes on Lower Cambrian Archaeostraca from Shensi and Yunnan. *Acta Palaeontologica Sinica 4*, 425–445 (in Chinese with English summary).

Huo Shi-cheng 1965: Additional notes on Lower Cambrian Archaeostraca from Shensi and Yunnan. *Acta Palaeontologica Sinica 13*(2), 291–307.

Huo Shi-cheng & Cui Zhi-lin 1989: On the ages of the Tsun llid-bearing strata in China. *Geological Review 35*, 72–85 (in Chinese).

Huo Shi-cheng & Shu Degan 1982: Notes on Lower Cambrian Bradoriida (Crustacea) from western Sichuan and southern Shaanxi. *Acta Palaeontologica Sinica 21*(3), 322–329 (in Chinese with English summary).

Huo Shi-cheng & Shu Degan 1983: On the phylogeny and ontogeny of Bradoriida with discussions of the origin of Crustacea. *Journal of Northwest University 38*(1), 82–88 (in Chinese with English summary).

Huo Shi-cheng & Shu Degan 1985: *Cambrian Bradoriida of South China.* Northwest University Publishing House, Xi'an (in Chinese with English summary).

Huo Shi-cheng, Shu Degan & Cui Zhilin 1991: *Cambrian Bradoriida of China.* Geological Publishing House, Beijing (in Chinese with English summary).

Huo Shi-cheng, Shu Degan, Zhang Xi-guang, Cui Zhilin & Tong Jiaowen 1983: Notes on Cambrian bradoriids from Shaanxi, Yunnan, Sichuan, Guizhou, Hubei and Guangdong. *Journal of Northwest University 13*, 56–75 (in Chinese with English summary).

Huo Shi-cheng, Shu Degan & Zhao Jing-zhou 1986: Research on Cambrian Bradoriids. *Acta Geologica Sinica 60*(1), 18–30.

Huys, R. 1991: Tantulocarida (Crustacea: Maxillopoda): a new taxon from the temporary meiobenthos. *P.S.Z.N.I.: Marine Ecology 12*(1), 1–34.

Itô, T. 1989: Origin of the basis in copepod limbs, with reference to remipedian and cephalocarid limbs. *Journal of Crustacean Biology 9*(1), 85–103.

Jaanusson, V. 1957: Middle Ordovician Ostracodes of central and southern Sweden. *Bulletin of the Geological Institutions of the University of Uppsala 37*(3/4), 173–442.

Jiang Zhiwen & Xiao Shuhai 1985: Entdeckung und Bedeutung frühkambrischer Phosphatocopida aus Waye, Provinz Xinjiang. *In* Huo Shicheng & Shu Degan (eds): *Cambrian Bradoriida of South China*, 179–186, 250. Northwest University Publishing House, Xi'an (in Chinese).

Jones, P.J. & McKenzie, K.G. 1980: Queensland Middle Cambrian Bradoriida (Crustacea): new taxa, palaeobiogeography and biological affinities. *Alcheringa 4*, 203–225.

Jones, P.J. & McKenzie, K.G. 1981: *Flemingopsis*, a new name for the Cambrian phosphatocopine ostracode genus *Flemingia* Jones & McKenzie 1980. *Alcheringa 5*(3–4), 310.

Jones, T.R. 1872: Note on the Entomostraca from the Cambrian rocks of St. David's. *Quarterly Journal of the Geological Society of London (I) 28*(2), 183–185.

Kaestner, A. 1967: *Lehrbuch der Speziellen Zoologie, Band I Wirbellose, 2. Teil*, I–VIII, 849–1242, 2nd edn. Gustav Fischer, Stuttgart.

Kempf, E.K. 1986a: Index and bibliography of marine Ostracoda 1, Index A. *Sonderveröffentlichungen des Geologischen Instituts der Universität zu Köln 50*, 1–762.

Kempf, E.K. 1986b: Index and bibliography of marine Ostracoda 2, Index B. *Sonderveröffentlichungen des Geologischen Instituts der Universität zu Köln 51*, 1–712.

Kempf, E.K. 1987: Index and bibliography of marine Ostracoda 3, Index C. *Sonderveröffentlichungen des Geologischen Instituts der Universität zu Köln 52*, 1–774.

Kingsley, J.S. 1894: The classification of the Arthropoda. *American Naturalist 28*, 118–135, 220–235.

Kitching, I.J., Forey, P.L., Humphreys, C.J. & Williams, D.M. 1998: *Cladistics. The Theory and Practice of Parsimony Analysis*, 2nd edn,

1–228. The Systematics Association Publication, Oxford University Press, Oxford.

Kobayashi, T. 1954: Fossil Estherians and allied fossils. *Journal of the Faculty of Science of the University of Tokyo (2) 9*(1), 1–192.

Kobayashi, T. & Kato, F. 1951: On the ontogeny and the ventral morphology of *Redlichia chinenesis* with description of *Alutella nakamurai*, new gen. and sp. *Journal of the Faculty of Science University of Tokyo (2) 8*(3), 99–143.

Kozur, H. 1974: Die Bedeutung der Bradoriida als Vorläufer der postkambrischen Ostracoden. *Zeitschrift der geologischen Wissenschaft Berlin 2*(7), 823–830.

Kruizinga, P. 1918: Bijdrage tot de kennis der sedimentaire zwerfsteenen in Nederland. (Zweerfsteenen van Baltischen oorsprong, uitgezondered die, welke in en bij de stad Groningen en bij Maarn zijn gevonden). *Verhandelingen van het Geologisch-Mijnbouwkundig Genostschap voor Nederland en Kolonien Geologische Serie 4*, I–XII, 1–271.

Kukalová-Peck, J. 1992: The "Uniramia" do not exist: the ground plan of the Pterygota as revealed by Permian Diaphanopterodea from Russia (Insecta: Palaeodictyopteroidea). *Canadian Journal of Zoology 69*, 236–255.

Kukalová-Peck, J. 1998: Arthropod phylogeny and "basal" morphological structures. *In* Fortey, R.A. & Thomas, R.H. (eds.): *Arthropod Relationships*. Systematics Association Special Volume Series 55, 249–268. Chapman & Hall, London.

Kummerow, E. 1924: Beiträge zur Kenntnis der Ostracoden und Phyllocariden aus nordischen Diluvialgeschieben. *Jahrbuch der Preußischen Geologischen Landesanstalt zu Berlin 44*, 405–448.

Kummerow, E. 1928: Beiträge zur Kenntnis der Fauna und der Herkunft der Diluvialgeschiebe. *Jahrbuch der Preussischen Geologischen Landesanstalt zu Berlin 48*(1), 1–59.

Kummerow, E. 1931: Über die Unterschiede zwischen Phyllocariden und Ostracoden. *Centralblatt für Mineralogie, Geologie und Paläontologie, Abteilung B 1931*(5), 242–257.

Landing, E. 1980: Late Cambrian–Early Ordovician macrofaunas and phosphatic microfaunas, St. John-Group, New Brunswick. *Journal of Paleontology 54*(4), 752–761.

Landing, E., Bowring, S.A., Davidek, K.L., Rushton, A.W.A., Fortey, R.A. & Wimbledon, W.A.P. 2000: Cambrian–Ordovician boundary age and duration of the lowest Ordovician Tremadoc Series based on U-Pb zircon dates from Avalonian Wales. *Geological Magazine 137*, 485–494.

Langer, W. 1973: Zur Ultrastruktur, Mikromorphologie und Taphonomie des Ostracoda-Carapax. *Palaeontographica 114*, Abt. A, 1–54.

Lauterbach, K.-E. 1973: Schlüsselereignisse in der Evolution der Stammgruppe der Euarthropoda. *Zoologische Beiträge 19*(NF), 251–299.

Lauterbach, K.-E. 1980: Schlüsselereignisse in der Evolution des Grundplans der Arachnata (Arthropoda). *Abhandlungen und Verhandlungen des naturwissenschaftlichen Vereins Hamburg* (NF) 2, 163–327.

Lauterbach, K.-E. 1983: Synapomorphien zwischen Trilobiten- und Cheliceratenzweig der Arachnata. *Zoologischer Anzeiger 210*(3/4), 213–238.

Lauterbach, K.-E. 1988: Zur Position angeblicher Crustacea aus dem Ober-Kambrium im Phylogentischen System der Mandibulata (Arthropoda). *Verhandlungen des naturwissenschaftlichen Vereins in Hamburg (NF) 30*, 409–467.

Li Yu-wen 1975: On the Cambrian Ostracoda with new materials from Sichuan, Yunnan and southern Shaanxi, China. *Professional Papers of Stratigraphy and Palaeontology 2*, 37–72 (in Chinese).

Li Yu-wen 1983: "Bradoriida". *In: Paleontological Atlas of Southwest China, Volume of Microfossils*, 7–22. Geological Publishing House, Beijing (in Chinese).

Lindström, G. 1888: *List of the Fossil Fauna of Sweden, 1, Cambrian and Lower Silurian*, 1–145. G., Stockholm.

Linnarsson, J.G.O. 1869a: Om Vestergötlands cambriska och siluriska aflagringar. *Kongliga Svenska Vetenskaps–Akademiens Handlingar 8*(2), 1–89.

Linnarsson, J.G.O. 1869b: Diagnoses specierum novarum e classe Crustaceorum in depositis Cambricis et Siluricis Vestrogotiae Sueciae repertarum. *Öfversigt af Kongliga Vetenskaps-Akademiens Förhandlingar 2*, 191–196.

Linnarsson, J.G.O. 1875: Öfversigt af Nerikes öfvergångsbildningar. *Öfversigt af Kongliga Vetenskaps-Akademiens Förhandlingar 32*(5), 3–47.

Lochman, C. & Hu Chung-Hung 1960: Upper Cambrian faunas from the Northwest Wind River Mountains, Wyoming. Part I. *Journal of Palaeontology 34*(5), 793–834.

Maas, A. & Waloszek, D. 2001a: Cambrian derivatives of the early arthropod stem lineage, Pentastomids, Tardigrades and Lobopodians – an "Orsten" perspective. *Zoologischer Anzeiger 240*(3/4), 451–459.

Maas, A. & Waloszek, D. 2001b: Larval development of *Euphausia superba* Dana, 1852 and a phylogenetic analysis of the Euphausiacea. *Hydrobiologia 448*(1/3), 143–169.

Maddison, W.P. & Maddison, D.R. 1992: MacClade Version 3. Sinauer Associates, Sunderland, MA.

Maddocks, R.F. 1982: Evolution within the Crustacea, Part 4: Ostracoda. *In* Abele, L.G. (ed.): *The Biology of Crustacea Volume 1: Systematics, the Fossil Record, and Biogeography*, 221–239. Academic Press, New York.

Malz, H. 1990: Nomenclatorial annotations on 'Cambrian and early Ordovician "Ostracoda" (Bradoriida) in China' by Shu, 1990. *Courier Forschungs–Institut Senckenberg 123*, 331–332.

Manuel, M., Kruse, M., Müller, W.E.G. & Le Parco, Y. 2000: The comparison of beta-thymosin homologues among Metazoa supports an arthropod–nematode clade. *Journal of Molecular Evolution 51*, 378–381.

Martin, J.W. & Davis, G.E. 2001: An updated classification of the Recent Crustacea. *National History Museum of Los Angeles County, Science Series 39*, 1–124.

Martinsson, A. 1974: The Cambrian of Norden. *In* Holland, C.H. (ed.): *Lower Palaeozoic Rocks of the World, Volume 2: Cambrian of the British Isles, Norden, and Spitsbergen*, 185–283. J. Wiley, London.

Matthews, S.C. 1973: Notes on open nomenclature and on synonymy lists. *Palaeontology 16*(4), 713–719.

McKenzie, K.G. 1970: [Book Review] Review of Moore, R.C. (ed.): *Treatise on Invertebrate Paleontology, Part R, Arthropoda 4*, R1–R651. University of Kansas and Geological Society of America, Boulder, CO [1969]. *Crustaceana 19*(1), 110–112.

McKenzie, K.G., Angel, M.V., Becker, G., Hinz-Schallreuter, I., Kontrovitz, M., Parker, A.R., Schallreuter, R.E.L. & Swanson, K.M. 1999: Ostracods. *In* Savazzi, E. (ed.): *Functional Morphology of the Invertebrate Skeleton*, 459–507. John Wiley & Sons, Chichester.

McKenzie, K.G., Müller, K.J. & Gramm, M.N. 1983: Phylogeny of Ostracoda. *In* Schram, F.R. (ed.): *Crustacean Phylogeny*, 29–46. A. A. Balkema, Rotterdam.

Melnikova, L.M. 1988: Some Bradoriids (Crustacea) from the Botomian Stage of the East Trans–Baikal Region. *Paleontologicheskii Zhurnal 1987*, 128–131 (in Russian).

Melnikova, L.M. 1990: Early and Late Cambrian Bradoriida (ostracods) from north-eastern central Kazakhstan. *In* Repina, L.N. (ed.): *Cambrian Biostratigraphy and Palaeontology of Northern Asia*, 170–176. Transactions of the Institute of Geology and Geophysics 765. Nauka Siberian Branch, Novosibirsk (in Russian).

Melnikova, L.M. & Mambetov, A.M. 1990: Nizhnekembriyskiye ostrakody Severnogo Tjan'-Shanya. *Paleontologiceskij Zhurnal 1990*(3), 57–62.

Melnikova, L.M. & Mambetov, A.M. 1991: Lower Cambrian Ostracodes of Northern Tyan'-Shan. *Paleontological Journal 24*(3), 56–61 (English translation of Melnikova & Mambetov 1990).

Melnikova, L.M., Siveter, Da.J. & Williams, M. 1997: Cambrian Bradoriida and Phosphatocopida (Arthropoda) from the former Soviet Union. *Journal of Micropalaeontology 16*, 179–191.

Moore, R.C. (ed.). 1961: *Treatise on Invertebrate Palaeontology, Part Q: Ostracoda*. Kansas University Press, Lawrence, KS.

Moritz, M. 1993: Unterstamm Arachnata. *In* Gruner, H.-E. (ed.): *Band I: Wirbellose Tiere, 4. Teil: Arthropoda (ohne Insecta)*, 64–442. Gustav Fischer, Jena.

Müller, A.H. 1989: *Lehrbuch der Paläozoologie. Band II: Invertebraten, Teil 3: Arthropoda 2 – Hemichordata*. VEB Gustav Fischer, Jena.

Müller, K.J. 1964a: Ostracoda (Bradorina) mit phosphatischen Gehäusen aus dem Oberkambrium von Schweden. *Neues Jahrbuch der Geologie und Paläontologie, Abhandlungen 121*(1), 1–46.

Müller, K.J. 1964b: Phosphatisierte Gehäuse bei Ostracoda aus dem oberen Kambrium und die Bedeutung von Apatit als Schalenbildner altpaläozoischer Metazoa. *Die Naturwissenschaften 2*, 1–2.

Müller, K.J. 1973: Late Cambrian and Early Ordovician Conodonts from northern Iran. *Geological Survey of Iran Report 30*, 1–80.

Müller, K.J. 1975: »Heraultia« *varensalensis* (Crustacea) aus dem Unteren Kambrium, der älteste Fall von Geschlechtsdimorphismus. *Paläontologische Zeitschrift 49*(1/2), 168–180.

Müller, K.J. 1979a: Phosphatocopine ostracodes with preserved appendages from the Upper Cambrian of Sweden. *Lethaia 12*(1), 1–27.

Müller, K.J. 1979b: Ostracoden mit erhaltenen Gliedmaßen aus einem oberkambrischen Stinkkalk-Geschiebe. *Der Geschiebesammler 13*(2), 91–94.

Müller, K.J. 1981a: Arthropods with phosphatized soft parts from the Upper Cambrian "Orsten" of Sweden. *Short Papers for the Second International Symposium on the Cambrian System* 1981, Open-File Report, 81–743, 147–151. United States Department of the Interior Geological Survey.

Müller, K.J. 1981b: Weichteile von Fossilien aus dem Erdaltertum. *Forschung, Mitteilungen der DFG 2/1981*, 6–9.

Müller, K.J. 1982a: *Hesslandona unisulcata* sp. nov. with phosphatised appendages from Upper Cambrian "Orsten" of Sweden. *In* Bate, R.H., Robinson, E. & Sheppard, L.M. (eds.): *Fossil and Recent Ostracods*, 276–304. Ellis Horwood, Chichester.

Müller, K.J. 1982b: Phosphatisation of soft tissue on Cambrian Crustacea. *In*: Academy of Sciences of the USSR, Siberian Branch, Institute of Geology and Geophysics (ed.) *Geology of Phosphorite Deposits and Problems of Phosphoritegenesis*, 65–66. Novosibirsk.

Müller, K.J. 1982c: Weichteile von Fossilien aus dem Erdaltertum. *Die Naturwissenschaften 69*, 249–254.

Müller, K.J. 1983: Crustacea with preserved soft parts from the Upper Cambrian of Sweden. *Lethaia 16*(2), 93–109.

Müller, K.J. 1985: Exceptional preservation in calcareous nodules. *Philosophical Transactions of the Royal Society of London B 311*, 67–73.

Müller, K.J. 1990: Upper Cambrian "Orsten". *In* Briggs, D.E.G. &

Crowther, P.R. (eds.): *Palaeobiology, a Synthesis*, 274–277. Blackwell Scientific Publications, Oxford.

Müller, K.J. & Hinz, I. 1991: Upper Cambrian conodonts from Sweden. *Fossils and Strata 28*, 1–153.

Müller, K.J. & Walossek, D. 1985a: A remarkable arthropod fauna from the Upper Cambrian "Orsten" of Sweden. *Transactions of the Royal Society of Edinburgh, Earth Sciences 76*, 161–172.

Müller, K.J. & Walossek, D. 1985b: Skaracarida, a new order of Crustacea from the Upper Cambrian of Västergötland, Sweden. *Fossils and Strata 17*, 1–65.

Müller, K.J. & Walossek, D. 1986a: *Martinssonia elongata* gen. et sp. n., a crustacean-like euarthropod from the Upper Cambrian "Orsten" of Sweden. *Zoologica Scripta 15*(1), 73–92.

Müller, K.J. & Walossek, D. 1986b: Arthropod larvae from the Upper Cambrian of Sweden. *Transactions of the Royal Society of Edinburgh, Earth Sciences 77*, 157–179.

Müller, K.J. & Walossek, D. 1987: Morphology, ontogeny and life habit of *Agnostus pisiformis* from the Upper Cambrian of Sweden. *Fossils and Strata 19*, 1–124.

Müller, K.J. & Walossek, D. 1988: External morphology and larval development of the Upper Cambrian maxillopod *Bredocaris admirabilis*. *Fossils and Strata 23*, 1–70.

Müller, K.J. & Walossek, D. 1991a: Ein Blick durch das <Orsten>-Fenster in die Arthropodenwelt vor 500 Millionen Jahren. *Verhandlungen der Deutschen Zoologischen Gesellschaft 84*, 281–294.

Müller, K.J. & Walossek, D. 1991b: "Orsten" arthropods – small in size but of great impact on biological and phylogenetic interpretations. *Geologiska Föreningens i Stockholm Förhandlingar 113*, 87–90.

Müller, K.J., Walossek, D. & Zakharov, A. 1995: "Orsten" type phosphatized soft-integument preservation and a new record from the Middle Cambrian Kuonamka Formation in Siberia. *Neues Jahrbuch für Geologie und Paläontologie Abhandlungen 197*(1), 101–118.

Newman, W.D. & Knight, M.D. 1984: The carapace and crustacean evolution – a rebuttal. *Journal of Crustacean Biology 4*(4), 682–687.

Nielsen, C. 1995: *Animal Evolution. Interrelationships of the living Phyla*. Oxford University Press, New York.

Olesen, J. 2001: External morphology and larval development of *Derocheilocaris remanei* Delamare-Deboutteville & Chappuis 1951 (Crustacea, Mystacocarida), with a comparison of crustacean segment and tagmosis patterns. *Biologiske Skrifter udgivet Det Kongelige Danske Videnskabernes Selskab 53*, 1–59.

Öpik, A.A. 1968: Ordian (Cambrian) Crustacea Bradoriida of Australia. *Bureau of Mineral Resources, Geology and Geophysics Bull. 103*, I–V, 1–45.

Pennant, T. 1777: *The British Zoology, vol. 4, 4th edn. Crustacea, Mollusca, Testacea*. B. White, London.

Pokorny, V. 1978: Ostracodes. *In* Haq, B.U. & Boersma, A. (eds.): *Introduction to Marine Micropaleontology*, 109–149. Elsevier, New York.

Poulsen, C. 1923: Bornholms Olenuslag og deres Fauna. *Danmarks Geologiske Undersøgelse, 2. Raekke 40*, 5–83.

Rabien, A. 1954: Zur Taxionomie und Chronologie der Oberdevonischen Ostracoden. *Abhandlungen des Hessischen Landesamtes für Bodenforschung 9*, 1–268.

Ramsköld, L. & Edgecombe, G.D. 1996: Trilobite appendage structure – *Eoredlichia* reconsidered. *Alcheringa 20*, 269–276.

Resser, C.E. 1945: Cambrian history of the Grand Canyon region. Part II: Cambrian fossils of the Grand Canyon. *Carnegie Institution of Washington Publication 563*, 169–220.

Reyment, R.A. 1983: Professor Klaus Müller, Bonn. *Terra Cognita 3*(1), 5–6.

Richter, R. 1948: *Einführung in die zoologische Nomenklatur durch Erläuterung der Internationalen Regeln.* Senckenbergische Naturforschende Gesellschaft, Frankfurt.

Richter, S. & Scholtz, G. 2001: Phylogenetic analysis of the Malacostraca (Crustacea). *Journal of Zoological Systematics and Evolutionary Research 39*, 1–23.

Ritterbush, L.A. 1983: Position of the labrum in agnostid trilobites. *Lethaia 16*(4), 309–310.

Rushton, A.W.A. 1969: *Cyclotron,* a new name for *Polyphyma* Groom. *Geological Magazine 106,* 216–217.

Rushton, A.W.A. 1974: The Cambrian of Wales and England. *In* Holland, C.H. (ed.): *Lower Palaeozoic Rocks of the World, Volume 2: Cambrian of the British Isles, Norden, and Spitsbergen,* 43–121. J. Wiley, London.

Rushton, A.W.A. 1978: Fossils from the Middle–Upper Cambrian transition in the Nuneaton district. *Palaeontology 21,* 245–283.

Sanders, H.L. 1957: The Cephalocarida and Crustacean phylogeny. *Systematic Zoology 6,* 112–128.

Sanders, H.L. 1963: The Cephalocarida. Functional morphology, larval development, comparative external anatomy. *Memoirs of the Connecticut Academy of Arts & Sciences 15,* 1–80.

Schallreuter, R.E.L. 1984: Geschiebe-Ostrakoden I. *Neue Jahrbücher der Geologie und Paläontologie, Abhandlungen 169*(1), 1–40.

Schmidt-Rhaesa, A. & Bartolomaeus, T. 2001: Fortschritte in der zoologischen Systematik – Von der Systema Naturae zum phylogenetischen System. *Naturwissenschaftliche Rundschau 54*(3), 121–131.

Schmidt-Rhaesa, A., Bartolomaeus, T., Lemburg, C., Ehlers, U. & Garey, J.R. 1998: The position of the arthropoda in the phylogenetic system. *Journal of Morphology 238,* 263–285.

Schminke, H.K. 1996: Crustacea, Krebse. *In* Westheide, W. & Rieger, R. (ed.): *Spezielle Zoologie, Teil 1: Einzeller und Wirbellose,* 501–581. Gustav Fischer, Stuttgart.

Schmitt, M. 1994: *Wie sich das Leben entwickelte. Die faszinierende Geschichte der Evolution.* Mosaik Verlag, München.

Scholtz, G. 1995: Head segmentation in Crustacea – an immuno-cytochemical study. *Zoology 98,* 104–114.

Scholtz, G. 1998: Cleavage, germ band formation and head segmentation: the ground pattern of the Euarthropoda. *In* Fortey, R.A. & Thomas, R.H. (eds.): *Arthropod Relationships,* 317–322. Chapman & Hall, London.

Scholtz, G. 2000: Evolution of the nauplius stage in malacostracan crustaceans. *Journal of Zoological Systematic and Evolutionary Research 38,* 175–187.

Schram, F.R. 1982: The fossil record and evolution of Crustacea. *In* Abele, L.G. (ed.): *The Biology of Crustacea Volume 1: Systematics, the Fossil Record, and Biogeography,* 93–147. Academic Press, New York.

Schram, F.R. 1986: *Crustacea.* Oxford University Press, New York.

Schram, F.R. & Emerson, M.J. 1990: The origin of Crustacean biramous appendages and the evolution of Arthropoda. *Science 250*(4981), 667–669.

Schram, F.R. & Hof, C.H.J. 1998: Fossils and the interrelationships of major Crustacean groups. *In* Edgecombe, G.D. (ed.): *Arthropod Fossils and Phylogeny,* 233–302. Columbia University Press, New York.

Schram, F.R. & Koenemann, S. 2001: Developmental genetics and arthropod evolution: part I, on legs. *Evolution & Development 3*(5), 343–354.

Schrank, E. 1973: Fauna und Kontakt Mittelkambrium/Oberkambrium

in einem Geschiebe. *Zeitschrift für geologische Wissenschaften 1,* 85–99.

Scott, H.W. 1961: Shell morphology of Ostracoda. *In* Moore, R.C. (ed.): *Treatise on Invertebrate Paleontology, Part Q, Arthropoda 3,* Q21–37. Geological Society of America and University of Kansas Press, Lawrence, KS.

Seilacher, A. 2001: Concretion morphologies reflecting diagenetic and epigenetic pathways. *Sedimentary Geology 143,* 41–57.

Shu Degan 1990a: *Cambrian and Lower Ordovician Bradoriida from Zhejiang, Hunan and Shaanxi Provinces.* Northwest University Press, Xi'an (in Chinese with English summary).

Shu Degan 1990b: Cambrian and Early Ordovician "Ostracoda" (Bradoriida) in China. *Courier Forschungsinstitut Senckenberg 123,* 315–330.

Shu Degan & Chen Ling 1994: Cambrian palaeobiogeography of Bradoriida. *Journal of Southeast Asian Earth Sciences 9*(3), 289–299.

Shu Degan, Vannier, J., Luo Huilin, Chen Ling, Zhang Xing-liang & Hu Shixue 1999: Anatomy and lifestyle of *Kunmingella* (Arthropoda, Bradoriida) from the Chengjiang fossil Lagerstätte (Lower Cambrian; Southwest China). *Lethaia 32,* 279–298.

Shu Degan, Zhang Xing-liang & Geyer, G. 1995: Anatomy and systematic affinities of the Lower Cambrian bivalved arthropod *Isoxys auritus. Alcheringa 19,* 333–342.

Siveter, D.J., Waloszek, D. & Williams, M. 2003: *Klausmuelleria salopensis* gen. et sp. nov., a phosphatocopid crustacean with preserved appendages from the lower Cambrian, England. *Special Paper in Palaeontology 70.*

Siveter, D.J. & Williams, M. 1997: Cambrian bradoriid and phosphatocopid arthropods of North America. *Special Papers in Palaeontology 57,* 1–69.

Siveter, D.J., Williams, M. & Rushton, A.W.A. 1995: Distribution and affinities of British Cambrian Bradoriids. *In* Ríha, J. (ed.): *Ostracoda and Biostratigraphy,* 416. A. A. Balkema, Rotterdam.

Siveter, D.J., Williams, M. & Waloszek, D. 2001: A Phosphatocopid Crustacean with appendages from the Lower Cambrian. *Science 293,* 479–481.

Snodgrass, R.E. 1956: Crustacean metamorphoses. *Smithsonian Miscellaneous Collections 131*(10), 1–79.

Snodgrass, R.E. 1958: Evolution of arthropod mechanisms. *Smithsonian Miscellaneous Collections 138*(2), 1–77.

Spjeldnaes, N. 1966: N. P. Angelin's work on fossil ostracodes. *Geologiska Föreningens i Stockholm Förhandlingar 88,* 407–409.

Steusloff, A. 1895: Neue Ostrakoden aus Diluvialgeschieben von Neu-Brandenburg. *Zeitschrift der Deutschen Geologischen Gesellschaft 46*(4), 775–787.

Storch, V. & Welsch, U. 1997: *Systematische Zoologie.* Gustav Fischer, Stuttgart.

Størmer, L. 1939: Studies on trilobite morphology. Part I. The thoracic appendages and their phylogenetic significance. *Norsk Geologisk Tidsskrift, utgitt av Norsk Geologisk Forening 19*(2–3), 143–273.

Straelen, V. van & Schmitz, G. 1934: Crustacea Phyllocarida (= Archaeostraca). *In* Quenstedt, W. (ed.): *Fossilium Catalogus (I: Animalia), Part 64,* 1–246. W. Junk, Berlin.

Struve, W. 1966: Beiträge zur Kenntnis devonischer Brachiopoden, 15: Einige Atrypinae aus dem Silurium und Devon. *Senckenbergiana Lethaia 47,* 123–163.

Swanson, K.M. 1989a: Ostracod phylogeny and evolution – a manawan perspective. *Courier des Forschungsinstitutes Senckenberg 113,* 11–20.

Swanson, K.M. 1989b: *Manawa staceyi* n. sp. (Punciidae, Ostracoda):

soft anatomy and ontogeny. *Courier des Forschungsinstitutes Senckenberg 113*, 235–249.

Swanson, K.M. 1990: The punciid ostracod – a new crustacean evolutionary window. *Courier des Forschungsinstitutes Senckenberg 123*, 11–18.

Swofford, D.L. 1990: PAUP: Phylogenetic analysis using parsimony, version 3.1. (Computer program distributed by Illinois State National History Survey, Champaign, IL).

Sylvester-Bradley, P.C. 1961: Archaeocopida. *In* Moore, R.C. (ed.): *Treatise on Invertebrate Paleontology, Part Q, Arthropoda 3*, Q100–105. Geological Society of America and University of Kansas Press, Lawrence, KS.

Taylor, A. & Rushton, A.W.A. 1972: The pre-Westphalian geology of the Warwickshire coalfield. *Bulletin of the Geological Survey of Great Britain 35*(13), I–VII, 1–152.

Terfelt, F. 2000: Upper Cambrian trilobite faunas and biostratigraphy at Kakeled on Kinnekulle, Västergötland, Sweden. *Instituto Superior de Correlación Geológica (Insugeo), Miscelánea 6*, 137–139.

Thomas, A.T., Owens, R.M. & Rushton, A.W.A. 1984: Trilobites in British stratigraphy. *Special Report of the Geological Society 16*, 1–78.

Tong Haowen 1987: Fossil Phosphatocopida from Lower Cambrian of China. *Acta Micropalaeontologica Sinica 4*(4), 427–437.

Ulrich, E.O. & Bassler, R.S. 1931: Cambrian bivalved Crustacea of the order Conchostraca. *Proceedings of the United States National Museum 78*(4), 1–130.

Valentine, J.W. & Collins, A.G. 2000: The significance of moulting in Ecdysozoan evolution. *Evolution & Development 2*(3), 152–156.

Vannier, J. & Walossek, D. 1998: Cambrian bivalved arthropods [Review of Siveter & Williams 1997: "Cambrian bradoriid and phosphatocopid arthropods of North America"]. *Lethaia 31*, 97–98.

Vannier, J., Wang Shang Qi & Coen, M. 2001: Leperditicopid arthropods (Ordovician–Late Devonian): functional morphology and ecological range. *Journal of Paleontology 75*(1), 75–95.

Walcott, C.D. 1890: The fauna of the Lower Cambrian or *Olenellus* Zone. *Tenth Annual Report of the United States Geological Survey, 1888/89*(1), 509–763.

Wallerius, I.D. 1895: *Undersökningar öfver Zonen med Agnostus lœvigatus i Vestergötland. Jämte en inledande öfversikt af Vestergötlands samtliga Paradoxideslager.* Gleerupska Universitets-Bokhandeln, Hjalmar Möller, Lund.

Walossek, D. 1993: The Upper Cambrian *Rehbachiella* and the phylogeny of Branchiopoda and Crustacea. *Fossils and Strata 32*, 1–202.

Walossek, D. 1999: On the Cambrian diversity of Crustacea. *In* Schram, F.R. & von Vaupel Klein, J.C. (eds.): *Crustaceans and the Biodiversity Crisis.* Proceedings of the Fourth International Crustacean Congress, Amsterdam, The Netherlands, 20–24 July 1998, vol. 1, 3–27. Brill Academic Publishers, Leiden.

Waloszek, D. 2002: Cambrian "Orsten"-type arthropods and the phylogeny of Crustacea. *In* Legakis, A., Sfenthourakis, S., Polymeni, R. & Thessalou-Legaki, M. (eds.): *The New Panorama of Animal Evolution.* Proceedings of the 18th International Congress of Zoology, Athens 2000. Pensoft Publishers, Sofia.

Walossek, D., Hinz-Schallreuter, I., Shergold, J.H. & Müller, K.J. 1993: Three-dimensional preservation of arthropod integument from the Middle Cambrian of Australia. *Lethaia 26*(1), 7–15.

Walossek, D., Høeg, J.T. & Shirley, T.C. 1996: Larval development of the rhizocephalan cirripede *Briarosaccus tenellus* (Maxillopoda: Thecostraca) reared in the laboratory: a scanning electron microscopy study. *Hydrobiologia 328*, 9–47.

Waloszek, D. & Maas, A. 2001: Phosphatocopida: sister group of the Eucrustacea. *Zoology 104* (Suppl. IV).

Walossek, D. & Müller, K.J. 1990: Upper Cambrian stem-lineage crustaceans and their bearing upon the monophyletic origin of Crustacea and the position of *Agnostus*. *Lethaia 23*(4), 409–427.

Walossek, D. & Müller, K.J. 1992: The "Alum Shale Window" – contribution of "Orsten" arthropods to the phylogeny of Crustacea. *Acta Zoologica 73*(5), 305–312.

Walossek, D. & Müller, K.J. 1994: Pentastomid parasites from the Lower Palaeozoic of Sweden. *Transactions of the Royal Society of Edinburgh, Earth Sciences 85*, 1–37.

Walossek, D. & Müller, K.J. 1998a: Cambrian "Orsten"-type arthropods and the phylogeny of Crustacea. *In* Fortey, R.A. & Thomas, R.H. (eds.): *Arthropod Relationships.* Systematics Association Special Volume Series 55, 139–153. Chapman & Hall, London.

Walossek, D. & Müller, K.J. 1998b: Early arthropod phylogeny in the light of the Cambrian "Orsten" fossils. *In* Edgecombe, G.D. (ed.): *Arthropod Fossils and Phylogeny*, 185–231. Columbia University Press, New York.

Walossek, D. & Szaniawski, H. 1991: *Cambrocaris baltica* n. gen. n. sp., a possible stem-lineage crustacean from the Upper Cambrian of Poland. *Lethaia 24*(4), 363–378.

Weitschat, W. 1983a: Ostracoden (O. Myodocopida) mit Weichkörper-Erhaltung aus der Unter-Trias von Spitzbergen. *Paläontologische Zeitschrift 57*, 314–322.

Weitschat, W. 1983b: On *Triadocypris spitzbergensis* Weitschat. *Stereo-Atlas of Ostracod Shells 10*, 127–138.

Westergård, A.H. 1910: Index to N. P. Angelin's Palaeontologia Scandinavica with notes. *Lunds Univ. Årsskr. N. F. Avd. 2 6*(1), 1–48 (also in: *Kungliga Fysiografiska Sällskapets Handlingar N. F. 21*(2); *Meddelanden från Lunds Geologiska Fältklubb Ser. B 5*).

Westergård, A.H. 1922: Sveriges olenidskiffer. *Sveriges Geologiska Undersökning Ser. C 18*, 1–241.

Westergård, A.H. 1947: Supplementary notes on the Upper Cambrian trilobites of Sweden. *Sveriges Geologiska Undersökning, Ser. C 489*, 1–34.

Westergård, A.H. 1953: Non-Agnostidean trilobites of the Middle Cambrian of Sweden. III. *Sveriges Geologiska Undersökning Ser. C 526*, 1–58.

Whatley, R.C., Siveter, Da.J. & Boomer, I.D. 1993: Arthropoda (Crustacea: Ostracoda). *In* Benton, M.J. (ed.): *The Fossil Record 2*, 343–356. Chapman & Hall, London.

Whatley, R.C., Siveter, D.J. & Boomer, I.D. 1999: Arthropoda (Crustacea: Ostracoda). *In* Benton, M.J. (ed.): *The Fossil Record 3*, 343–356. Chapman & Hall, London.

Whittington, H.B. 1975: Trilobites with appendages from the Middle Cambrian, Burgess Shale, British Columbia. *Fossils and Strata 4*, 97–136.

Whittington, H.B. & Almond, J.E. 1987: Appendages and habits of the Upper Ordovician trilobite *Thriarthrus eatoni*. *Philosophical Transactions of the Royal Society of London B 317*, 1–46.

Williams, M. & Siveter, D.J. 1998: *British Cambrian and Tremadoc Bradoriid and Phosphatocopid Arthropods.* The Palaeontographical Society, London.

Williams, M., Siveter, D.J., Berg-Madsen, V. & Hinz-Schallreuter, I. 1994a: On *Vestrogothia longispinosa*. *Stereo-Atlas of Ostracod Shells 21*, 21–26.

Williams, M., Siveter, D.J., Rushton, A.W.A. & Berg-Madsen, V. 1994b: The Upper Cambrian bradoriid ostracod *Cyclotron lapworthi* is a hesslandonid. *Transactions of the Royal Society of Edinburgh, Earth Sciences 85*, 123–130.

Williams, M., Siveter, D.J. & Peel, J.S. 1996: *Isoxys* (Arthropoda) from the early Cambrian Sirius Passet Lagerstätte, North Greenland. *Journal of Paleontology 70*(6), 947–954.

Williamson, D.I. 1982: Larval morphology and diversity. *In* Abele, L.G. (ed.): *The Biology of Crustacea 2. Embryology, Morphology and Genetics*, 43–110. Academic Press, New York.

Wiman, C. 1905: Studien über das Nordbaltische Silurgebiet. I. Olenellussandstein, Obolussandstein und Ceratopygeschiefer. *Bulletin of the Geological Institution of the University of Upsala 6*, 12–76.

Xiao Bing & Zhao Jing zhou 1986: Lower Cambrian bradoriids from Aksu – Wushi region of Xinjiang. *Journal of Northwest University 16*(3), 73–89.

Yager, J. & Humphreys, W.F. 1996: *Lasionectes exleyi*, sp. nov., the first Remipede Crustacean record from Australia and the Indian Ocean, with a key to the world species. *Invertebrate Taxonomy 10*, 171–187.

Zhang Tan-rong 1981. Bradoriida. *In: Palaeontological Atlas of Northwest China (Xinjiang branch 1)*, 213–214. Science Press, Beijing. [In Chinese].

Zhang Wen-tang 1974: Bradoriida. *In A Handbook of the Stratigraphy and Palaeontology in Southwest China*, 107–111. Science Press, Beijing (in Chinese).

Zhang Wen-tang & Hou Xian-guang 1985: Preliminary notes on the occurrence of the unusual trolobite *Naraoia* in Asia. *Acta Palaeontologica Sinica 24*(6), 591–595.

Zhang Xi-guang 1987: Moult stages and dimorphism of Early Cambrian bradoriids from Xichuan, Henan, China. *Alcheringa 11*, 1–19.

Zhang Xi-guang & Pratt, B.R. 1993: Early Cambrian Ostracode larvae with a univalved carapace. *Science 262*, 93–94.

Zhao Jing-zhou 1989a: Cladistics and classification of Cambrian Bradoriids from North China. *Acta Palaeontologica Sinica 28*(4), 463–473.

Zhao Jing-zhou 1989b: Problems on species and genera of Cambrian Bradoriida from north China. *Acta Micropalaeontologica Sinica 6*(4), 409–418 (in Chinese with English summary).

Zhao Yuhong & Tong Haowen 1989: On *Dabashanella retroswinga* Huo, Shu & Fu. *Stereo-Atlas of Ostracod Shells 16*(3), 13–16.

Zhou Ben-he 1985: Discovery of Early Cambrian ostracods on North Anhui and its significance. *Bulletin of Chengdu Institute of Geology and Mineral Resources, Chinese Academy of Geological Sciences 6*, 95–101.

Ziegler, B. 1998: *Einführung in die Paläobiologie, Teil 3, Spezielle Paläontologie*. E. Schweizerbart, Stuttgart.

Appendix A – Data for the phylogenetic analysis

PAUP settings

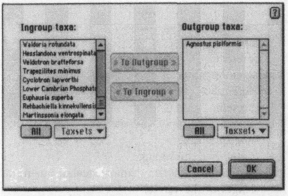

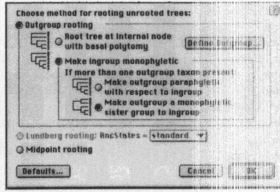

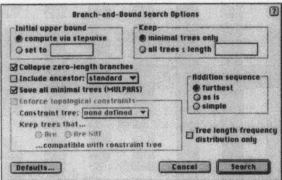

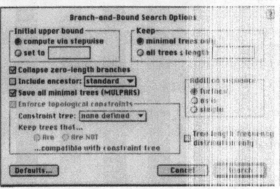

List of all coded characters

This list contains all coded characters referred to as (chX:Y) in the text, where X is the character and Y is the coded character state. A character applied to a structure, which can itself be absent or present, is coded as "not applicable". See Appendix (Character matrix) for the data matrix. Characters are assigned to superunits to provide easier access.

Character group A: shield

Character 1: shield
  0: not applicable
  1: not bivalved
  2: bivalved.
Character 2: hinge line
  0: not applicable
  1: short, dorsal area curved
  2: long, dorsal area straight.
Character 3: location of the maximum length of the shield
  1: ventral to the midline or on the midline
  2: on the dorsal rim
  3: between the dorsal rim and the midline.
Character 4: location of the maximum height of the valves

  0: not applicable
  1: anterior to the antero-posterior midline (pre plete)
  2: on the antero-posterior midline (amplete)
  3: posterior to the antero-posterior midline (ost-plete).
Character 5: free margin of the valves curve back dorsally
  0: not applicable
  1: more anteriorly
  2: more posteriorly
  3: anteriorly and posteriorly more or less equal.
Character 6: anterior margin of the valves
  0: not applicable
  1: rather straight
  2: curved.
Character 7: median/ventral margin of the valve
  0: not applicable
  1: rather straight
  2: curved.
Character 8: posterior margin of the valves
  0: ot applicable
  1: straight
  2: curved.
Character 9: valves leave gap
  0: not applicable

1: no, close tightly
2: yes.
Character 10: lobe L_1
 0: absent
 1: present.
Character 11: lobe L_2
 0: absent
 1: present.
Character 12: lobe L_3
 0: absent
 1: present.
Character 13: lobe L_4
 0: absent
 1: present.
Character 14: lobe L_5
 0: absent
 1: present.
Character 15: lobe L_6
 0: absent
 1: present.
Character 16: surface of the shield with a spine
 0: no
 1: yes.
Character 17: location of the spines
 0: not applicable
 1: centrally
 2: antero-centrally.
Character 18: direction of the spines
 0: not applicable
 1: lateral
 2: postero-lateral.
Character 19: outgrowths of the right shield margin
 0: not present
 1: present, short triangular outgrowth
 2: present, distinct spine.
Character 20: outgrowths of the left shield margin
 0: not present
 1: present, short triangular outgrowth
 2: present, distinct spine.
Character 21: groove at the outer rim of the shield
 0: absent
 1: present.

Character group B: dorsal area

Character 22: dorsal area of the bivalve shield
 0: not applicable
 1: without any structures
 2: with anterior and posterior plates
 3: with a complete dorsal bar (interdorsum).
Character 23: width of the interdorsum
 0: not applicable
 1: between 1/14 and 1/25 the length of the valve
 2: less than 1/25 the length of the valve
 3: more than 1/12 the length of the valve.

Character 24: interdorsum in cross-section
 0: not applicable
 1: flat
 2: convex.
Character 25: ornamentation on the median part of the interdorsum
 0: not applicable
 1: absent
 2: present.
Character 26: anterior dorsal loop
 0: absent
 1: present.
Character 27: shape of the anterior dorsal loop
 0: not applicable
 1: dome-like thickening
 2: short spine (shorter than or equal to one ninth the length of the valves)
 3: long spine (longer than one eighth the length of the valves).
Character 28: direction of the anterior cardinal spine
 0: not applicable
 1: directed dorsally
 2: directed antero-dorsally
 3: directed anteriorly.
Character 29: posterior dorsal loop
 0: absent
 1: present.
Character 30: shape of the posterior dorsal loop
 0: not applicable
 1: dome-like thickening
 2: short spine (shorter than or equal to one ninth the length of the valves)
 3: long spine (longer than one sixth the length of the valves).
Character 31: direction of the posterior cardinal spine
 0: not applicable
 1: directed dorsally
 2: directed postero-dorsally
 3: directed posteriorly
 4: directed postero-ventrally.
Character 32: outgrowths on the anterior spine
 0: not applicable
 1: not present
 2: present.
Character 33: outgrowths on the posterior cardinal spine
 0: not applicable
 1: not present
 2: present.

Character group C: doublure

Character 34: doublure
 0: absent
 1: present.

Character 35: location of the minimum width of the doublure
0: not applicable
1: anteriorly
2: ventrally
3: posteriorly
4: anteriorly and posteriorly equally narrow
5: anteriorly and ventrally equally narrow.

Character 36: location of the maximum width of the doublure
0: not applicable
1: anteriorly
2: medially
3: posteriorly
4: postero-ventrally
5: antero-ventrally.

Character 37: maximum width of the doublure relative to the length of the valves
0: not applicable
1: less than or equal to 1/10 the length of the valves
2: between one sixth and one ninth the length of the valves
3: more than one fifth the length of the valves.

Character 38: pits on the doublure
0: not applicable
1: not present
2: present.

Character 39: small outgrowth structures on the doublure
0: not applicable
1: not present
2: present.

Character 40: shape of the outgrowths
0: not applicable
1: conical, dome-like
2: bottle-like.

Character 41: pores on the doublure
0: not applicable
1: absent
2: present.

Character 42: parallel lines on the doublure
0: not applicable
1: absent
2: present.

Character group D: body and soft parts

Character 43: extension of the body to the lateral side of the shield
1: almost not recognisable
2: obvious, far towards the ventral side.

Character 44: fusion of the body to the shield
1: including the four anterior-most limb-bearing segments

2: including the five anterior-most limb-bearing segments
3: including the six anterior-most limb-bearing segments
4: including the seven anterior-most limb-bearing segments
5: including the eight anterior-most limb-bearing segments
6: including the nine anterior-most limb-bearing segments
7: including 10 or more anterior-most limb-bearing segments.

Character 45: antennula segments
1: less than 14 annuli/segment
2: more than 14 annuli/segment.

Character 46: antennula size
1: less than one quarter the head length
2: equal to or more than one third the head length.

Character 47: furca
0: not applicable
1: absent
2: present.

Character 48: setation of furca
0: not applicable
1: absent
2: present.

Character 49: location of the setation of each furcal ramus
0: not applicable
1: only terminally
2: medially
3: laterally
4: medially and laterally.

Character 50: limb stem composition of the antenna
0: only basis
1: basis with median outgrowth underneath (proximal endite)
2: basis with additional element underneath (coxa).

Character 51: suture between the coxa and the basis of the antenna of early stages
0: not applicable, no coxa developed
1: present, coxa and basis separate
2: absent, coxa and basis fused.

Character 52: suture between the coxa and the basis of the antenna of late stages
0: not applicable
1: present, coxa and basis separate
2: absent, coxa and basis fused.

Character 53: endopods of the antenna
0: not applicable
1: composed of seven segments
2: composed of five segments
3: composed of four segments
4: composed of three segments
5: composed of two segments.

Character 54: limb stem composition of the mandible
  0: only basis
  1: basis with median outgrowth underneath (proximal endite)
  2: basis with additional element underneath (coxa).
Character 55: suture between the coxa and the basis of the mandible of early stages
  0: not applicable, no coxa developed
  1: present, coxa and basis separate
  2: absent, coxa and basis fused.
Character 56: suture between the coxa and the basis of the mandible of late stages
  0: not applicable, no coxa developed
  1: present, coxa and basis separate
  2: absent, coxa and basis fused.
Character 57: endopodites of the second pair of post-antennular limbs (mandible)
0: not applicable
  1: composed of seven segments
  2: composed of five segments
  3: composed of four segments
  4: composed of three segments
  5: composed of two segments.
Character 58: at least the first and second post-mandibular limbs are of the same shape
  0: no, first post-mandibular limb different from the second
  1: yes.

Character 59: limb stem composition of the post-mandibular limbs
  0: only basis
  1: basis with median outgrowth underneath (proximal endite)
  2: basis with additional element underneath (coxa).
Character 60: endopodites of the third pair of post-mandibular limbs (thoracopods I)
  1: composed of seven segments
  2: composed of five segments
  3: composed of four segments
  4: composed of three segments.
Character 61: hypostome with a posterior outgrowth (labrum)
  0: absent
  1: present.
Character 62: single sternal plate, the sternum:
  0: absent
  1: present.
Character 63: sternites of the segment of the mandible with paired humps (paragnaths)
  0: absent
  1: present.
Character 64: first larval stage
  1: with antennula and three additional pairs of limbs ("head-larva")
  2: with antennula and two additional pairs of limbs (orthonauplius)
  3: different, more advanced stage.

Character matrix

Species	Character															
	1	2	3	4	5	6	7	8	9	10	11	12	13	14	15	16
Hesslandona unisulcata	2	2	3	1	2	2	2	2	1	1	0	0	0	0	0	0
Hesslandona necopina	2	2	2	1	0	1	1	1	1	0	0	0	0	0	0	0
Hesslandona kinnekullensis	2	2	3	1	3	2	2	2	1	0	0	0	0	0	0	0
Hesslandona trituberculata	2	2	3	1	2	2	2	1	2	1	1	1	0	0	0	0
Hesslandona ventrospinata	2	2	3	3	2	1	2	2	1	1	1	1	1	1	1	0
Hesslandona suecica n. sp.	2	2	3	2	3	2	2	2	1	0	0	0	0	0	0	0
Hesslandona angustata n. sp.	2	2	3	2	3	2	2	2	1	0	0	0	0	0	0	0
Hesslandona curvispina n. sp.	2	2	2	1	0	1	2	1	1	1	1	1	0	0	0	0
Hesslandona toreborgensis n. sp. 2	2	3	1	3	2	2	2	1	0	0	0	0	0	0	0	0
Trapezilitesminimus	2	2	1	2	2	2	2	2	1	1	0	0	0	0	0	0
Waldoria rotundata	2	2	3	3	3	2	2	2	1	1	1	1	1	1	0	1
Veldotron bratteforsa	2	2	3	1	2	2	1	2	2	1	1	1	1	1	1	0
Falites fala	2	2	3	3	2	2	2	2	1	1	1	1	0	0	0	0
Vestrogothia spinata	2	2	3	1	2	1	2	1	1	0	0	0	0	0	0	1
Cyclotron lapworthi	2	2	2	1	0	2	2	2	1	1	1	1	1	1	1	0
"*Phosphatocopida sp.*"	2	?	2	?	0	?	?	?	?	0	0	0	0	0	0	0
Euphausia superba	1	0	3	0	0	0	0	0	0	0	0	0	0	0	0	0
†*Rehbachiella kinnekullensis*	1	0	2	0	0	0	0	0	0	0	0	0	0	0	0	0
†*Martinssonia elongata*	1	0	2	0	0	0	0	0	0	0	0	0	0	0	0	0
†*Agnostus pisiformis*	1	0	2	0	0	0	0	0	0	0	0	0	0	0	0	0

Species	Character														
	17	18	19	20	21	22	23	24	25	26	27	28	29	30	31
Hesslandona unisulcata	0	0	0	0	0	3	1	1	1	1	1	0	1	?	0
Hesslandona necopina	0	0	0	0	1	3	1	2	1	1	3	2	1	3	2
Hesslandona kinnekullensis	0	0	0	0	0	3	1	2	1	1	3	1	1	3	1
Hesslandona trituberculata	0	0	0	0	0	3	1	2	1	1	2	3	1	2	3
Hesslandona ventrospinata	0	0	1	1	0	3	1	2	1	1	3	2	1	3	2
Hesslandona suecica n. sp.	0	0	0	0	0	3	3	1	1	1	2	0	1	2	2
Hesslandona angustata n. sp.	0	0	0	0	0	3	2	1	1	0	0	0	0	0	0
Hesslandona curvispina n. sp.	0	0	0	0	0	3	1	2	1	1	3	2	1	3	4
Hesslandona toreborgensis n. sp.	0	0	0	0	0	3	1	2	1	1	1	0	1	2	2
Trapezilitesminimus	0	0	0	0	0	3	1	1	1	1	1	0	1	?	0
Waldoria rotundata	1	1	0	0	0	3	2	1	1	0	0	0	0	0	0
Veldotron bratteforsa	0	0	0	0	0	3	1	0	1	0	0	0	0	0	0
Falites fala	0	0	0	0	0	2	0	0	0	0	0	0	0	0	0
Vestrogothia spinata	2	2	0	2	0	2	0	0	0	0	0	0	1	2	3
Cyclotron lapworthi	0	0	0	0	?	3	2	?	1	0	0	0	?	?	?
"Phosphatocopida sp."	0	0	?	?	?	1	0	0	0	0	0	0	0	0	0
Euphausia superba	0	0	0	0	0	0	0	0	0	0	0	0	0	0	0
†*Rehbachiella kinnekullensis*	0	0	0	0	0	0	0	0	0	0	0	0	0	0	0
†*Martinssonia elongata*	0	0	0	0	0	0	0	0	0	0	0	0	0	0	0
†*Agnostus pisiformis*	0	0	0	0	0	0	0	0	0	0	0	0	0	0	0

Species	Character														
	33	34	35	36	37	38	39	40	41	42	43	44	45	46	47
Hesslandona unisulcata	0	1	1	4	1	2	2	2	2	1	1	6	1	?	2
Hesslandona necopina	2	1	2	3	1	1	1	1	1	1	1	7	1	?	?
Hesslandona kinnekullensis	1	1	2	3	1	1	1	0	1	1	?	?	?	?	?
Hesslandona trituberculata	1	1	5	3	2	1	2	1	1	1	?	?	?	?	?
Hesslandona ventrospinata	1	1	2	3	1	1	2	1	1	1	1	5	1	1	?
Hesslandona suecica n. sp.	1	1	2	3	1	1	1	0	1	1	1	5	1	1	?
Hesslandona angustata n. sp.	0	1	2	3	1	1	1	0	1	1	1	3	1	1	?
Hesslandona curvispina n. sp.	2	1	2	3	1	1	2	1	1	1	1	6	?	?	?
Hesslandona toreborgensis n. sp.	1	1	2	3	1	1	1	0	1	1	?	?	?	?	?
Trapezilitesminimus	0	1	4	2	2	1	1	0	1	1	1	4	1	1	?
Waldoria rotundata	0	1	1	4	2	1	2	1	2	1	1	6	1	1	2
Veldotron bratteforsa	0	1	1	4	2	1	2	1	1	1	?	?	?	?	?
Falites fala	0	1	1	4	3	1	1	0	1	2	1	5	1	1	?
Vestrogothia spinata	1	1	2	3	2	1	2	1	1	1	1	?	?	?	?
Cyclotron lapworthi	?	?	?	?	?	?	?	?	?	?	?	?	?	?	?
"Phosphatocopida sp."	0	1	1	3	1	0	0	0	0	0	?	?	?	?	?
Euphausia superba	0	0	0	0	0	0	0	0	0	0	2	7	2	2	2
†*Rehbachiella kinnekullensis*	0	0	0	0	0	0	0	0	0	0	2	1	2	2	2
†*Martinssonia elongata*	0	0	0	0	0	0	0	0	0	0	2	1	2	2	0
†*Agnostus pisiformis*	0	0	0	0	0	0	0	0	0	0	2	1	2	2	0

Species	Character														
	49	50	51	52	53	54	55	56	57	58	59	60	61	62	63
Hesslandona unisulcata	4	2	2	2	5	2	2	2	5	1	1	4	1	1	1
Hesslandona necopina	?	2	2	2	5	2	2	2	5	1	1	4	1	1	1
Hesslandona kinnekullensis	?	?	?	?	?	?	?	?	?	?	?	?	?	?	?
Hesslandona trituberculata	?	?	?	?	?	?	?	?	?	?	?	?	?	?	?
Hesslandona ventrospinata	?	2	?	2	5	2	?	2	5	1	1	4	1	1	1
Hesslandona suecica n. sp.	?	2	2	2	5	2	2	2	5	1	1	4	1	1	1
Hesslandona angustata n. sp.	?	2	2	2	5	2	2	2	5	1	1	4	1	1	1
Hesslandona curvispina n. sp.	?	2	2	2	5	2	2	2	5	1	1	4	1	1	1
Hesslandona toreborgensis n. sp.	?	?	?	?	?	?	?	?	?	?	?	?	?	?	?
Trapezilitesminimus	?	2	?	2	5	2	?	2	5	1	1	4	1	1	1
Waldoria rotundata	4	2	?	2	5	2	?	2	5	1	1	4	1	1	1
Veldotron bratteforsa	?	?	?	?	?	?	?	?	?	?	?	?	?	?	?
Falites fala	?	2	2	2	5	2	2	2	5	1	1	4	1	1	1

Vestrogothia spinata	?	2	2	2	5	2	1	2	5	1	1	4	1	1	1	1
Cyclotron lapworthi	?	?	?	?	?	?	?	?	?	?	?	?	?	?	?	?
"Phosphatocopida sp."	?	2	1	?	4	2	1	?	4	?	1	?	1	1	1	1
Euphausia superba	0	2	1	2	4	2	1	2	5	0	2	2	1	1	1	2
†*Rehbachiella kinnekullensis*	4	2	1	1	4	2	1	1	0	0	1	3	1	1	1	2
†*Martinssonia elongata*	0	1	0	0	2	1	0	0	2	1	1	2	0	0	0	1
†*Agnostus pisiformis*	0	0	0	0	1	0	0	0	1	1	0	1	0	0	0	1

Appendix B – Synonymy of Phosphatocopine taxa

An alphabetical list of all taxa referred to Phosphatocopina with a complete list of their records in the literature and some remarks. It is unknown whether all species are in fact phosphatocopines. Most of them are poorly known and the question of inclusion or non-inclusion of species to Phosphatocopina remains part of a future investigation.

Possible Phosphatocopine taxa

Alutella Kobayashi & Kato, 1951 (type species: *Alutella nakamurai* Kobayashi & Kato, 1951 by original designation)

1951   *Alutella* Kobayashi & Kato, gen. nov. – Kobayashi & Kato, p. 139.

1968   *Alutella* Kobayashi & Kato, 1951 – Öpik, p. 31 (referred to Svealutidae nov.).

1986a   *Alutella* Kobayashi & Kato, 1951 – Kempf, p. 45.

1986b   *Alutella* Kobayashi & Kato, 1951 – Kempf, p. 657.

1987   *Alutella* Kobayashi & Kato, 1951 – Kempf, p. 259.

1989b   *Alutella* – Zhao, p. 410.

1990b   *Alutella* Kobayashi & Kato 1951 – Shu, p. 323.

Alutella duplicata Shu, 1990 (type locality and horizon: Lower Cambrian, Shaanxi, China)

1990b   *Alutella duplicata* Shu 1987 (unpublished manuscript) – Shu, p. 328, pl. 3, figs. 39–41.

Alutella nakamurai Kobayashi & Kato, 1951 (type locality and horizon: Lower and Middle (?) Cambrian, Sanshihiplu Station, Liaotung, South Manchuria)

1951   *Alutella nakamurai* Kobayashi & Kato, new species – Kobayashi & Kato, p. 139, pl. III, fig. 15.

1968   *Alutella nakamurai* Kobayashi & Kato, 1951 – Öpik, p. 26.

1986a   *Alutella nakamurai* Kobayashi & Kato, 1951 – Kempf, p. 45.

1986b   *Alutella nakamurai* Kobayashi & Kato, 1951 – Kempf, p. 389.

1987   *Alutella nakamurai* Kobayashi & Kato, 1951 – Kempf, p. 259.

1989b   *Alutella nakamurai luonanensis* (Yi, Cui & Huo) – Zhao, p. 410, table 1, pl. 1, fig. 1.

1989b   *Liella luonanensis* – Zhao, table 1 (synonymised with *Alutella nakamurai luonanensis*).

Aparchona Hinz-Schallreuter, 1993 (type species:

Aparchona klafacki Hinz-Schallreuter, 1993 by original designation)

1993c   *Aparchona* n. gen. – Hinz-Schallreuter, p. 412.

1996   *Aparchona* Hinz-Schallreuter, 1993 – Hinz-Schallreuter & Koppka, pp. 27, 30.

1998   ?*Aparchona* Hinz-Schallreuter, 1993 – Hinz-Schallreuter, p. 116.

Aparchona grandispinosa Hinz-Schallreuter & Koppka, 1996 (type locality and horizon: Middle Cambrian, erratic boulders of Nienhagen, Mecklenburg, Germany)

1996   *Aparchona grandispinosa* sp. nov. – Hinz-Schallreuter & Koppka, p. 30.

Aparchona klafacki Hinz-Schallreuter, 1993 (type locality and horizon: Late Middle Cambrian, erratic boulders of Gislövshammar, Skåne, Sweden)

1993c   *Aparchona klafacki* n. sp. – Hinz-Schallreuter, pp. 392, 412, figs. 10.2, 10.3.

1996   *Aparchona klafacki* Hinz-Schallreuter, 1993 – Hinz-Schallreuter & Koppka, p. 30.

Bidimorpha Hinz-Schallreuter, 1993 (type species: *Bidimorpha bidimorpha* Hinz-Schallreuter, 1993)

1993b   *Bidimorpha* n. g. – Hinz-Schallreuter, pp. 329, 332–335, 341, 343–345.

1993c   *Bidimorpha* Hinz-Schallreuter, 1993 – Hinz-Schallreuter, pp. 395, 402, 405, 408, 409.

1994b   *Bidimorpha* [Hinz-Schallreuter, 1993] – Williams et al., p. 128.

1997   *Bidimorpha* – Hinz-Schallreuter, p. 13.

1998   *Bidimorpha* Hinz-Schallreuter, 1993 – Hinz-Schallreuter, pp. 103, 107, 118, 132, text-figs. 3, 4.

1998   *Bidimorpha* Hinz-Schallreuter, 1993 – Williams & Siveter, p. 34.

Bidimorpha arator Hinz-Schallreuter, 1998 (type locality and horizon: Middle Cambrian, Isle of Bornholm)

1993c   *Bidimorpha inversa* n. sp. – Hinz-Schallreuter, pp. 400, 410 (*partim*), figs. 7.4a, b, 10.1.

1998   *Bidimorpha arator* n. sp. – Hinz-Schallreuter, pp. 103, 106, 107, 118, 132, pl. 5, fig. 7; l. 6, fig. 6; pl. 7, figs. 5, 7; text-fig. 1; table 5.

Bidimorpha bidimorpha Hinz-Schallreuter, 1993 (type locality and horizon: Middle Cambrian, erratic boulders of Gislövshammar, Skåne, Sweden)

1902   "*Beyrichia*" *angelini* Barr., var. *armata* i. var. – Grönwall, pp. 162, 169, 227, pl. 4, 27 (*non Beyrichia? armata* Richter 1863, p. 12).

1929   "*Beyrichia*" *angelini* Barr. var. *armata*, – Gürich, p. 43, fig. 6; text-fig. 2 (cop. Grönwall, pl. 4, fig. 27).

1929   *Polyphyma armata* (Grönwall) als var. – Gürich, p. 44, text-pl. 2, fig. 6.

1931   *Polyphyma armata* (Grönwall) – Ulrich &

Bassler, pp. 8, 11, 67, 68, 121, 129, pl. 8, fig. 31 (cop. Grönwall 1902, pl. 4, fig. 27).

1934   *Beyrichia angelini armata* Grönwall = *Polyphyma armata* – Bassler & Kellett, p. 185.

1934   *Polyphyma angelini* var. *armata* (Grönwall), 1902 – van Straelen & Schmitz, pp. 197, 211, 229, 237, 245.

1969   *C.* [*Cyclotron*] *armatum* (Groenwall) – Rushton, p. 216.

1973   *C.* [*Cyclotron*] *armatum* (Grönwall, 1902) – Schrank, p. 90.

1986a   *Beyrichia angelini armata* Groenwall, 1902 – Kempf, p. 110.

1986a   *Cyclotron armatum* (Groenwall, 1902) Rushton, 1969 – Kempf, p. 203.

1986a   *Polyphyma armata* (Groenwall, 1902) Ulrich & Bassler, 1931 – Kempf, p. 604.

1986b   *Beyrichia angelini armata* Groenwall, 1902 – Kempf, p. 50.

1986b   *Cyclotron armatum* (Groenwall, 1902) Rushton, 1969 – Kempf, p. 64.

1986b   *Polyphyma armata* (Groenwall, 1902) Ulrich & Bassler, 1931 – Kempf, p. 64.

1987   *Beyrichia angelini armata* Groenwall, 1902 – Kempf, p. 120.

1987   *Polyphyma armata* (Groenwall, 1902) Ulrich & Bassler, 1931 – Kempf, p. 168.

1987   *Cyclotron armatum* (Groenwall, 1902) Rushton, 1969 – Kempf, p. 523.

1993b   *Bidimorpha bidimorpha* n. sp. – Hinz-Schallreuter, pp. 329, 331, 332, 335, 342, 345, figs. 1A, 3–6 (pro *Beyrichia armata* Grönwall, 1902 non Richter, 1863).

1993c   *Bidimorpha bidimorpha* – Hinz-Schallreuter, pp. 396, 402, 409.

1994b   *Beyrichia armata* Grönwall, 1902 – Williams et al., p. 128 (referred to *Bidimorpha*).

1996   *Bidimorpha bidimorpha* – Hinz-Schallreuter & Koppka, p. 29.

1997   *Bidimorpha bidimorpha* Hinz-Schallreuter, 1993 – Hinz-Schallreuter, p. 13.

1998   *Beyrichia angelini* Barrande var. *armata* – Hinz-Schallreuter, p. 104 (referred to Grönwall 1902).

1998   *Bidimorpha bidimorpha* Hinz-Schallreuter, 1993 – Hinz-Schallreuter, pp. 104, 106, 107, 118, 132.

1999   *Bidimorpha bidimorpha* Hinz-Schallreuter – McKenzie et al., p. 464, fig. 33.7.

Bidimorpha inversa Hinz-Schallreuter, 1993 (type locality and horizon: Middle Cambrian, Gislövshammar, Skåne, Sweden)

1993c   *Bidimorpha inversa* sp. n. – Hinz-Schallreuter, pp. 392, 410, figs. 7.2–7.4, 10.1.

1998   *Bidimorpha inversa* Hinz-Schallreuter, 1993 – Hinz-Schallreuter, pp. 106, 118.

Bidimorpha labiator Hinz-Schallreuter, 1998 (type locality and horizon: Middle Cambrian, Isle of Bornholm)

1998   *Bidimorpha labiator* n. sp. – Hinz-Schallreuter, pp. 103, 105, 106, 118, 122, pl. 6, figs. 1–7; pl. 8, figs. 2, 3; pl. 10, fig. 3; text-figs. 1, 6, 7; table 6.

Bidimorpha sexspinosa Hinz-Schallreuter, 1998 (type locality and horizon: Middle Cambrian, Isle of Bornholm)

1998   *Bidimorpha sexspinosa* n. sp. – Hinz-Schallreuter, pp. 103, 118, 124, pl. 5, fig. 8a–c; table 7.

"*Bythocypris*" *polita* Steusloff, 1895 (type locality and horizon: Cambrian, erratic boulders of Neu-Brandenburg, Germany)

1895   *Bythocypris polita* n. sp. – Steusloff, p. 775, pl. 58, fig. 31.

1924   *Bythocypris polita* Steusloff – Kummerow, pp. 406, 408.

1931   *Lepiditta? polita* (Steusloff) – Ulrich & Bassler, pp. 11, 95, 128, 129, pl. 7, fig. 28.

1934   *Bythocypris polita* Steusloff = *Lepiditta polita* – Bassler & Kellett, p. 231.

1954   *Bythocypris polita* Steusloff – Kobayashi, p. 129.

1982a   *Bythocypris polita* Steusloff – Müller, p. 279.

1984   *Bythocypris polita* Steusloff, 1895 – Schallreuter, p. 2.

1986a   *Bythocypris polita* Steusloff, 1894 – Kempf, p. 148.

1986a   *Lepiditta? polita* (Steusloff, 1894) Ulrich & Bass, 1931 – Kempf, p. 460.

1986b   *Bythocypris polita* Steusloff, 1894 – Kempf, p. 456.

1986b   *Lepiditta? polita* (Steusloff, 1894) Ulrich & Bass, 1931 – Kempf, p. 456.

1987   *Bythocypris polita* Steusloff, 1894 – Kempf, p. 109.

1987   *Lepiditta? polita* (Steusloff, 1894) Ulrich & Bass, 1931 – Kempf, p. 168.

1993c   *Bythocypris polita* Steusloff, 1895 – Hinz-Schallreuter, p. 396.

1998   *Bythocypris polita* [Steusloff, 1895] – Hinz-Schallreuter, pp. 104, 115.

Remarks. – Kummerow (1924) assumed that *Bythocypris polita* is a juvenile form of *Aristozoë primordialis* (= *Hesslandona minima*). Schallreuter (1984) synonymised "*Bythocypris*" *polita* with *Hesslandona necopina* Müller, 1964. However, the species described by Müller (1964a) differs from the species described by Steusloff (1895) in lacking lobes on the valves. Thus, a synonymy of both species cannot be the case. However, as the material of

Steusloff (1895) is lost (Hinz-Schallreuter 1993c), statements on the true relationship of "*Bythocypris*" *polita* cannot be made herein.

Comleyopsis Hinz, 1993 (type species: *Comleyopsis schallreuteri* Hinz, 1993 by original designation)
1987   *Hesslandona*? Müller – Hinz, p. 59.
1993   *Comleyopsis* n. g. – Hinz, pp. 5, 11, 12, fig. 2b1.
1993b   *Comleyopsis* Hinz, 1993 – Hinz-Schallreuter, pp. 341–343.
1993c   *Comleyopsis* Hinz, 1993 – Hinz-Schallreuter, pp. 386, 410.
1996   *Comleyopsis* Hinz, 1993 – Hinz-Schallreuter & Koppka, pp. 27–29.
1998   *Comleyopsis* Hinz, 1993 – Hinz-Schallreuter, p. 115.
1998   *Comleyopsis* Hinz, 1993 – Williams & Siveter, p. 30.

Comleyopsis iecta Hinz-Schallreuter & Koppka, 1996 (type locality and horizon: Middle Cambrian, erratic boulders of Nienhagen, Mecklenburg, Germany)
1996   *Comleyopsis iecta* sp. nov. – Hinz-Schallreuter & Koppka, pp. 28, 29.

Comleyopsis schallreuteri Hinz, 1993 (type locality and horizon: Lower Cambrian, Atdabanian, Comley, England)
1987   *Hesslandona*? n. sp. B – Hinz, p. 59, pl. 3, figs. 8, 15; table 1.
1991a   *Hesslandona*? n. sp. B – Hinz, p. 233.
1993   *Comleyopsis schallreuteri* n. sp. – Hinz, pp. 3, 12, fig. 4E, F.
1993   *Hesslandona*? n. sp. of Huiz, 1987 – Whatley et al., p. 345.
1998   *Comleyopsis schallreuteri* Hinz, 1993 – Williams & Siveter, p. 30, pl. 6, figs. 11, 12.
1999   *Hesslandona*? n. sp. of Hinz, 1987 – Whatley et al., p. 345.

Cyclotron Rushton, 1969 (type species: *Polyphyma lapworthi* Groom, 1902 by original designation)
1902   *Polyphyma* gen. nov. – Groom, pp. 83, 84, 86, 87, *non Polyphyma* Jakovlev, 1877, p. 73 [Insecta], *nec Polyphyma* Hamm, 1881, p. 38 [Bryozoa].
1913   *Polyphyma* Groom – Eastman, p. 735.
1924   *Polyphyma* Groom – Broili, p. 629.
1931   *Polyphyma* Groom – Ulrich & Bassler, pp. 11, 51, 64, 65, 129.
1934   *Polyphyma* Groom – Bassler & Kellett, p. 435.
1947   *Polyphyma* – Westergård, p. 19.
1961   *Polyphyma* Groom, 1902 – Sylvester-Bradley *in* Moore, p. Q103.
1968   *Polyphyma* Groom – Öpik, pp. 6, 26.
1969   *Polyphyma* – Andres, p. 179.

1969   *Cyclotron* nom. nov. – Rushton, p. 216 pro *Polyphyma* Groom, 1902 *non* Jakovlev, [187]7).
1972   *Cyclotron* Rushton, 1969 – Taylor & Rushton, pp. 13, 18, 25.
1973   *Cyclotron* Rushton, 1969 – Schrank, p. 8.
1974   *Cyclotron* Rushton, 1969 (= *Polyphyma* Groom, 1902) – Kozur (referred to Hipponicharionidae Sylvester-Bradley, 1962).
1975   *Polyphyma* Groom, 1902 – Li, p. 43.
1978   *Cyclotron* Rushton, 1969 – Rushton, p. 273.
1979   *Cyclotron* Rushton, 1969 – Bednarczyk, p. 117.
1980   *Cyclotron* – Jones & McKenzie, p. 219.
1981   *Cyclotron* Rushton 1969 – Gründel, pp. 60, 61, 64.
1986a   *Cyclotron* 〈*Polyphyma*〉 Rushton, 1969 – Kempf, p. 203.
1986a   *Polyphyma* Groom, 1902 – Kempf, p. 603.
1986b   *Cyclotron* 〈*Polyphyma*〉 Rushton, 1969 – Kempf, p. 667.
1986b   *Polyphyma* Groom, 1902 – Kempf, p. 691.
1987   *Polyphyma* Groom, 1902 – Kempf, p. 121.
1987   *Cyclotron* 〈*Polyphyma*〉 Rushton, 1969 – Kempf, p. 523.
1987   *Cyclotron* – Zhang, p. 9.
1989   *Polyphyma* Groom, 1902 – A. H. Müller, p. 54.
1993b   *Cyclotron* Rushton, 1969 (pro *Polyphyma* Groom, 1902) – Hinz-Schallreuter, pp. 332, 333.
1993c   *Cyclotron* Rushton, 1969 (*nom. nov.* pro *Polyphyma* Groom, 1902 *non* Jakovlev, 1877 *non* Hamm, 1881) – Hinz-Schallreuter, pp. 386, 391, 402, 405, 408.
1994   *Cyclotron* – Hinz & Jones, p. 368.
1994b   *Cyclotron* Rushton – Williams et al., pp. 123, 124, 126.
1995   *Cyclotron* – Siveter et al., p. 416.
1997   *Cyclotron* (pro *Polyphyma*) – Hinz-Schallreuter, p. 13.
1997   *Cyclotron* Rushton – Siveter & Williams, p. 59.
1998a   *Cyclotron* – Clarkson et al., pp. 257, 261, 263, 264.
1998b   *Cyclotron* – Clarkson et al., p. 27.
1998   *Cyclotron* Rushton, 1869 – Hinz-Schallreuter, pp. 106, 115, 118 (as member of Vestrogothiinae).
1998   *Cyclotron* Rushton – Williams & Siveter, p. 32.

Remarks. – Taxa which have been referred to *Cyclotron* include specimens of *Beyrichia angelini* Barrande, 1872 of Linnarsson (1875). Barrande (1872) introduced the name *Beyrichia angelini*, but without description or illustration, for the specimen in fig. 36 on plate A of Angelin's unpublished Palaeontologia Scandinavica Part III (Spjeldnaes 1966). According to Westergård (1947, p. 19), this specimen is lost. The first combined published description and figure of *Beyrichia angelini* was by Linnarsson (1875). Linnarsson's specimen was re-figured by Westergård who, in the absence of Angelin's original, considered it to be the type of *Beyrichia angelini*.

Cyclotron Angelini (Linnarsson, 1875) Rushton, 1969 (type locality and horizon: Upper Cambrian, southern Sweden)

1875 *Beyrichia Angelini* Barr. – Linnarsson, pp. 16, 29, 45, pl. 5, fig. 11.

1888 *Beyrichia Angelini* Barr. – Lindström, p. 6.

1893 *Beyrichia Angelini* Barr. – Holm, p. 110 (footnote).

1902 *Beyrichia Angelini*, Barr. – Groom, pp. 83, 87, 88.

1910 *Beyrichia Angelini* – Westergård, p. 5.

1918 *Beyrichia Angelini* Barr. – Kruizinga, p. 61.

1922 *Polyphyma Angelini* – Westergård, pp. 9, 10.

1923 *Polyphyma Angelini* (Barrande) – Poulsen, pp. 52, 61, 81, fig. 19.

1929 "*Beyrichia*" *Angelini* – Gürich, pp. 38, 44.

1929 *Polyphyma Angelini* – Gürich, p. 44.

1931 *Polyphyma Angelini* – Kummerow, pp. 254, 255, text-fig. 17.

1931 *Polyphyma Angelini* (Barrande) – Ulrich & Bassler, pp. 11, 67, 127, 129, pl. 8, fig. 30.

1934 *Polyphyma Angelini* (Barrande), 1872 – van Straelen & Schmitz, pp. 196, 211, 229, 237, 245.

1947 *Polyphyma Angelini* (Barrande, 1872) – Westergård, p. 18, (on p. 19 referred to *Beyrichia angelini* of Linnarsson 1875), pl. 1, fig. 15 [re-figured from Linnarsson (1875)].

1964a *Polyphyma Angelini* (Barrande, 1872) – Müller, pp. 4, 5, 38.

1966 *Beyrichia Angelini* [Barrande] – Spjeldnaes, p. 407.

1968 *Polyphyma Angelini* – Öpik, p. 6.

1969 *Cyclotron Angelini* (Barrande) – Rushton, p. 216.

1972 *C. Angelini* – Taylor & Rushton, p. 18.

1973 *C. Angelini* (Barrande) – Schrank, p. 90.

1978 *C. Angelini* (Barrande) – Rushton, p. 278.

1974 *Polyphyma Angelini* – Martinsson, p. 208.

1986a *Beyrichia Angelini* Barrande, 1872 – Kempf, p. 110.

1986a *Beyrichia Angelini angelini* Barrande, 1872 – Kempf, p. 110.

1986a *Cyclotron Angelini* (Barrande, 1872) Rushton, 1969 – Kempf, p. 203.

1986a *Polyphyma Angelini* (Barrande, 1872) Ulrich & Bassler, 1931 – Kempf, p. 604.

1986b *Beyrichia Angelini* Barrande, 1872 – Kempf, p. 50.

1986b *Beyrichia Angelini angelini* Barrande, 1872 – Kempf, p. 50.

1986b *Cyclotron Angelini* (Barrande, 1872) Rushton, 1969 – Kempf, p. 50.

1986b *Polyphyma Angelini* (Barrande, 1872) Ulrich & Bassler, 1931 – Kempf, p. 50.

1987 *Beyrichia Angelini* Barrande, 1872 – Kempf, p. 62.

1987 *Beyrichia Angelini angelini* Barrande, 1872 – Kempf, p. 62.

1987 *Polyphyma Angelini* (Barrande, 1872) Ulrich & Bassler, 1931 – Kempf, p. 168.

1987 *Cyclotron Angelini* (Barrande, 1872) Rushton, 1969 – Kempf, p. 523.

1991a *Beyrichia Angelini* Barrande (1872) – Hinz, p. 231.

1993b *Beyrichia Angelini* – Hinz-Schallreuter, p. 332.

1993b *Cyclotron Angelini* (Barrande, 1872) – Hinz-Schallreuter, p. 333.

1993c *Cyclotron Angelini* [Barrande, 1872] – Hinz-Schallreuter, p. 396.

1993c *Beyrichia Angelini* Barrande, 1872 – Hinz-Schallreuter, p. 408.

1994b *Beyrichia Angelini* Linnarsson, 1875 – Williams *et al.*, pp. 126, 128 (referred to *Cyclotron*).

1994b *Cyclotron Angelini* (Linnarsson, 1875) – Williams *et al.*, p. 129, fig. 6q.

1997 *Cyclotron Angelini* – Hinz-Schallreuter, p. 13.

1998b *Cyclotron Angelini* – Clarkson *et al.*, p. 28.

1998 *Beyrichia Angelini* Barrande, 1872 – Hinz-Schallreuter, p. 104.

1998 *Cyclotron Angelini* (Linnarsson, 1875) – Williams & Siveter, p. 33.

Cyclotron cambricum Gründel *in* Gründel & Buchholz, 1981 (Upper Cambrian, Zone 2, erratic boulders from the Isle of Rügen, Germany)

1981 *Cyclotron cambricum* n. sp. – Gründel *in* Gründel & Buchholz, p. 64, pl. II, fig. 13.

1986a *Cyclotron cambricum* Gruendel, 1981 – Kempf, p. 203.

1986b *Cyclotron cambricum* Gruendel, 1981 – Kempf, p. 112.

1987 *Cyclotron cambricum* Gruendel, 1981 – Kempf, p. 710.

1993b *Cyclotron cambricum* Gründel *in* Gründel & Buchholz, 1981 – Hinz-Schallreuter, p. 333.

1993c *Cyclotron cambricum* Gründel *in* Gründel & Buchholz, 1981 – Hinz-Schallreuter, p. 408.

1994b *C. cambrium* [Gründel *in* Gründel & Buchholz (1981)] – Williams *et al.*, p. 128 (sic!).

1999 *Cyclotron cambricum* Gründel, 1981 – Whatley *et al.*, p. 343 (as Hipponicharionidae).

Cyclotron furcatocostatum Gründel *in* Gründel &

Buchholz, 1981 (Upper Cambrian, Zone 2, erratic boulders from the Isle of Rügen, Germany)

1979　*Cyclotron nodomarginatum* Schrank, 1973 – Bednarczyk, pp. 215, 217, 218, plate, figs. 1a–b, 2–4?, 7–8.

1981　*Cyclotron furcatocostatum* n. sp. – Gründel *in* Gründel & Buchholz, p. 64, pl. II, figs. 11, 12.

1986a　*Cyclotron furcatocostatum* Gruendel, 1981 – Kempf, p. 203.

1986b　*Cyclotron furcatocostatum* Gruendel, 1981 – Kempf, p. 233.

1987　*Cyclotron furcatocostatum* Gruendel, 1981 – Kempf, p. 710.

1993b　*Cyclotron furcatocostatum* Gründel *in* Gründel & Buchholz, 1981 – Hinz-Schallreuter, p. 333.

1993c　*Cyclotron furcatocostatum* Gründel *in* Gründel & Buchholz, 1981 – Hinz-Schallreuter, pp. 393, 408.

1994b　*C. furcatocostatum* [Gründel *in* Gründel & Buchholz (1981)] – Williams *et al.*, pp. 128, 129.

1996c　*Cyclotron furcatocostatum* – Hinz-Schallreuter, p. 28.

Cyclotronidae Gründel *in* Gründel & Buchholz, 1981

1981　Cyclotronidae n. f. – Gründel *in* Gründel & Buchholz, p. 64.

1993c　Cyclotronidae Gründel *in* Gründel & Buchholz, 1981 – Hinz-Schallreuter, p. 402 (as synonymous with Vestrogothiidae).

1998　Cyclotronidae Gründel *in* Gründel & Buchholz, 1981 – Hinz-Schallreuter, pp. 116, 118 [as synonymous, with Vestrogothiinae *sensu* Hinz-Schallreuter (1998)].

Cyclotron lapworthi (Groom, 1902) Rushton, 1969 (type locality and horizon: Upper Cambrian, Zone 1, Malvern Hills, England)

1902　*Polyphyma lapworthi* sp. nov. – Groom, pp. 83, 84, 87, 88, pl. 3, figs. 1–8.

1923　*Polyphyma lapworthi* – Poulsen, p. 52.

1931　*Polyphyma lapworthi* Groom – Kummerow, p. 254, fig. 16.

1931　*Polyphyma lapworthi* Groom – Ulrich & Bassler, pp. 11, 65, 66, 68, 129, pl. 8, figs. 26, 27.

1931　*Polyphyma marginata*, new species – Ulrich & Bassler, pp. 11, 66, 67, 68, 129, pl. 8, figs. 28, 29.

1934　*Polyphyma lapworthi* Groom, 1902 – van Straelen & Schmitz, pp. 197, 211, 229, 237, 245.

1961　*P. lapworthi* – Sylvester-Bradley *in* Moore, p. Q103, fig. 39.2.

1968　"*Polyphyma*" *lapworthi* Groom – Allen & Rushton, p. 39.

1969　*Cyclotron lapworthi* (Groom) – Rushton, p. 216.

1969　*C. marginatum* (Ulrich & Bassler) – Rushton, p. 216.

1971　*Cyclotron* [*Polyphyma*] *lapworthi* (Groom) – Earp & Hains, p. 34.

1972　*Cyclotron lapworthi* (Groom) – Taylor & Rushton, pp. 13, 18, pls. 2, 4 (borehole records).

1973　*Polyphyma lapworthi* Groom, 1902 – Schrank, pp. 89, 90 (as type species of *Cyclotron*)

1973　*Cyclotron marginatum* (Ulrich & Bassler) [1931] – Schrank, p. 90.

1974　*Cyclotron* [*Polyphyma*] *lapworthi* (Groom) – Rushton, p. 104.

1978　*Cyclotron lapworthi* (Groom, 1902) – Rushton, p. 278.

1978　*C. marginatum* (Ulrich & Bassler, 1931) – Rushton, p. 278.

1981　*Cyclotron lapworthi* – Gründel, p. 65.

1986a　*Cyclotron lapworthi* (Groom, 1902) Rushton, 1969 – Kempf, p. 203.

1986a　*Cyclotron marginatum* (Ulrich & Bassler, 1931) Rushton, 1969 – Kempf, p. 203.

1986a　*Polyphyma lapworthy* Groom, 1902 – Kempf, p. 604 (sic!).

1986a　*Polyphyma marginata* Ulrich & Bassler, 1931 – Kempf, p. 604.

1986b　*Cyclotron lapworthi* (Groom, 1902) Rushton, 1969 – Kempf, p. 322.

1986b　*Cyclotron marginatum* (Ulrich & Bassler, 1931) Rushton, 1969 – Kempf, p. 356.

1986b　*Polyphyma lapworthy* Groom, 1902 – Kempf, p. 322 (sic!).

1986b　*Polyphyma marginata* Ulrich & Bassler, 1931 – Kempf, p. 356.

1987　*Polyphyma lapworthy* Groom, 1902 – Kempf, p. 120 (sic!).

1987　*Polyphyma marginata* Ulrich & Bassler, 1931 – Kempf, p. 168.

1987　*Cyclotron lapworthi* (Groom, 1902) Rushton, 1969 – Kempf, p. 523.

1987　*Cyclotron marginatum* (Ulrich & Bassler, 1931) Rushton, 1969 – Kempf, p. 523.

1989　*Polyphyma lapworthi* Groom – Müller, p. 54, fig. 46b.

1993b　*Cyclotron lapworthi* (Groom, 1902) – Hinz-Schallreuter, p. 332, fig. 1D.

1993c　*Polyphyma lapworthi* Groom, 1902 – Hinz-Schallreuter, p. 408.

1994b　*Cyclotron lapworthi* (Groom, 1902) – Williams *et al.*, pp. 123–126, 128, figs. 1, 2, 6a–o, r

1994b　*Polyphyma marginata* Ulrich & Bassler, 1931 – Williams *et al.*, p. 128, (= *C. lapworthi*)

1997　*Cyclotron lapworthi* (Groom) – Siveter & Williams, p. 59, pl. 8, figs. 8, 9.

1998　*Polyphyma marginata* Ulrich & Bassler, 1931 – Hinz-Schallreuter, p. 104.

1998　*Polyphyma lapworthi* (Groom, 1902) – Hinz-Schallreuter, p. 118.

1998　*Cyclotron lapworthi* (Groom) – Williams & Siveter, pp. 10, 32, pl. 6, figs. 2–5, 6(?).

Cyclotron cf. *lapworthi* Williams, Siveter, Rushton & Berg-Madsen, 1994 (locality and horizon: Upper Cambrian, Zone 2, England, Newfoundland, Canada)

1994b　*Cyclotron* cf. *lapworthi* – Williams *et al.*, p. 129, fig. 6p.

Cyclotron nodomarginatum Schrank, 1973 (type locality and horizon: Upper Cambrian, Zone 2, Niederfinow, Kreis Frankfurt/Oder, Germany)

1973　*Cyclotron nodomarginatum* n. sp. – Schrank, pp. 89, 92, pl. 3, figs. 2, 2a.

1978　*C. nodomarginatum* Schrank – Rushton, p. 278.

1981　*Cyclotron nodomarginatum* – Gründel, p. 65, pl. II, figs. 13, 14.

1986a　*Cyclotron nodomarginatum* Schrank, 1973 – Kempf, p. 203.

1986b　*Cyclotron nodomarginatum* Schrank, 1973 – Kempf, p. 397.

1987　*Cyclotron nodomarginatum* Schrank, 1973 – Kempf, p. 593.

1993b　*Cyclotron nodomarginatum* Schrank, 1973 – Hinz-Schallreuter, p. 333.

1993c　*Cyclotron nodomarginatum* Schrank, 1973 – Hinz-Schallreuter, p. 408.

1994b　*C. nodomarginatum* Schrank, 1973 – Williams *et al.*, p. 128.

Cyclotron poulseni Gründel *in* Gründel & Buchholz, 1981 (type locality and horizon: Upper Cambrian, Zone 2, Oleå, Isle of Bornholm, Denmark)

1923　*Polyphyma angelini* Barrande – Poulsen, p. 52, fig. 19.

?1978　*Cyclotron* sp. – Rushton, p. 278, pl. 26, fig. 15.

1981　*Cyclotron poulseni* n. sp. – Gründel *in* Gründel & Buchholz, p. 65, pl. III, figs. 1, 2.

1986a　*Cyclotron poulseni* Gruendel, 1981 – Kempf, p. 203.

1986b　*Cyclotron poulseni* Gruendel, 1981 – Kempf, p. 462.

1987　*Cyclotron poulseni* Gruendel, 1981 – Kempf, p. 710.

1993b　*Cyclotron poulseni* Gründel *in* Gründel & Buchholz, 1981 – Hinz-Schallreuter, p. 333.

1993c　*Cyclotron poulseni* Gründel *in* Gründel & Buchholz, 1981 – Hinz-Schallreuter, p. 408.

1994b　*C. poulseni* [Gründel *in* Gründel & Buchholz (1981)] – Williams *et al.*, p. 128.

1998　*Polyphyma angelini* – Hinz-Schallreuter, p. 104 [referred to Poulsen (1923)].

1998　*Cyclotron poulseni* n. sp. – Hinz-Schallreuter, p. 104 [referred to Gründel (1981)].

Cyclotron ventrocurvatum Gründel *in* Gründel & Buchholz, 1981 (type locality and horizon: Upper Cambrian, Zone 2, erratic boulders of Dwasieden, Isle of Rügen, Germany)

1981　*Cyclotron ventrocurvatum* n. sp. – Gründel *in* Gründel & Buchholz, p. 66, pl. III, fig. 6.

1986a　*Cyclotron ventrocurvatum* Gruendel, 1981 – Kempf, p. 203.

1986b　*Cyclotron ventrocurvatum* Gruendel, 1981 – Kempf, p. 634.

1987　*Cyclotron ventrocurvatum* Gruendel, 1981 – Kempf, p. 710.

1993b　*Cyclotron ventrocurvatum* Gründel *in* Gründel & Buchholz, 1981 – Hinz-Schallreuter, p. 333.

1993c　*Cyclotron ventrocurvatum* Gründel *in* Gründel & Buchholz, 1981 – Hinz-Schallreuter, pp. 393, 408, 409.

1994b　*C. ventrocurvatum* [Gründel *in* Gründel & Buchholz (1981)] – Williams *et al.*, pp. 128, 129.

Cyclotron n. sp. 1 Gründel *in* Gründel & Buchholz, 1981 (locality and horizon: ?Upper Cambrian, ?Zone 2, northeastern Germany)

1981　*Cyclotron* n. sp. 1 – Gründel *in* Gründel & Buchholz, p. 66, pl. III, fig. 4.

Cyclotron n. sp. 2 Gründel *in* Gründel & Buchholz, 1981 (locality and horizon: Upper Cambrian, Zone 1, northeastern Germany)

1981　*Cyclotron* n. sp. 2 – Gründel *in* Gründel & Buchholz, p. 66, pl. III, fig. 5.

Cyclotron sp. cf. *poulseni* n. sp. Gründel *in* Gründel & Buchholz, 1981 (locality and horizon: Middle Cambrian, Zone B2, northeastern Germany)

1981　*Cyclotron* sp. cf. *poulseni* n. sp. – Gründel *in* Gründel & Buchholz, pl. III, fig. 3.

1993b　*C.* sp., cf. *poulseni* [Gründel *in* Gründel & Buchholz (1981)] – Hinz-Schallreuter, p. 334.

Cyclotron sp. C Williams & Siveter, 1998 (locality and horizon: Upper Cambrian, Zone 1, Warwickshire, England)

1972　*C.* aff. *angelini* (Barrande) – Taylor & Rushton, pp. 18, 25, pl. 4 (borehole record).

1978　*Cyclotron* sp. – Rushton, p. 278, pl. 26, fig. 15.

1998　*Cyclotron* sp. C – Williams & Siveter, p. 33, pl. 6, fig. 1.

Cyclotron sp. D Williams & Siveter, 1998 (locality and horizon: Upper Cambrian, Zone 1, Pembrokeshire, South Wales)

1998   *Cyclotron* sp. D – Williams & Siveter, p. 34, pl. 5, fig. 13.

Cyclotron sp. Clarkson, Ahlberg & Taylor, 1998 (locality and horizon: Upper Cambrian, Zone 2, Andrarum, Skåne, Sweden)
1998a   *Cyclotron* sp. – Clarkson *et al.*, fig. 3J.

Cyclotron sp. nov. Taylor & Rushton, 1972 (locality and horizon: Upper Cambrian, Zone 5, Warwickshire, England)
1972   *C.* sp. nov. – Taylor & Rushton, p. 25.

Dabashanella Huo, Shu & Fu *in* Huo, Shu, Zhang, Cui & Tong, 1983 (type species: *Dabashanella hemicyclica* Huo, Shu & Fu *in* Huo, Shu, Zhang, Cui & Tong, 1983 by original designation)
1983   *Dabashanella* Huo, Shu & Fu gen. nov. – Huo, Shu & Fu *in* Huo *et al.*, p. 68.
1986   *Dabashanella* – Huo *et al.*, p. 25.
1987   *Dabashanella* Huo, Shu & Fu, 1983 – Tong, p. 433.
1987   *Dabashanella* Huo, Shu & Fu – Zhang, p. 16.
1989a   *Dabashanella* Huo, Shu & Fu, 1983 – Zhao, p. 472.
1989b   *Dabashanella* – Zhao, p. 412.
1990   *Dabashanella* Huo, Shu & Fu, 1983 – Abushik *et al.*, p. 46.
1990   *Xianjiangella* Jiang & Xiao, 1985 – Abushik *et al.*, p. 46 (synonymised with *Dabashanella*) (sic!).
1990   *Dabashanella* Huo, Shu & Fu, 1983 – Melnikova & Mambetov, pp. 57, 58, 60.
1990   *Xinjangella* – Melnikova & Mambetov, pp. 57, 58, 60 (synonymised with *Dabashanella*).
1990a   *Dabashanella* Huo, Shu & Fu – Shu, pp. 67, 76.
1990b   *Dabashanella* Huo, Shu & Fu 1983 – Shu, p. 323.
1991   *Dabashanella* Huo, Shu & Fu, 1983 – Huo *et al.*, pp. 178, 212.
1991   *Dabashanella* Huo, Shu & Fu, 1983 – Melnikova & Mambetov, pp. 56, 57, 59.
1991   *Xinjangella* – Melnikova & Mambetov, pp. 56, 57, 59 (synonymised with *Dabashanella*) (sic!).
1992c   *Dabashanella* Shu – Hinz, p. 7.
1993b   *Dabashanella* – Hinz-Schallreuter, p. 344.
1996   *Dabashanella* – Hinz-Schallreuter & Koppka, pp. 27, 28, 36, 38, 40, 41.
1998   *Dabashanella* Huo, Cui & Fu, 1983 – Williams & Siveter, p. 35 (sic!).

Dabashanella erecta Shu, 1990 (type locality and horizon: Lower Cambrian, Shaanxi, China)
1990a   *Dabashanella erecta* sp. nov. – Shu, pp. 68, 92, pl. 10, figs. 11, 12.

Dabashanella hemicyclica Huo, Shu & Fu *in* Huo, Shu,

Zhang, Cui & Tong, 1983 (type locality and horizon: Lower Cambrian, Shaanxi, China)
1973   *Indiana sipa* – Fleming, p. 6, pl. 1, figs. 3.
1980   *Indiana? sipa* Fleming 1973 – Jones & McKenzie, pp. 217, 218, 220.
1983   *Dabashanella hemicyclica* Huo, Shu & Fu sp. nov. – Huo, Shu & Fu *in* Huo *et al.*, 68, 69, 75, pl. 5, figs. 18–20, text-fig. II-26.
1985   *Dabashanella hemicyclica* Huo, Shu & Fu 1983 – Huo & Shu, p. 175, pl. 28, figs. 18–20
1986   *Dabashanella hemicyclica* – Huo *et al.* text-figs. 3–5.
1986a   *Indiana sipa* Fleming, 1973 – Kempf, p. 3.
1986b   *Indiana sipa* Fleming, 1973 – Kempf, p. 9.
1986   *Dabashanella conica* sp. nov. – Zhao & Xiao *in* Xiao & Zhao, p. 86.
1987   *Indiana sipa* Fleming, 1973 – Kempf, p. 3.
1987   *Dabashanella hemicyclica* Huo, Shu & Fu 1983 – Tong, p. 434, pl. 1, figs. 11–16.
1987   *Dabashanella hemicyclica* – Zhang, p. 1.
1987   *Indiana sipa* Fleming – Zhang, p. 1.
1989a   *Dabashanella hemicyclica* Huo, Shu & Fu 1983 – Zhao, pl. 1, figs. 1, 2.
1989b   *Dabashanella hemicyclica* Huo, Shu & Fu – Zhao, pp. 411, 412, table 1, pl. 1, figs. 5, 16.
1989b   *Dabashanella conica* Zhao & Xiao – Zhao, p. 411, table 1 (synonymised with *Dabashanella hemicyclica*).
1989b   *Mononotella lentiformis* – Zhao, table 1 synonymised with *Dabashanella hemicyclica*)
1989   *Dabashanella hemicyclica* Huo, Shu & Fu 1983 – Zhao & Tong, p. 15.
1990   *D. hemicyclica* Huo, Shu & Fu, 1983 – Melnikova & Mambetov, pp. 57, 60.
1990a   *Dabashanella hemicyclica* Huo, Shu & Fu 1983 – Shu, pp. 67, 68, 75, 92, pl. 11, figs. 9–
1990b   *Dabashanella hemicyclica* Huo, Shu & Fu *in* Huo *et al.* 1983 – Shu, p. 328, pl. 3, figs. 1–46.
1991   *Dabashanella hemicyclica* Huo, Shu & Fu 1983 – Huo *et al.*, p. 179, pl. 38, figs. 13–16.
1991   *D. hemicyclica* Huo, Shu & Fu, 1983 – Melnikova & Mambetov, pp. 56, 59.
1996   *Dabashanella hemicyclica* Huo & Shu *in* Huo *et al.*, 1983 – Hinz-Schallreuter & Koppka, p. 36.

Dabashanella qilianensis Cui & Wang, 1991 *in* Huo, Shu & Cui, 1991 (type locality and horizon: Middle Cambrian, Maojiagou in Datong, Qinghai, China)
1991   *Dabashanella qilianensis* sp. nov. – Cui & Wang *in* Huo *et al.*, pp. 180, 181, 212, p. 45, figs. 10–12; text-fig. 8-101.

Dabashanella qinghaiensis Cui & Wang *in* Huo, Shu & Cui, 1991 (type locality and horizon: Middle Cambrian, Maojiagou in Datong, Qinghai, China)

1991 *Dabashanella qinghaiensis* sp. nov. – Cui & Wang in Huo *et al.*, pp. 179, 212, pl. 45, figs. 7–9; text-fig. 8-100.

Dabashanella reniformis Jiang & Xiao, 1985 (type locality and horizon: Lower Cambrian, Zhenba, Shaanxi, China)

1985 *Dabashanella reniformis* n. sp. – Jiang & Xiao, p. 250.

1985 *Xinjiangella reniformis* sp. nov. – Huo & Shu, p. 185, pl. 1, figs. 7, 8.

1987 *Paraphaseolella renoformis* (Jiang & Xiao) comb. nov. – Tong, p. 434.

1989a *Dabashanella reniformis* Jiang & Xiao, 1985 – Zhao, pl. 1, fig. 6.

1989b *Xinjiangella reniformis* Jiang & Xiao – Zhao, p. 411, table 1; pl. 1, figs. 10–12.

1989b *Selliformella reniformis* (Jiang & Xiao) – Zhao, p. 411, table 1 (synonymised with *Dabashanella reniformis*).

1990 *Xinjangella reniformis* – Melnikova & Mambetov, pp. 57, 61 (synonymised with *Dabashanella retroswinga*) (sic!).

1991 *Xinjangella reniformis* – Melnikova & Mambetov, pp. 56, 60 (synonymised with *Dabashanella retroswinga*) (sic!).

Dabashanella retroswinga Huo, Shu & Fu *in* Huo, Shu, Zhang, Cui & Tong, 1983 (type locality and horizon: Lower Cambrian, Shaanxi, China)

1983 *Dabashanella retroswinga* Huo, Shu & Fu sp. nov. – Huo, Shu & Fu *in* Huo *et al.*, pp. 68, 75, pl. 5, figs. 14–17; text-fig. II-27.

1985 *Dabashanella retroswinga* – Huo & Shu, p. 175, pl. 28, figs. 13–15.

1985 *Xinjiangella venustois* n. sp. – Jiang & Xiao *in* Huo & Shu, pp. 184, 185, pl. 36, figs. 1–4.

1985 *Xinjiangella venustois* Jiang & Xiao – Huo & Shu, p. 184, pl. 1, figs. 6, 9.

1987 *Dabashanella retroswinga* Huo, Shu & Fu – Tong, p. 433, pl. 1, figs. 3–10; pl. 2, figs. 1–18.

1987 *D. retroswinga* – Zhang, p. 1.

1989 *Dabashanella retroswinga* Huo *et al.* – Huo & Cui, p. 80, pl. 2, figs. 11–13.

1989b *Xinjiangella venustois* Jiang & Xiao – Zhao, p. 411, table 1 (synonymised with *D. hemicyclica*).

1989b *Xinjiangella dorsonodis* – Zhao, table 1 (synonymised with *D. hemicyclica*).

1989b *Palaeomeishucunella xibeiensis* – Zhao, table 1 (synonymised with *Dabashanella hemicyclica*).

1989 *Dabashanella retroswinga* – Zhao & Tong, pp. 13–16, pl. 16, figs. 1–3.

1990a *Dabashanella retroswinga* Huo, Shu & Fu – Shu, pp. 67, 70, 93, pl. 14, fig. 10.

1990 *D. retroswinga* – Abushik *et al.*, p. 46.

1990 *Dabashanella retroswinga* Huo, Shu & Fu, 1983 – Melnikova & Mambetov, pp. 57, 60, 61, pl. 7, fig. 2a, b.

1990 *Xinjangella venustois* – Melnikova & Mambetov, pp. 57, 61 (synonymised with *Dabashanella retroswinga*) (sic!).

1991 *Dabashanella retroswinga* – Huo *et al.*, p. 178, pl. 38, figs. 1–12; text-figs. 4–7.

1991 *Dabashanella retroswinga* Huo, Shu & Fu, 1983 – Melnikova & Mambetov, pp. 56, 59, 60, pl. 7, figs. 1–8.

1991 *Xinjangella venustois* – Melnikova & Mambetov, pp. 56, 60 (synonymised with *Dabashanella retroswinga*) (sic!).

1993b *Dabashanella retroswinga* – Hinz-Schallreuter, p. 346.

1996 *Dabashanella retroswinga* Huo & Shu *in* Huo *et al.*, 1983 – Hinz-Schallreuter & Koppka, pp. 34, 36, 41.

1997 *Dabashanella retroswinga* – Melnikova *et al.*, pp. 181, 187, pl. 1, fig. 9, text-fig. 3.

Dabashanella squamiformis (Xiao & Zhao, 1986) Zhao, 1989 (type locality and horizon: Lower Cambrian, Xichuan, Henan, Shaanxi, China)

1986 *? squamiformis* n. sp. – Xiao & Zhao, p. 80.

1989a *Dabashanella squamiformis* (Xiao & Zhao, 1986) – Zhao, pl. 1, fig. 7.

Dabashanella tenuis Shu, 1990 (type locality and horizon: Lower Cambrian, China)

1990a *Dabashanella tenuis* sp. nov. – Shu, pp. 67, 92, pl. 11, figs. 1–8, 25.

Dabashanellidae Zhao, 1989

1989b Dabashanellidae fam. nov. – Zhao, p. 472.

1990a Dabashanellina suborder new – Shu, pp. 40, 67.

1990a Dabashanellidae fam. nov. – Shu, p. 67.

1990b Dabashanellina, n. suborder – Shu, p. 323.

1990 Dabashanellidae – Malz, p. 331 [referred to Shu (1990b)].

1990 Dabashanellina – Malz, p. 331 [referred to Shu (1990b)].

1990 Dabashanellida, order nov. – Melnikova *in* Abushik *et al.*, p. 46.

1990 Dabashanellidae, fam. nov. – Melnikova *in* Abushik *et al.*, p. 46.

1990 Dabashanellida Melnikova, ordo nov. – Melnikova *in* Melnikova & Mambetov, p. 60.

1990 Dabashanellidae Melnikova, fam. nov. – Melnikova *in* Melnikova & Mambetov, p. 60.

1991 Dabashanellidae Huo, Cui, Wang & Zhang (fam. nov.) – Huo *et al.*, pp. 178, 211.

1991 Dabashanellidae Huo, Cui, Wang & Zhang (fam. nov.) – Huo *et al.*, pp. 178, 211.

1991   Dabashanellida Melnikova, ordo nov. –
        Melnikova *in* Melnikova & Mambetov, p. 59.
1991   Dabashanellidae Melnikova, fam. nov. –
        Melnikova *in* Melnikova & Mambetov, p. 59.
1992c   ("Dabashanellids" – Hinz, p. 7.)
1996   Dabashanellina Shu, 1990 – Hinz-Schallreuter
        & Koppka, pp. 27, 34, 36, 38.
1996   ("dabashanellid phosphatocopids" – Hou
        et al., p. 1141.)
1997   Dabashanellina – Hinz-Schallreuter, p. 13.
1999   (dabashanellids – McKenzie *et al.*, pp. 461,
        463.)

Dielymella Ulrich & Bassler, 1931 (assignment to
Phosphatocopina very uncertain)
1931   *Dielymella*, n. gen. – Ulrich & Bassler, pp. 9,
        85, 86, 128.
1945   *Dielymella* Ulrich & Bassler, 1931 – Resser,
        p. 216.
1960   *Dielymella* – Lochman & Hu, p. 826.
1961   *Dielymella* Ulrich & Bassler, 1931 – Sylvester-
        Bradley *in* Moore, p. Q103, fig. 39.5.
1974   *Dielymella* Ulrich & Bassler, 1931 – Kozur,
        p. 826 (referred to Indianidae Ulrich &
        Bassler, 1931).
1980   *Dielymella* Ulrich & Bassler, 1931 – Jones &
        McKenzie, p. 217.
1983   *Dielymella* Ulrich & Bassler – Huo *et al.*, p. 65.
1986a   *Dielymella* Ulrich & Bassler, 1931 – Kempf,
        p. 316.
1986b   *Dielymella* Ulrich & Bassler, 1931 – Kempf,
        p. 670.
1987   *Dielymella* Ulrich & Bassler, 1931 – Kempf,
        p. 168.
1987   *Dielymella* – Zhang, p. 16.
1990   *Dielymella* Ulrich & Bassler, 1931 –
        Melnikova, p. 175.
1997   *Dielymella* Ulrich & Bassler – Siveter &
        Williams, p. 62.

Dielymella? brevis Ulrich & Bassler, 1931 (type locality
and horizon: Lower Cambrian, Sunset Hill, Salisbury,
Vermont, USA)
1890   *Nothozoe? vermontana* Whitfield – Walcott,
        p. 628, pl. 80, fig. 4a, b, ?fig. 4.
1931   *Dielymella brevis* new species – Ulrich &
        Bassler, p. 89, pl. 10, figs. 12, 13.
1986a   *Dielymella brevis* Ulrich & Bassler, 1931 –
        Kempf, p. 316.
1986b   *Dielymella brevis* Ulrich & Bassler, 1931 –
        Kempf, p. 103.
1987   *Dielymella brevis* Ulrich & Bassler, 1931 –
        Kempf, p. 168.
1996c   *Dielymella brevis* – Hinz-Schallreuter, p. 28.
1997   *Dielymella brevis* Ulrich & Bassler – Siveter &
        Williams, p. 63, pl. 9, fig. 9.

Dielymella? dubia Jones & McKenzie, 1980 (type locality
and horizon: Middle Cambrian, Georgina Basin,
Queensland, Australia)
1980   *Dielymella? dubia* sp. nov. – Jones & McKenzie,
        p. 217, fig. 8E–I.
1986a   *Dielymella? dubia* Jones & McKenzie, 1980 –
        Kempf, p. 316.
1986b   *Dielymella? dubia* Jones & McKenzie, 1980 –
        Kempf, p. 191.
1987   *Dielymella? dubia* Jones & McKenzie, 1980 –
        Kempf, p. 697.
1987   *Dielymella? dubia* Jones & McKenzie – Zhang,
        pp. 8, 16, 18.

Dielymella recticardinalis Ulrich & Bassler, 1931 (locality
and horizon: Lower Cambrian, Nankoweap Valley,
Grand Canyon, Colorado, Arizona, USA)
1931   *Dielymella recticardinalis*, new species – Ulrich
        & Bassler, pp. 9, 85–89, pl. 10, figs. 3–7.
1931   *Dielymella recticardinalis angustata*, new variety
        – Ulrich & Bassler, p. 87, pl. 10, fig. 8.
1931   *Dielymella nasuta*, new species – Ulrich &
        Bassler, p. 88, pl. 10, figs. 10, 11.
1931   *Dielymella appressa*, new species – Ulrich &
        Bassler, p. 88, pl. 10, fig. 9.
1931   *Dielymella dorsalis*, new species – Ulrich &
        Bassler, p. 89, pl. 10, fig. 1.
1945   *Dielymella appressa* Ulrich & Bassler – Resser,
        p. 216, pl. 27, figs. 10, 21.
1945   *Dielymella dorsalis* Ulrich & Bassler – Resser,
        p. 216, pl. 26, fig. 20.
1945   *Dielymella nasuta* Ulrich & Bassler – Resser,
        p. 216, pl. 27, figs. 11, 12, 23.
1945   *Dielymella recticardinalis* Ulrich & Bassler –
        Resser, p. 216, pl. 27, figs. 13, 15, 19, 20.
1945   *Dielymella recticardinalis angustata* Ulrich &
        Bassler – Resser, p. 216, pl. 27, fig. 24.
1961   *Dielymella recticardinalis* Ulrich & Bassler –
        Sylvester-Bradley, p. Q103, fig. 39.5.
1975   *Dielymella recticardinalis* Ulrich & Bassler, 1931
        – Li, p. 44, pl. III, fig. 14 (sic!).
1980   *D. [Dielymella] recticardinalis* Ulrich & Bassler,
        1931 – Jones & McKenzie, p. 217.
1986a   *Dielymella appressa* Ulrich & Bassler, 1931 –
        Kempf, p. 316.
1986a   *Dielymella dorsalis* Ulrich & Bassler, 1931 –
        Kempf, p. 316.
1986a   *Dielymella nasuta* Ulrich & Bassler, 1931 –
        Kempf, p. 316.
1986a   *Dielymella recticardinalis* Ulrich & Bassler, 1931
        – Kempf, p. 316.
1986a   *Dielymella recticardinalis angustata* Ulrich &
        Bassler, 1931 – Kempf, p. 316.
1986a   *Dielymella recticardinalis recticardina* Ulrich &
        Bassler, 1931 – Kempf, p. 316 (name too long).

1986b *Dielymella appressa* Ulrich & Bassler, 1931 –
Kempf, p. 59.

1986b *Dielymella dorsalis* Ulrich & Bassler, 1931 –
Kempf, p. 188.

1986b *Dielymella nasuta* Ulrich & Bassler, 1931 –
Kempf, p. 391.

1986b *Dielymella recticardinalis* Ulrich & Bassler, 1931
– Kempf, p. 494.

1986b *Dielymella recticardinalis angustata* Ulrich &
Bassler, 1931 – Kempf, p. 494.

1986b *Dielymella recticardinalis recticardina* Ulrich &
Bassler, 1931 – Kempf, p. 494 (name too long).

1987 *Dielymella appressa* Ulrich & Bassler, 1931 –
Kempf, p. 168.

1987 *Dielymella dorsalis* Ulrich & Bassler, 1931 –
Kempf, p. 168.

1987 *Dielymella nasuta* Ulrich & Bassler, 1931 –
Kempf, p. 168.

1987 *Dielymella recticardinalis* Ulrich & Bassler, 1931
– Kempf, p. 168.

1987 *Dielymella recticardinalis angustata* Ulrich &
Bassler, 1931 – Kempf, p. 168.

1987 *Dielymella recticardinalis recticardina* Ulrich &
Bassler, 1931 – Kempf, p. 168 (name too long).

1990 *Dielymella recticardinalis* – Melnikova, p. 176.

1997 *Dielymella recticardinalis* Ulrich & Bassler –
Siveter & Williams, p. 62, pl. 9, figs. 1–8.

"*Dielymella*" sp. Melnikova, Siveter & Williams, 1997
(locality and horizon: Late Cambrian, area of the
former Soviet Union)

1997 "*Dielymella*" sp. – Melnikova *et al.*, text-fig. 3.

Dielymella? sp. Melnikova, 1990 (locality and horizon:
Late Cambrian, northeastern central Kazakhstan)

1990 *Dabashanella*? sp. – Melnikova, p. 171.

1990 *Dielymella*? sp. – Melnikova, p. 175, pl. 31,
fig. 4.

1997 "undetermined   phosphatocopid"   (=
Dielymella sp. of Melnikova, 1990) –
Melnikova *et al.*, p. 185, pl. 3, fig. 3.

Eohesslandona Shu, 1990 (type species: *Eohesslandona
usualis* Shu, 1990 by original designation)

1990a *Eohesslandona* gen. nov. – Shu, p. 65.

1990b *Eohesslandona* Shu 1987 (unpublished manuscript) – Shu, p. 323.

Eohesslandona? sp. Hinz, 1993 (locality and horizon:
Middle Cambrian, Mount Murray, Queensland,
Australia)

1993 *Eohesslandona*? sp. – Hinz, fig. 4D.

Eohesslandonà usualis Shu, 1990 (type locality and
horizon: Cambrian, China)

1990a *Eohesslandona usualis* sp. nov. – Shu, pp. 65,
66, 92, pl. 13, figs. 1–3.

1990b *Eohesslandona usualis* Shu 1987 (unpublished
manuscript) – Shu, p. 328, pl. 3 figs. 42, 43.

1990 *Eohesslandona usualis* sp. nov. – Malz, table 1
[referred to Shu (1990b)].

Epactridion Bengtson *in* Bengtson, Conway Morris,
Cooper, Jell & Runnegar, 1990 (type species:
Epactridion portax Bengtson *in* Bengtson, Conway
Morris, Cooper, Jell & Runnegar, 1990; assignment
uncertain)

1990 *Epactridion* Bengston, gen. nov. (in press) –
Abushik *et al.*, p. 45 (sic!).

1990 *Epactridion* Bengtson, gen. nov. – Bengtson *in*
Bengtson *et al.*, pp. 322, 323.

1991a *Epactridion* Bengtson *in* Bengtson *et al.* 1990 –
Hinz, p. 233.

Epactridion portax Bengtson *in* Bengtson, Conway
Morris, Cooper, Jell & Runnegar, 1990 (type locality
and horizon: Lower Cambrian, South Australia)

1990 *Epactridion portax* Bengtson, sp. nov. –
Bengtson *in* Bengtson *et al.*, p. 323,
figs. 204–206.

Euphosphatocopina new name

This taxon is defined herein by a set of autapomorphies
(see p. 165)

Falites Müller, 1964 (type species: *Falites fala* Müller,
1964 by original designation, see p. 131)

1964a *Falites* n. g. – Müller, p. 25.

1965 *Falites* Müller – Adamczak, p. 32.

1972 *Falites* Mueller – Taylor & Rushton, pp. 13,
18, 25.

1974 *Falites* Müller, 1964 – Kozur, p. 826.

1978 *Falites* Müller, 1964 – Rushton, p. 276.

1980 *Falites* Müller, 1964 – Landing, p. 757.

1981a *Falites* – Müller, p. 147.

1983 *Falites* – Briggs, pp. 9, 10.

1983 *Falites* – Müller, p. 94.

1986 *Falites* – Huo *et al.*, p. 23.

1986 *Falites* K.J. Müller, 1964 – Schram, p. 415.

1986a *Falites* Mueller, 1964 – Kempf, p. 354.

1986b *Falites* Mueller, 1964 – Kempf, p. 673.

1987 *Falites* Mueller, 1964 – Kempf, p. 436.

1987 *Falites* – Zhang, pp. 5, 9.

1990 *Falites* – Bengtson *in* Bengtson *et al.*, p. 323.

1990a *Falites* Müller, 1964 – Shu, pp. 66, 77.

1990b *Falites* Müller 1964 – Shu, pp. 318, 323.

1991 *Falites* Müller, 1964 – Huo *et al.*, p. 181 (as
Falies), p. 212.

1993c *Falites* Müller, 1964 – Hinz-Schallreuter,
pp. 386, 395, 399–403.

1995 *Falites* – Siveter *et al.*, p. 416.

1996a *Falites* Müller, 1964 – Hinz-Schallreuter, p. 85.

1996b *Falites* Müller, 1964 – Hinz-Schallreuter,
pp. 89, 91.

1998 *Falites* Müller, 1964 – Hinz-Schallreuter, pp. 104, 106–108, 112, 115.

Falites angustiduplicatus Müller, 1964 (type locality and horizon: Upper Cambrian, Zone 5d–f, Trolmen, Kinnekulle, Sweden)

1964a *Falites angustiduplicata* n. sp. – Müller, p. 28, pl. 4, figs. 1–4, text-fig. 1.

1967 *Falites angustiduplicata* Müller (1964) – Bate et al., p. 538.

1974 *F. angustiduplicata* – Martinsson, p. 212.

1978 *F. angustiduplicatus* Müller – Rushton, p. 277, text-fig. 2.

1979a *Falites angustiduplicata* Müller, 1964 – Müller, figs. 32, 34E.

1986a *Falites angustiduplicatus* Mueller, 1964 – Kempf, p. 354.

1986b *Falites angustiduplicatus* Mueller, 1964 – Kempf, p. 53.

1987 *Falites angustiduplicatus* Mueller, 1964 – Kempf, p. 436.

1993c *Falites angustiduplicatus* Müller, 1964 – Hinz-Schallreuter, p. 400.

1996b *F. [Falites] angustiduplicatus* [Müller, 1964] – Hinz-Schallreuter, p. 91.

1998 *Falites angustiduplicatus* Müller, 1964 – Williams & Siveter, pp. 28, 29.

1999 *Falites angustiduplicata* Müller, 1964 – McKenzie et al., p. 344 (sic!).

Falites aff. *angustiduplicatus* Shu, 1990 (locality and horizon: Cambrian, China)

1990a *Falites* aff. *angustiduplicata* Müller, 1964 – Shu, pp. 66, 92, pl. 12, figs. 4, 5.

1990b *Falites* aff. *angustiduplicata* Müller, 1964 – Shu, p. 328, pl. 3, fig. 38.

Falites cycloides Müller, 1964 (type locality and horizon: Upper Cambrian, Zone 5c, Stenåsen, Falbygden, Sweden)

1964a *Falites cycloides* n. sp. – Müller, p. 26, pl. 5, figs. 2–5.

1974 *F. cycloides* – Martinsson, p. 212.

1978 *F. cycloides* Müller – Rushton, p. 277.

1986a *Falites cycloides* Mueller, 1964 – Kempf, p. 355.

1986b *Falites cycloides* Mueller, 1964 – Kempf, p. 165.

1987 *Falites cycloides* Mueller, 1964 – Kempf, p. 436.

1993c *Falites cycloides* Müller, 1964 – Hinz-Schallreuter, p. 400.

1996b *F. [Falites] cycloides* [Müller, 1964] – Hinz-Schallreuter, p. 91.

1998 *Falites cycloides* – Hinz-Schallreuter, p. 114.

1998 *Falites cycloides* Müller, 1964 – Williams & Siveter, pp. 28, 29.

Falites fala Müller, 1964 (type locality and horizon: Upper Cambrian, Zone 5c, Stenåsen, Falbygden, Sweden)

*v 1964a *Falites fala* n. sp. – Müller, p. 25, pl. 3, figs. 3–10; pl. 5, fig. 6; text-fig. 2.

. 1965 *Falites fala* Müller – Adamczak, pp. 28, 29, pl. 1, figs. 4, 5a–c; text-fig. 1.

. 1972 *Falites* sp. – Taylor & Rushton, pp. 18, 25.

. 1974 *Falites fala* Müller – Martinsson, p. 211.

p. 1978 *Falites fala* Müller – Rushton, p. 277 [partim: specimens of Monks Park Formation non specimens from Outwoods Formation and text-fig. 2. (= *Hesslandona unisulcata*)].

. 1979a *Falites fala* Müller, 1964 – Müller, pp. 20, figs. 1, 10, 11, 21, 25.

. 1980 *Falites fala* Müller, 1964 – Landing, p. 7.

. 1981 *Falites fala* Müller – Gründel, pp. 63, 69, pl. 2, figs. 6, 7.

. 1982b *Falites fala* Müller – Müller, fig. 2.

. 1983 *Falites fala* Müller – McKenzie et al., 36, fig. 6.

. 1983 *Falites fala* Müller, 1964 – Reyment, fig. 1.

. 1986a *Falites fala* Mueller – Kempf, p. 355.

. 1986b *Falites fala* Mueller – Kempf, p. 216.

. 1987 *Falites fala* Mueller – Kempf, p. 436.

. 1987 *Falites fala* Müller – Tong, p. 433.

. 1987 *Falites fala* – Zhang, p. 5.

. 1989 *Falites fala* Müller, 1979 – Zhao & Tong, 15 [referred to Müller (1979a, fig. 10a–c)].

. 1991 *Falites fala* Müller – Huo et al., p. 181.

. 1993b *Falites fala* – Hinz-Schallreuter, p. 346.

. 1993c *Falites fala* Müller, 1964 – Hinz-Schallreuter, p. 400.

. 1996a *Falites fala* Müller, 1964 – Hinz-Schallreuter, p. 85.

. 1996b *Falites fala* Müller, 1964 – Hinz-Schallreuter, pp. 89, 91, pl. 23; text-figs. 1, 2.

· 1998 *Falites fala* Müller, 1964 – Hinz-Schallreuter, pp. 112, 114, 132.

. 1998 *Falites fala* Müller, 1964 – Williams & Siveter, pp. 28, 29, pl. 5, figs. 1–4.

. 1999 *Falites fala* Müller – McKenzie et al., 64.

· 1999 *Falites fala* Müller, 1964 – Whatley et al., p. 344.

Falites insula Hinz-Schallreuter, 1998 (type locality and horizon: Middle Cambrian, Isle of Bornholm)

1998 *Falites insula* n. sp. – Hinz-Schallreuter, pp. 103, 112, pl. 2; pl. 3, fig. 8; text-fig. 2; table 3.

Falites marsupiata Cui & Wang, 1991 in Huo, Shu & Cui, 1991 (type locality and horizon: Middle Cambrian, Maojiagou in Datong, Qinghai, China)

1991 *Falites marsupiata* sp. nov. – Cui & Wang in

Huo *et al.*, pp. 181, 212, pl. 45, figs. 14; text-fig. 8-102.

"*Falites*" *pateli* Landing, 1980 (locality and horizon: Upper Cambrian, New Brunswick, Canada)
1980   *Falites pateli* n. sp. – Landing, p. 758, fig. 4A–E.
1986a *Falites pateli* Landing, 1980 – Kempf, p. 355.
1986b *Falites pateli* Landing, 1980 – Kempf, p. 435.
1987   *Falites pateli* Landing, 1980 – Kempf, p. 697.
1993c ?*Falites pateli* Landing – Hinz-Schallreuter, p. 400.
1996b *F.* [*Falites*] *pateli* Landing, 1980 – Hinz-Schallreuter, p. 91.
1997   "*Falites*" *pateli* Landing – Siveter & Williams, p. 60.
1998   "*Falites*" *pateli* Landing, 1980 – Williams & Siveter, p. 29.

Falitidae Müller, 1964
1964a Falitidae n. f. – Müller, p. 24.
1980   Falitidae – Jones & McKenzie, p. 219.
1980   Falitidae Müller, 1964 – Landing, p. 757.
1982a Falitidae Müller – Müller, p. 285.
1983   Falitidae – Briggs, p. 9, table 1.
1986   Falitidae K.J. Müller, 1964 – Schram, p. 417.
1990a Falitidae Müller, 1964 – Shu, p. 66.
1991   Falitidae Müller, 1964 – Huo *et al.*, pp. 178, 181, 212.
1993c Falitidae Müller, 1964 – Hinz-Schallreuter, pp. 391, 396, 399.
1996b Falitidae Müller, 1964 – Hinz-Schallreuter, p. 89.
1998   Falitinae Müller, 1964 – Hinz-Schallreuter, pp. 104, 106, 110, 112 (subfamily of Hesslandonidae).
1998   Falitidae Müller, 1964 – Williams & Siveter, p. 27.
1999   Falitidae Müller, 1982 – Whatley *et al.*, p. 344.

Flemingopsis Jones & McKenzie, 1980 [type species: *Flemingia duo* Jones & McKenzie, 1980 (= *Flemingopsis duo*) by original designation]
1980   *Flemingia* gen. nov. – Jones & McKenzie, p. 214 (*non* Johnston, 1832).
1981   *Flemingopsis* n. gen. – Jones & McKenzie, p. 310 (*pro Flemingia non* Johnston, 1832 fide Johnston 1846, p. 447 [Polychaeta]; *non* de Koninck, 1881, p. 93 [Gastropoda]; non Jeffreys, 1884, p. 116 [Gastropoda]).
1986   *Flemingopsis* Jones & McKenzie – Huo *et al.*, p. 22.
1986a *Flemingia* Jones & McKenzie, 1980 – Kempf, p. 358.
1986a *Flemingopsis* ⟨*Flemingia*⟩ Jones & McKenzie, 1981 – Kempf, p. 358.

1986b *Flemingia* Jones & McKenzie, 1980 – Kempf, p. 673.
1986b *Flemingopsis* ⟨*Flemingia*⟩ Jones & McKenzie, 1981 – Kempf, p. 673.
1987   *Flemingia* Jones & McKenzie, 1980 – Kempf, p. 697.
1987   *Flemingopsis* ⟨*Flemingia*⟩ Jones & McKenzie, 1981 – Kempf, p. 712.
1987   *Flemingopsis* Jones & McKenzie – Zhang, pp. 8, 10, 12.
1989b *Flemingopsis* – Zhao, pp. 410, 411.
1990   *Flemingopsis* Jones & McKenzie, 1981 – Abushik *et al.*, p. 43.
1993a *Flemingopsis* – Hinz-Schallreuter, p. 321.
1993c *Flemingopsis* – Hinz-Schallreuter, pp. 395, 440.

Flemingopsis duo Jones & McKenzie, 1980 (type locality and horizon: Middle Cambrian, Georgina Basin, Queensland, Australia)
1973   *Zepaera* sp. – Fleming, p. 8, pl. 1, figs. 15–17; text-figs. A10, A11.
1980   *Flemingia duo* sp. nov. – Jones & McKenzie, p. 215.
1981   *Flemingia duo* Jones & McKenzie, 1980 – Jones & McKenzie, p. 310 (referred to *Flemingopsis*).
1986a *Flemingia duo* Jones & McKenzie, 1980 – Kempf, p. 358.
1986a *Flemingopsis duo* (Jones & McKenzie, 1980a) Jones & McKenzie, 1981 – Kempf, p. 358.
1986b *Flemingia duo* Jones & McKenzie, 1980 – Kempf, p. 193.
1986b *Flemingopsis duo* (Jones & McKenzie, 1980) Jones & McKenzie, 1981 – Kempf, p. 193.
1987   *Flemingia duo* Jones & McKenzie, 1980 – Kempf, p. 697.
1987   *Flemingopsis duo* (Jones & McKenzie, 1980) Jones & McKenzie, 1981 – Kempf, p. 712.
1987   *Flemingopsis duo* Jones & McKenzie – Zhang, p. 14.
1993a *Flemingopsis duo* Jones & McKenzie – Hinz-Schallreuter, pp. 320, 321.

Flemingopsis ventrospinata (Hinz-Schallreuter, 1993) (type locality and horizon: Middle Cambrian, Late Templetonian, Mount Murray, Queensland, Australia)
1993a *Flemingopsis ventrospinata* – Hinz-Schallreuter, p. 320, fig. 4.1.
1993c *Flemingopsis ventrospinata* – Hinz-Schallreuter, p. 395 (footnote).
1999   *Flemingopsis ventrospinata* Hinz-Schallreuter – McKenzie *et al.*, p. 463.

Gladioscutum Hinz-Schallreuter & Jones, 1994 (type species *Gladioscutum lauriei* Hinz-Schallreuter & Jones, 1994)

1994   *Gladioscutum* n. gen. – Hinz-Schallreuter &
          Jones, pp. 361, 370, 371.

Gladioscutum lauriei Hinz-Schallreuter & Jones, 1994
(type locality and horizon: Middle Cambrian,
Georgina Basin, Queensland, Australia)
1994   *Gladioscutum lauriei* n. sp. – Hinz-Schallreuter
          & Jones, pp. 361, 370, figs. 2–5.
1999   *Gladioscutum lauriei* Hinz-Schallreuter & Jones
          – McKenzie *et al.*, p. 463.

Hesslandona Müller, 1964 (type species: *Hesslandona necopina* Müller, 1964 by original designation, see p. 21)
1964a   *Hesslandona* n. g. – Müller, p. 21.
1965   *Hesslandona* Müller – Adamczak, pp. 29, 33.
1972   *Hesslandona* Mueller – Taylor & Rushton,
          p. 13.
1974   *Hesslandona* – Kozur, p. 823.
1975   *Hesslandona* – Müller, p. 177.
1978   *Hesslandona* Müller, 1964 – Rushton, p. 279.
1979   *Hesslandona* Müller, 1964 – Bednarczyk,
          p. 218.
1981a   *Hesslandona* – Müller, p. 147.
1982a   *Hesslandona* Müller, 1964 – Müller, p. 279ff.
1983   *Hesslandona* – Briggs, pp. 9–11.
1983   *Hesslandona* – Müller, p. 94.
1986   *Hesslandona* – Huo *et al.*, pp. 23–25.
1986   *Hesslandona* – Schram, p. 415.
1986a   *Hesslandona* Mueller, 1964 – Kempf, p. 400.
1986b   *Hesslandona* Mueller, 1964 – Kempf, p. 677.
1987   *Hesslandona* Mueller, 1964 – Kempf, p. 436.
1990   *Hesslandona* Müller, 1964 – Abushik *et al.*,
          p. 41.
1990   *Hesslandona* Müller, 1964 – Melnikova &
          Mambetov, p. 57.
1990a   *Hesslandona* – Shu, pp. 66, 76.
1991   *Hesslandona* Müller, 1964 – Melnikova &
          Mambetov, p. 56.
1992a   *Hesslandona* – Hinz, p. 15.
1993   *Hesslandona* Müller, 1964 – Hinz, p. 12,
          fig. 2b3.
1993b   *Hesslandona* Müller, 1964 – Hinz-Schallreuter,
          pp. 333, 343, 344, 347.
1993c   *Hesslandona* Müller, 1964 – Hinz-Schallreuter,
          pp. 386, 395, 396, 402.
1994   *Hesslandona* Müller – Hinz-Schallreuter, p. 13.
1998   *Hesslandona* – Cohen *et al.*, pp. 251, 253.
1998   *Hesslandona* Müller, 1964 – Hinz-Schallreuter,
          pp. 106–108, 110, 115, 116, 132; text-fig. 1.
1998   *Hesslandona* Müller, 1964 – Williams &
          Siveter, p. 30.
1998   *Hesslandona* – Ziegler, p. 223.
2001   *Hesslandona* – Chen *et al.*, fig. 4.

Hesslandona abdominalis Hinz-Schallreuter, 1998 (type

locality and horizon: Middle Cambrian, [ ] of
Bornholm)
1985b   Hesslandonid ostracode (new species ?) – erg-
          Madsen, p. 140, fig. 5A–D.
1993c   *Hesslandona reichi* ssp. n. A – nz-
          Schallreuter, p. 399.
1998   *Hesslandona abdominalis* n. sp. – nz-
          Schallreuter, pp. 103, 112, 115, 116, 11 , 22,
          124, 126, pl. 1, figs. 1, 2; pl. 9, fig. 3; 10,
          fig. 1; table 4.

Hesslandona angustata n. sp. (see p. 89, type locali and
horizon: Upper Cambrian, Zone 1, m,
Kinnekulle, Sweden)

Hesslandona curvispina n. sp. (see p. 93, type locali and
horizon: Upper Cambrian, Zone 1, Back org,
Kinnekulle, Sweden)
p. 1972 Undeterminable species of *Vestrogo* –
          Taylor & Rushton, p. 18.
. 1978 *Hesslandona trituberculata* (Lochmann Hu,
          1960) – Rushton, p. 279, pl. 26, fig. 11 ext-
          fig. 2.
   1979 *Hesslandona necopina* Müller, 19 –
          Bednarczyk, pp. 215, 218, plate, figs. 5, 9a,
          b, 10.
. 1981 *Hesslandona trituberculata* (Lochmann Hu)
          – Gründel, p. 63, pl. 3, fig. 9 (non fig. .
. 1986a *Hesslandona trituberculata* (Lochman Hu,
          1960) Rushton, 1978 – Kempf, p. 400.
. 1986b *Hesslandona trituberculata* (Lochman Hu,
          1960) Rushton, 1978 – Kempf, p. 610.
. 1987 *Hesslandona trituberculata* (Lochman Hu,
          1960) Rushton, 1978 – Kempf, p. 670.
p. 1992a *Hesslandona trituberculata* Gründel – inz,
          pp. 13, 15 [only fig. 9 of Gründel (198

Hesslandona kinnekullensis Müller, 1964 (see p. 6 , type
locality and horizon: Upper Cambrian, Zc 2,
Brattefors, Kinnekulle, Sweden)
v* 1964a *Hesslandona kinnekullensis* n. sp. – M ller,
          p. 23, pl. 1, figs. 7–9, ?11.
v. 1964a *Hesslandona necopina* – Müller, pl. 1, s. 3,
          4.
.   1974 *H. kinnekullensis* – Martinsson, p. 20
.   1978 *H. kinnekullensis* Müller – Rushton, 279.
non 1979a *Hesslandona kinnekullensis* Müller, 4 –
          Müller, fig. 36 (= *Vestrogotkia* sp. or lites
          sp.).
.   1985a *Hesslandona kinnekullensis* Müller, 4 –
          Müller & Walossek, fig. 5g.
.   1986a *Hesslandona kinnekullensis* Mueller, 4 –
          Kempf, p. 400.
.   1986b *Hesslandona kinnekullensis* Mueller, 4 –
          Kempf, p. 308.
.   1987 *Hesslandona kinnekullensis* Mueller, 4 –
          Kempf, p. 436.

. 1989b *Hesslandona kinnekullensis* Müller – Zhao, p. 412.

. 1993c *Hesslandona kinnekullensis* Müller, 1964 – Hinz-Schallreuter, p. 396.

. 1998 *Hesslandona kinnekullensis* Müller, 1964 – Hinz-Schallreuter, pp. 115, 116.

Hesslandona necopina Müller, 1964 (see p. 58, type locality and horizon: Upper Cambrian, Zone 1, Gydhem, Falbygden, Sweden)

v* 1964a *Hesslandona necopina* n. sp. – Müller, p. 22, pl. 1, fig. 6.

non 1964a *Hesslandona necopina* n. sp. – Müller, pl. 1, figs. 1, 2 (= *Hesslandona trituberculata*).

non 1964a *Hesslandona necopina* n. sp. – Müller, pl. 1, figs. 3, 4 (= *Hesslandona kinnekullensis*).

non 1964a *Hesslandona necopina* n. sp. – Müller, pl. 1, fig. 5 (= *Hesslandona suecica* n. sp.).

. 1974 *Hesslandina necopina* – Martinsson, p. 208 (sic!).

. 1975 *Hesslandona necopina* Müller, 1964 – Müller, pl. 19, fig. 1 (st. 339).

. 1978 *H. necopina* – Rushton, p. 279.

non 1979 *Hesslandona necopina* Müller, 1964 – Bednarczyk, pp. 215, 218, plate, figs. 5, 6, 9a, b, 10.

v. 1979a *Hesslandona necopina* Müller, 1964 – Müller, fig. 7a–c.

. 1981 *H. necopina* K.J. Müller (1964) – Gründel *in* Gründel & Buchholz, p. 63.

v. 1983 *Hesslandona necopina* Müller, 1964 – McKenzie *et al.*, fig. 1.

. 1986 *Hesslandona necopina* – Huo *et al.*, text-fig. 4b–e.

. 1986a *Hesslandona necopina* Mueller, 1964 – Kempf, p. 400.

. 1986b *Hesslandona necopina* Mueller, 1964 – Kempf, p. 392.

. 1987 *Hesslandona necopina* Mueller, 1964 – Kempf, p. 436.

. 1987 *Hesslandona necopina* Müller – Tong, p. 433.

. 1989 *Hesslandona necopina* Müller, 1979 – Zhao & Tong, p. 15 [cop. Müller (1979a, fig. 7a–c)].

. 1964a *Hesslandona necopina* Müller – Hinz, p. 15 (cop. Müller, pl. 1, fig. 6).

. 1993b *Hesslandona necopina* Müller, 1964 – Hinz-Schallreuter, p. 333.

. 1993c *Hesslandona necopina* Müller, 1964 – Hinz-Schallreuter, p. 396.

. 1998 *Hesslandona necopina* Müller, 1964 – Hinz-Schallreuter, pp. 104, 115.

. 1998 *Hesslandona necopina* Müller, 1964 – Williams & Siveter, p. 30.

. 1999 *Hesslandona necopina* Müller – McKenzie *et al.*, fig. 33.5A.

Hesslandona reichi Hinz-Schallreuter, 1993 (type locality and horizon: Middle Cambrian, Zone C2, Gislövshammar, Skåne, Sweden)

1993c *Hesslandona reichi* n. sp. – Hinz-Schallreuter, pp. 392, 396, 399, figs. 6.1–6.3.

1998 *Hesslandona reichi* Hinz-Schallreuter, 1993 – Hinz-Schallreuter, pp. 115, 116.

Hesslandona suecica n. sp. (see p. 85, type locality and horizon: Upper Cambrian, Zone 1, Gum, Kinnekulle, Sweden)

. 1964a *Hesslandona necopina* – Müller, pl. 1, fig. 5.

. 1982c *Hesslandona asulcata* – Müller, fig. 3 (nom. nud.).

. 1996 *Hesslandona* sp. nov. – Hou *et al.*, fig. 9a, d (UB 1629, 1627).

Hesslandona toreborgensis n. sp. (see p. 104, type locality and horizon: Upper Cambrian, Zone 2, Toreborg, Kinnekulle, Sweden)

Hesslandona trituberculata (Lochmann & Hu, 1960) Rushton, 1978 (see p. 179, type locality and horizon: Upper Cambrian, Northwest Wind River Mountains, Wyoming, USA)

* 1960 *Dielymella? trituberculata*, n. sp. – Lochmann & Hu, pp. 793, 826, pl. 98, fig. 56.

v. 1964a *Hesslandona* n. sp. a – Müller, p. 24, pl. 1, fig. 10a, b.

non 1978 *Hesslandona trituberculata* (Lochmann & Hu, 1960) – Rushton, p. 279, pl. 26, fig. 11; text-fig. 2 (= *Hesslandona curvispina* n. sp.).

non 1981 *Hesslandona trituberculata* (Lochmann & Hu) – Gründel, p. 63, pl. 3, fig. 9 (= *Hesslandona curvispina* n. sp.).

non 1981 *Hesslandona trituberculata* (Lochmann & Hu) – Gründel, p. 63, pl. 3, fig. 10 (= *Hesslandona kinnekullensis*).

. 1986a *Dielymella? trituberculata* Lochman & Hu, 1960 – Kempf, p. 316.

non 1986a *Hesslandona trituberculata* (Lochman & Hu, 1960) Rushton, 1978 – Kempf, p. 400 (= *Hesslandona curvispina* n. sp.].

. 1986b *Dielymella? trituberculata* Lochman & Hu, 1960 – Kempf, p. 610.

non 1986b *Hesslandona trituberculata* (Lochman & Hu, 1960) Rushton, 1978 – Kempf, p. 610 (= *Hesslandona curvispina* n. sp.).

. 1987 *Dielymella? trituberculata* Lochman & Hu, 1960 – Kempf, p. 362.

non 1987 *Hesslandona trituberculata* (Lochman & Hu, 1960) Rushton, 1978 – Kempf, p. 670 (= *Hesslandona curvispina* n. sp.).

. 1993b *Hesslandona* n. sp. a [Müller 1964] – Hinz-Schallreuter, p. 342.

. 1993c *Dielymella? trituberculata* Lochmann & Hu, 1960 – Hinz-Schallreuter, p. 396 (referred to *Hesslandona*).

. 1994a *Hesslandona trituberculata* Lochman & Hu – Williams *et al.*, p. 23.

. 1997 *Vestrogothia trituberculata* (Lochmann & Hu) – Siveter & Williams, p. 60, pl. 8, fig. 7.

. 1998 *Dielymella? trituberculata* Lochman & Hu, 1960 – Hinz-Schallreuter, pp. 104, 115 (referred to *Hesslandona*).

. 1998 *Hesslandona trituberculata* (Lochman & Hu, 1960) – Williams & Siveter, p. 31.

Hesslandona? ventrospinata Gründel *in* Gründel & Buchholz, 1981 (see p. 77, type locality and horizon: Upper Cambrian, Zone 1, erratic boulders from the Isle of Rügen, Germany)

* 1981 *Hesslandona? ventrospinata* n. sp. – Gründel, p. 63, pl. 2, fig. 15.

. *1986a Hesslandona? ventrospinata* Gruendel, 1981 – Kempf, p. 400.

. *1986b Hesslandona? ventrospinata* Gruendel, 1981 – Kempf, p. 635.

. *1987 Hesslandona? ventrospinata* Gruendel, 1981 – Kempf, p. 710.

. 1998 *Hesslandona? ventrospinata* Gründel *in* Gründel & Buchholz, 1981 – Hinz-Schallreuter, pp. 115, (126) (referred to *Cyclotron*).

Hesslandona unisulcata Müller, 1982 (see p. 16, type locality and horizon: Upper Cambrian, Zone 1, Gum, Kinnekulle, Sweden)

. *1974 Falites fala* – Martinsson, p. 208 (= *Hesslandona unisulcata*).

p 1978 *Falites fala* Müller – Rushton, pp. 276, 277 [*partim*: specimens BGS BDA 1820 (pl. 26, fig. 12), BDA 1844, BDA 1855, BDA 1863; non BGS BDA 1824 (= *Waldoria rotundata*)]; text-fig. 2 (= *Hesslandona unisulcata*).

v* 1982a *Hesslandona unisulcata* sp. nov. – Müller, p. 279, plates 1–8; text-figs. 1–5.

v. 1982c *Hesslandona unisulcata* – Müller, fig. 2.

. 1983 *Hesslandona unisulcata* Müller, 1982 – McKenzie *et al.*, figs. 2, 3.

v. 1983 *Hesslandona unisulcata* Müller, 1982 – Reyment, fig. 2 [cop. Müller (1982a, pl. 6, fig. 1b)].

v. 1985a *Hesslandona unisulcata* Müller, 1982 – Müller & Walossek, p. 161, fig. 2a.

. 1986 *Hesslandona unisulcata* – Schram, fig. 33-10A.

. *1986a Hesslandona unisulcata* Mueller, 1982 – Kempf, p. 400.

. *1986b Hesslandona unisulcata* Mueller, 1982 – Kempf, p. 625.

. *1987 Hesslandona unisulcata* Mueller, 1982 – Kempf, p. 730.

. 1990 *Hesslandona unisulcata* – Müller, p. 27 [(cf. p. 16, 199AQ54)].

. 1991a *Hesslandona unisulcata* Müller, 1982 – Müller & Walossek, figs. 1, 2.

. 1993c *Hesslandona unisulcata* Müller, 1982 – Hinz-Schallreuter, p. 400.

. 1993c *Falites unisulcatus* (Müller, 1982) – Hinz-Schallreuter, pp. 392, 400, 402, fig. 7.

. 1993 *Hesslandona unisulcata* Müller 1982 – Whatley *et al.*, p. 345.

. 1995 *Hesslandona unisulcata* Müller, 1982 – Müller *et al.*, fig. 4F.

. 1996 *Hesslandona unisulcata* Müller, 1982 – Hou *et al.*, fig. 9b, c (UB 1628, 658).

. 1998 *Hesslandona sulcata* Müller – Boxhall, p. 162 (sic!).

: 1998 *Falites unisulcatus* – Hinz-Schallreuter, p. 114.

v. 1998a *Hesslandona unisulcata* Müller, 1982 – Walossek & Müller, fig. 12.3a.

v. 1998a Isolated post-mandibular limb of a phosphatocopine – Walossek & Müller, fig. 12.3b.

. 1998 *Hesslandona unisulcata* Müller, 1982 – Williams & Siveter, p. 28.

. *1998 Hesslandona* – Ziegler, fig. 234.1 [cop. Müller (1982a, fig. 5)].

. 1999 *Hesslandona unisulcata* Müller, 1982 – Whatley *et al.*, p. 345.

. 2001 "einer der ältesten bekannten Krebse aus der Gruppe Phosphatocopina" – Schmidt-Rhaesa & Bartolomaeus, fig. 9A.

. *2001 Hesslandona unisulcata* – Siveter *et al.*, p. 481.

Hesslandona n. sp. b Müller, 1964 (locality and horizon: Upper Cambrian, Zone 5d–f, Trolmen, Kinnekulle, Sweden)

1964a *Hesslandona* n. sp. b – Müller, p. 24, pl. 1, fig. 12a, b.

Hesslandonidae Müller, 1964

1964a Hesslandonidae n. f. – Müller, p. 21.

1974 Hesslandonidae Müller, 1964 – Kozur, p. 323.

1975 Hesslandonidae – Müller, p. 177.

1979 Hesslandonidae Müller, 1964 – Bednarczyk, p. 218.

1980 Hesslandonidae – Jones & McKenzie, p. 19.

1983 Hesslandonidae – Briggs, p. 9, table 1.

1985a Hesslandonidae – Berg-Madsen, p. 140.

1985 Hesslandonidae Müller, 1964 – Huo & Shu, p. 184.

1985a   Hesslandonidae – Müller & Walossek, pp. 165, 169, 171.

1986   Hesslandonidae K.J. Müller, 1982 – Schram, p. 417.

1987   Hesslandonidae Müller 1964 – Hinz, p. 59.

1989b   Hesslandonidae – Zhao, p. 411.

1990   Hesslandonidae Müller, 1982 – Abushik *et al.*, p. 41.

1990   Hesslandonidae Müller – Melnikova & Mambetov, p. 57, (58).

1990a   Hesslandonidae Müller, 1964 – Shu, p. 65.

1991b   ("hesslandonids" – Hinz, p. 69.)

1991   Hesslandonidae Müller – Melnikova & Mambetov, p. 56, (57).

1992   ("Hesslandonid" – Hinz & Jones, pp. 9, 11.)

1993   Hesslandonidae Müller, 1964 – Hinz, pp. 3, (5, 10), 11, 12.

1993a   Hesslandonidae Müller, 1964 – Hinz-Schallreuter, pp. 308, 310.

1993b   Hesslandonidae Müller, 1964 – Hinz-Schallreuter, pp. 329–333, 343, 347.

1993c   Hesslandonidae Müller, 1964 – Hinz-Schallreuter, pp. 386, 391, 396, 402.

1993   Hesslandonidae Müller, 1964 – Whatley *et al.*, p. 345.

1995   Hesslandonidae – Hinz-Schallreuter, p. 4154.

1996b   Hesslandonidae – Hinz-Schallreuter, p. 89.

1997   Hesslandonidae – Hinz-Schallreuter, p. 13 (as possible synonym of Vestrogothiidae).

1998   Hesslandonidae Müller, 1964 – Hinz-Schallreuter, pp. 104, 110.

1998   Hesslandoninae Müller, 1964 – Hinz-Schallreuter, pp. 106, 110, 114, 117 (subfamily of Hesslandonidae).

1998   Hesslandonidae Müller, 1964 – Williams & Siveter, pp. 27, 30.

1999   Hesslandonidae – McKenzie *et al.*, p. 463.

1999   Hesslandonidae Müller, 1964 – Whatley *et al.*, p. 345.

Hesslandonid ostracode Berg-Madsen, 1985 (locality and horizon: Middle/Upper Cambrian, Læså, Isle of Bornholm, Denmark)

1985b   Hesslandonid ostracode – Berg-Madsen, p. 141 (*non* p. 140, fig. 5A–D).

Hesslandonina Müller, 1982

1982a   Hesslandonina nov. – Müller, p. 279.

1983   Hesslandonocopina Müller, 1982 – McKenzie *et al.*, pp. 36, 37.

1986   Hesslandonocopina K.J. Müller, 1982 – Schram, p. 417.

1987   Hesslandonocopina Müller 1982 – Hinz, p. 59.

1987   Hesslandonina – Zhang, p. 10.

1989b   Hesslandonina – Zhao, p. 411.

1990   Hesslandonida Müller, 1982 (nom. transl. hic) – Abushik *et al.*, p. 41.

1990   Hesslandonida Müller, 1982 – Melnikova & Mambetov, p. 58.

1990a   Hesslandonina Müller, 1982 – Shu, pp. 40, 65, 79.

1990b   Hesslandonina Müller 1982 – Shu, p. 323.

1991a   Hesslandonina – Hinz, p. 233.

1991a   Hesslandonocopina Müller, 1982 – Hinz, p. 233.

1991   Hesslandonida Müller, 1982 – Melnikova & Mambetov, p. 57.

1993b   Hesslandonina [Müller, 1982] – Hinz-Schallreuter, pp. 346, 347.

1993   Hesslandonidae Müller, 1982 – Whatley *et al.*, p. 345.

1994   Hesslandonida – Hinz-Schallreuter, p. 13.

1995   Hesslandonocopina – Hinz-Schallreuter, p. 415.

1996   Hesslandonocopina = Hesslandonina – Hinz-Schallreuter & Koppka, p. 38 (footnote).

1998   Hesslandonina – Williams & Siveter, p. 27.

1999   Hesslandonina Müller, 1982 – Whatley *et al.*, p. 345.

"*Leperdita*" *vexata* Hicks, 1871 (type locality and horizon: Cambrian, Longmynd Rocks, St. David's, North Wales)

1871   *Leperditia*? *vexata*, Hicks – Hicks, p. 401, pl. 15, figs. 15–17.

1872   *Leperditia*? *vexata*, Hicks – Jones, p. 184, pl. V, figs. 17, 18.

1934   *Leperditia*? *vexata* Hicks – Bassler & Kellett, p. 403.

1998   *Leperditia vexata* Hicks, 1871 – Hinz-Schallreuter, p. 104 (*Shergoldopsis vexata* on p. 104).

Remarks. – Jones (1872) illustrated *Leperditia vexata* but described it as a "larval trilobite". He erroneously interpreted the interdorsum of this species as the pleural segment of a trilobite. Bassler & Kellett (1934) regarded *Leperditia vexata* as a branchiopod or larval trilobite. Hinz-Schallreuter (1998) considered "*Leperditia*" *vexata* as a member of *Shergoldopsis*.

Liangshanella Huo, 1956 (type species: *Liangshanella liangshanensis* Huo, 1956 by original designation)

1956   *Liangshanella* n. g. – Huo, p. 427.

1965   *Liangshanella* Huo – Huo, pp. 294, 301.

1975   *Liangshanella* Huo, 1956 – Li, pp. 39, 43 (as *Lianshanella*), 45, 46.

1982   *Liangshanella* Huo, 1956 – Huo & Shu, p. 323.

1983   *Liangshanella* Huo, 1956 – Li, p. 15.

1985   *Liangshanella* Huo, 1956 – Huo & Shu, p. 56.

1985   *Liangshanella* – Zhang & Hou, pp. 592, 594.

1986 *Liangshanella* Huo – Huo *et al.*, pp. 25, 26.
1986a *Liangshanella* Huo, 1956 – Kempf, p. 464.
1986b *Liangshanella* Huo, 1956 – Kempf, p. 682.
1987 *Liangshanella* Huo, 1956 – Kempf, p. 305.
1987b *Liangshanella* Huo, 1956 – Hou, pp. 537–539.
1989a *Liangshanella* Huo, 1956 – Zhao, pp. 470, 472.
1989b *Liangshanella* – Zhao, pp. 412, 413.
1990 *Liangshanella* Huo, 1956 – Abushik *et al.*, p. 42.
1990a *Liangshanella* Huo, 1956 – Shu, p. 57.
1991 *Liangshanella* Huo, 1956 – Huo *et al.*, p. 210.
1996 *Liangshanella* Huo, 1956 – Hou *et al.*, p. 1140.
1997 *Liangshanella* Huo, 1956 – Melnikova *et al.*, p. 181.
1997 *Liangshanella* Huo, 1956 – Siveter & Williams, p. 53.

Liangshanella burgessensis Siveter & Williams, 1997 (type locality and horizon: Middle Cambrian, Burgess Shale, Walcott Query, British Columbia, Canada)
?1931 *Aluta rimulata*, new species – Ulrich & Bassler, p. 61, pl. 5, fig. 16.
1986a *Aluta*? *rimulata* Ulrich & Bassler, 1931 – Kempf, p. 45.
1986b *Aluta*? *rimulata* Ulrich & Bassler, 1931 – Kempf, p. 508.
1987 *Aluta*? *rimulata* Ulrich & Bassler, 1931 – Kempf, p. 167.
1997 *Liangshanella burgessensis* sp. nov. – Siveter & Williams, p. 53, pl. 8, figs. 3–5, ?6.
1999 *Liangshanella burgessensis* – Shu *et al.*, p. 292.

Liangshanella chiehi Huo, 1956 (type locality and horizon: Lower Cambrian, Shaanxi, China)
1956 *Liangshanella chiehi* – Huo, pp. 428, 445, pl. 1, fig. 4.
1965 *Liangshanella chiehi* Huo – Huo, pp. 294, 301, pl. 1, fig. 7.
1982 *Liangshanella chiehi* Huo – Huo & Shu, p. 323, pl. 1, fig. 1.
1986a *Liangshanella chiehi* Huo, 1956 – Kempf, p. 464.
1986b *Liangshanella chiehi* Huo, 1956 – Kempf, p. 125.
1987 *Liangshanella chiehi* Huo, 1956 – Kempf, p. 305.

Liangshanella kalpinensis Zhang, 1981 (type locality and horizon: Lower Cambrian, China)
1981 *Liangshanella kalpinensis* – Zhang, p. 214.
1986a *Liangshanella*? *kalpinensis* Zhang, 1981 – Kempf, p. 464.
1986b *Liangshanella*? *kalpinensis* Zhang, 1981 – Kempf, p. 303.
1987 *Liangshanella*? *kalpinensis* Zhang, 1981 – Kempf, p. 721.

1989b *Liangshanella kalpingensis* Zhang – Zhao, p. 413 (sic!).

Liangshanella liangshanensis Huo, 1956 (type locality and horizon: Lower Cambrian, Shaanxi, China)
1956 *Liangshanella liangshanensis* – Huo, pp. 427, 428, 445, pl. 1, figs. 1, 2.
1974 *Liangshanella liangshanensis* Huo – Zhang, p. 111, pl. 43, fig. 17.
1975 *Liangshanella Liangshanensis* Huo, 1956 – Li, p. 46 (sic!).
1986a *Liangshanella liangshanensis* Huo, 1956 – Kempf, p. 464.
1986b *Liangshanella liangshanensis* Huo, 1956 – Kempf, p. 331.
1987 *Liangshanella liangshanensis* Huo, 1956 – Kempf, p. 305.
1987b *Liangshanella liangshanensis* Huo, 1956 – Hou, p. 539, pl. 1, figs. 1–3.
1989b *Liangshanella liangshanensis* Huo – Zhao, p. 413.
1990a *Liangshanella liangshanensis* Huo, 1956 – Shu, pp. 57, 93, pl. 14, fig. 1.

Liangshanella minuta Shu, 1990 (type locality and horizon: Lower Cambrian, China)
1990a *Liangshanella minuta* Shu – Shu, p. 57 (referring to p. 93, pl. 14, figs. 2–5).
1990a *Liangshanella* cf. *minuta* Shu – Shu, p. 93, pl. 14, figs. 2–5.

Liangshanella obesa Huo, 1956 (type locality and horizon: Lower Cambrian, Shaanxi, China)
1956 *Liangshanella obesa* – Huo, pp. 428, 445, pl. 1, fig. 3.
1975 *Liangshanella obesa* Huo, 1956 – Li, p. 47.
1982 *Liangshanella obesa* Huo – Huo & Shu, p. 323, pl. 1, fig. 2.
1986a *Liangshanella obesa* Huo, 1956 – Kempf, p. 464.
1986b *Liangshanella obesa* Huo, 1956 – Kempf, p. 403.
1987 *Liangshanella obesa* Huo, 1956 – Kempf, p. 306.

Liangshanella orbicularis Li, 1975 (type locality and horizon: Lower Cambrian, Shaanxi, China)
1975 *Liangshanella orbicularis* Lee, sp. nov. – Li, pp. 44, 46, pl. IV, figs. 1–4.
1986a *Liangshanella orbicularis* Li(Y-W), 1975 – Kempf, p. 464.
1986b *Liangshanella orbicularis* Li(Y-W), 1975 – Kempf, p. 416.
1987 *Liangshanella orbicularis* Li(Y-W), 1975 – Kempf, p. 624.
1989b *Liangshanella orbicularis* Lee – Zhao, p. 413.

Liangshanella sayutinae Melnikova, 1988 (type locality

and horizon: Lower Cambrian, Lena-Aldan Region, Siberia, Russia)

1988　*Liangshanella sayutinae* sp. nov. – Melnikova, pp. 128, 129.

1997　*Liangshanella? sayutinae* – Melnikova *et al.*, p. 181, text-fig. 3.

Liangshanella similis Huo, Shu & Cui, 1991 (type locality and horizon: Lower Cambrian, Sugaitbulag in Wushi, Xinjiang, China)

1991　*Liangshanella similis* (sp. nov.) – Huo *et al.*, p. 210, pl. 15, figs. 19, 20, text-fig. 8-39.

Liangshanella sp. *sensu* Li (1975) (locality and horizon: Lower Cambrian, Shaanxi, China)

1975　*Liangshanella* sp. – Li, p. 45.

Liangshanella wushiensis Zhao & Xiao, 1986 (type locality and horizon: Lower Cambrian, Shaanxi, China)

1986　*Liangshanella wushiensis* n. sp. – Xiao & Zhao, p. 79, pl 2, fig. 4.

1989b　*Liangshanella wushiensis* Zhao & Xiao – Zhao, p. 413.

Liangshanella yunnanensis Zhang, 1974 (type locality and horizon: Lower Cambrian, Yunnan, China)

1974　*Liangshanella yunnanensis* (sp. nov.) – Zhang, p. 111, pl. 43, fig. 18.

1986a　*Liangshanella yunnanensis* Zhang, 1974 – Kempf, p. 464.

1986b　*Liangshanella yunnanensis* Zhang, 1974 – Kempf, p. 653.

1987　*Liangshanella yunnanensis* Zhang, 1974 – Kempf, p. 611.

Liangshanellidae Hou, 1987

1987b　Liangshanellidae fam. nov. – Hou, p. 538.

1989a　Liangshanellidae fam. nov. – Zhao, pp. 470, 472.

1989b　("Liangshanellids" – Zhao, p. 412.)

1991　Liangshanellidae fam. nov. – Huo *et al.*, p. 210.

Monasteriidae Jones & McKenzie, 1980

1980　Monasteriidae – Jones & McKenzie, p. 216.

1982a　Monasteriidae Jones & McKenzie – Müller, p. 285.

1986　Monasteridae Jones & McKenzie, 1980 – Schram, p. 417 (sic!).

1987　Monasteriidae Jones & McKenzie 1980 – Zhang, p. 10.

1990　Monasteriidae Jones & McKenzie, 1980 – Abushik *et al.*, p. 45.

1990　Monasteriidae Jones & McKenzie, 1980 – Bengtson *in* Bengtson *et al.*, pp. 322, 323.

1990　Monasteriidae Jones & McKenzie, 1980 – Melnikova, p. 174.

1990a　Monasteriidae Jones & McKenzie, 1979 (emended) – Shu, p. 60.

1991　Monasteriidae Jones & McKenzie, 1980 – Huo *et al.*, p. 177.

1992b　Monasteriidae [Jones & McKenzie, 1980] – Hinz, p. 123.

1993c　Monasteriidae – Hinz-Schallreuter, p. 396.

1998　Monasteriidae Jones & McKenzie, 1980 – Williams & Siveter, p. 27.

1999　Monasteriidae Müller, 1982 – Whatley *et al.*, p. 344.

Monasterium Fleming, 1973 (type species: *Monasterium oepiki* Fleming, 1973, to Monasteriidae)

1973　*Monasterium* – Fleming, p. 8.

1980　*Monasterium* Fleming, 1973 – Jones & McKenzie, p. 216.

1986a　Monasterium Fleming, 1973 – Kempf, p. 508.

1986b　Monasterium Fleming, 1973 – Kempf, p. 685.

1987　Monasterium Fleming, 1973 – Kempf, p. 583.

1987　*Monasterium* Fleming, 1973 – Zhang, pp. 8, 10.

1989a　*Xichuanella* gen. nov. – Zhao, p. 472, pl. 1, fig. 3.

1989b　*Xichuanella* Zhao, 1989 – Zhao, pp. 412, 415.

1990　*Monasterium* Fleming, 1973 – Abushik *et al.*, p. 45.

1990　*Monasterium* Fleming, 1973 – Bengtson *in* Bengtson *et al.*, pp. 322, 323.

1990　*Monasterium* Fleming, 1973 – Melnikova, pp. 171, 174, 175.

1990a　*Monasterium* Fleming – Shu, pp. 60–63.

1991　*Monasterium* Fleming, 1973 – Huo *et al.*, p. 177.

1992b　*Monasterium* Fleming, 1973 – Hinz, pp. 123, 125, 127.

1992b　*Xichuanella* Zhao – Hinz, pp. 123, 127 (as synonymous with *Monasterium*).

1993　*Monasterium* Fleming, 1973 – Hinz, pp. 1, 8.

1993a　*Monasterium* – Hinz-Schallreuter, p. 320.

1997　*Monasterium* – Melnikova *et al.*, p. 185.

1999　*Monasterium* – McKenzie *et al.*, p. 461.

Monasterium bucerum Zhang, 1987 (type locality and horizon: Lower Cambrian, Xichuan, Henan, China)

1987　*Monasterium bucerum* sp. nov. – Zhang, pp. 1, 11, figs. 2, 9A–E.

1989a　*Xichuanella bucerum* (Zhang, 1987) – Zhao, pp. 470, 472, pl. 1, fig. 3.

1989b　*Monasterium bucerum* – Zhao, p. 412, table 1; pl. 1, fig. 14 (referred to *Xichuanella*).

1990　*M. bucerum* Zhang, 1987 – Bengtson *in* Bengtson *et al.*, p. 323.

1990　*Monasterium bucerium* Zhang – Melnikova, p. 174 (sic!).

1991　*Monasterium bucerum* Zhang – Huo *et al.*, p. 177, pl. 37, figs. 9–13.

1992b　*M.* [*Monasterium*] *bucerum* Zhang – Hinz, pp. 123, 127.

1997   *Monasterium bucerum* Zhang, 1987 – Melnikova *et al.*, p. 185.

Monasterium dorium Fleming, 1973 (type locality and horizon: Middle Cambrian, Georgina Basin, Queensland, Australia)

1973   *Monasterium dorium* – Fleming, p. 6.

1986a *Monasterium dorium* Fleming, 1973 – Kempf, p. 508.

1986b *Monasterium dorium* Fleming, 1973 – Kempf, p. 188.

1987   *Monasterium dorium* Fleming, 1973 – Kempf, p. 583.

1990   *M. dorium* Fleming, 1973 – Bengtson *in* Bengtson *et al.*, p. 323.

1990   *Monasterium dorium* Fleming – Melnikova, p. 174.

1992b *Monasterium dorium* Fleming – Hinz, pp. 127, 129.

1993   *Monasterium dorium* – Hinz, p. 8.

1994   *Monasterium dorium* Fleming – Hinz & Jones, p. 364.

1999   *Monasterium dorium* Jones & McKenzie – McKenzie *et al.*, p. 463.

Monasterium ivshini Melnikova, 1990 (type locality and horizon: Late Cambrian, area of the former Soviet Union)

1990   *Monasterium ivshini* sp. nov. – Melnikova, pp. 171, 174, pl. 32, fig. 1.

1992b *Monasterium ivshini* Melnikova – Hinz, p. 127.

1997   *Monasterium ivshini* Melnikova, 1990 – Melnikova *et al.*, p. 185, pl. 3, fig. 4; text-fig. 3.

Monasterium oepiki Fleming, 1973 (type locality and horizon: Middle Cambrian, Georgina Basin, Queensland, Australia)

1971   Beyrichonidae gen. et sp. nov. A – Hill *et al.*, pl. Cm14, figs. 12–14.

1973   *Monasterium oepiki* – Fleming, p. 8, pl. 2, figs. 6–12; pl. 3, fig. 7; pl. 4, figs. 1–4; text-figs. A13, A14.

1980   *Monasterium oepiki* Fleming, 1973 – Jones & McKenzie, p. 217.

1986a *Monasterium opiki* Fleming, 1973 – Kempf, p. 508 (sic!).

1986b *Monasterium opiki* Fleming, 1973 – Kempf, p. 415 (sic!).

1987   *Monasterium opiki* Fleming, 1973 – Kempf, p. 583 (sic!).

1987   *Monasterium oepiki* Fleming, 1973 – Zhang, pp. 1, 11, pl. 2.

1989b *Monasterium oepiki* Fleming – Zhao, p. 412.

1990   *M. oepiki* – Abushik *et al.*, p. 45.

1990   *M. oepiki* Fleming, 1973 – Bengtson *in* Bengtson *et al.*, p. 323.

1990a *Monasterium oepiki* – Shu, p. 63.

1992b *Monasterium oepiki* Fleming, 1973 – Hinz, pp. 123, 125, 127, 129, pl. 19.

1993b *Monasterium oepiki* – Hinz-Schallreuter, pp. 329, 346.

1994   *Monasterium oepiki* Fleming – Hinz & Jones, pp. 364, 372.

1998   *Monasterium oepiki* Fleming, 1973 – Hinz-Schallreuter, p. 108.

1999   *Monasterium oepiki* Fleming – Whatley *et al.*, p. 344.

Monasterium seletinensis Melnikova, 1990 (type locality and horizon: Late Cambrian, area of the former Soviet Union)

1990   *Seletinella seletiensis* sp. nov. – Melnikova, p. 171.

1990   *Monasterium seletinensis* sp. nov. – Melnikova, p. 175, pl. 32, figs. 2–4.

1992b *Monasterium seletinensis* Melnikova – Hinz, p. 127.

1997   *Monasterium* (=gen. nov.?) *seletinensis* Melnikova, 1990 – Melnikova *et al.*, p. 85, pl. 3, fig. 2; text-fig. 3.

Monasterium sp. nov. Melnikova, Siveter & Williams, 1997 (locality and horizon: Upper Cambrian, Kazakhstan)

1997   *Monasterium* sp. nov. – Melnikova *et al.*, p. 4, fig. 11; text-fig. 3.

Monasteriidae gen. et sp. nov. Shu, 1990 (locality and horizon: Lower Cambrian, China)

1990a gen. et sp. nov. – Shu, p. 62 (refers to *Taociella? nodosa* gen. et sp. nov. on p. 93, pl. 14, fig. 11).

Naviformella Zhao & Xiao *in* Xiao & Zhao, 1986 (type species: *Naviformella antiquata* Zhao & Xiao *in* Xiao & Zhao, 1986 by original designation)

1986   *Naviformella* n. g. – Zhao & Xiao *in* Xiao & Zhao, p. 76.

Naviformella antiquata Zhao & Xiao *in* Xiao & Zhao, 1986 (type locality and horizon: Lower Cambrian, China)

1986   *Naviformella antiquata* Zhao & Xiao (gen. et sp. nov.) – Zhao & Xiao *in* Xiao & Zhao, p. 83, pl. 2, fig. 4.

1989b *Hesslandona antiquata* Zhao & Xiao – Zhao, p. 412, pl. 1, fig. 17; pl. 1, fig. 17.

Oepikaluta Jones & McKenzie, 1980 (type species: *Oepikaluta dissuta* Jones & McKenzie, 1980 by original designation)

1980   *Oepikaluta* gen. nov. – Jones & McKenzie, p. 211.

1986   *Oepikaluta* Jones & McKenzie – Huo *et al.*, p. 22.

1986a *Oepikaluta* Jones & McKenzie, 1980 – Kempf, p. 540.

1986b *Oepikaluta* Jones & McKenzie, 1980 – Kempf, p. 688.

1987   *Oepikaluta* Jones & McKenzie, 1980 – Kempf, p. 697.

1987   *Oepikaluta* Jones & McKenzie – Zhang, pp. 8, 10, 12.

1989b *Oepikaluta* – Zhao, pp. 410, 411.

1990   *Oepikaluta* Jones & McKenzie, 1980 – Abushik *et al.*, p. 43.

1990a *Oepikaluta* Jones & McKenzie, 1980 – Shu, pp. 49, 77.

1993   *Oepikaluta* – Hinz, p. 5.

Oepikaluta dissuta Jones & McKenzie, 1980 (type locality and horizon: Middle Cambrian, Georgina Basin, Queensland, Australia)

1980   *Oepikaluta dissuta* sp. nov. – Jones & McKenzie, p. 211, fig. 4A–I.

1982a *Oepikaluta dissuta* – Müller, pp. 281, 287.

1986a *Oepikaluta dissuta* Jones & McKenzie, 1980 – Kempf, p. 540.

1986b *Oepikaluta dissuta* Jones & McKenzie, 1980 – Kempf, p. 184.

1987   *Oepikaluta dissuta* Jones & McKenzie, 1980 – Kempf, p. 697.

1991   *Oepikaluta dissuta* Jones & McKenzie – Huo *et al.*, p. 208.

1993   *Oepikaluta dissuta* Jones & McKenzie, 1980 – Hinz, p. 8, fig. 4I, K.

1994   *Oepikaluta dissuta* Jones & McKenzie 1980 – Hinz & Jones, pp. 363, 364.

1996   *Oepikaluta dissuta* Jones & McKenzie – Hinz-Schallreuter & Koppka, p. 27.

1999   *Oepikaluta dissuta* Jones & McKenzie, 1980 – Whatley *et al.*, p. 344.

Oepikalutidae Jones & McKenzie, 1980

1980   Oepikalutidae nov. – Jones & McKenzie, p. 211.

1982a Oepikalutidae Jones & McKenzie – Müller, p. 285.

1986   Oepikalutidae Jones & McKenzie, 1980 – Schram, p. 417.

1987   Oepikalutidae Jones & McKenzie 1980 – Zhang, p. 11.

1989b ("Zepaerids" – Zhao, p. 410.)

1990   Houlongdongellinae Huo & Shu, 1985 – Abushik *et al.*, p. 42.

1990   Oepikalutidae Jones & McKenzie, 1980 – Abushik *et al.*, p. 43.

1990   Oepikalutidae Jones & McKenzie, 1980 – Bengtson *in* Bengtson *et al.*, p. 322.

1990a Oepikalutidae Jones & McKenzie, 1980 – Shu, p. 49.

1991   Houlongdongellidae (Jones & McKenzie, 1980) = Oepikalutidae Jones & McKenzie, 1980 – Huo *et al.*, p. 207.

1993c Oepikalutidae – Hinz-Schallreuter, pp. 396, (440).

1995   Oepikalutidae Jones & McKenzie – Hinz-Schallreuter, p. 415.

1998   Oepikalutidae – Hinz-Schallreuter, p. 109.

1998   Oepikalutidae Jones & McKenzie, 1980 – Williams & Siveter, p. 27.

1999   Oepikalutidae – McKenzie *et al.*, p. 463.

1999   Oepikalutidae Fleming – Whatley *et al.*, p. 344 (wrong assignment of author).

Paradabashanella Shu, 1990 (type species: *Paradabashanella elongata* Shu, 1990 by original designation)

1990a *Paradabashanella* gen. nov. – Shu, p. 70.

1990b *Paradabashanella* Shu, 1987 (unpublished manuscript) – Shu, p. 323.

1992c *Paradabashanella* – Hinz, p. 7.

Paradabashanella elongata Shu, 1990 (type locality and horizon: Lower Cambrian, China)

1990a *Paradabashanella elongata* sp. nov. – Shu, pp. 70, 92, pl. 10, figs. 13–17; text-fig. 45.

1990   *Paradabashanella elongata* – Malz, table 1 (referred to Shu 1990b).

1992c *Paradabashanella elongata* Shu, 1990 – Hinz, p. 5.

Paraphaseolella Tong, 1987 (type species: *Paraphaseolella typica* Tong, 1987 by original designation)

1987   *Paraphaseolella* gen. nov. – Tong, p. 434.

1990   *Paraphaseolella* – Melnikova & Mambetov, pp. 58–60 (synonymised with *Dabashanella*).

1991   *Paraphaseolella* – Melnikova & Mambetov, pp. 57–59 (synonymised with *Dabashanella*).

Paraphaseolella typica Tong, 1987 (type locality and horizon: Lower Cambrian, southern Shaanxi, China)

1987   *Paraphaseolella typica* sp. nov. – Tong, p. 434, pl. 1, figs, 1, 2; pl. 2, figs. 19–22.

1990   *Paraphaseolella typica* – Melnikova & Mambetov, pp. 58, 61 (synonymised with *Dabashanella retroswinga*).

1991   *Paraphaseolella typica* – Melnikova & Mambetov, pp. 57, 60 (synonymised with *Dabashanella retroswinga*).

Parashergoldopsis Hinz-Schallreuter, 1993 (type species: *Parashergoldopsis levis* Hinz-Schallreuter, 1993 by original designation)

1993a *Parashergoldopsis* n. g. – Hinz-Schallreuter, p. 310.

1993c *Parashergoldopsis* Hinz-Schallreuter, 1993 –
Hinz-Schallreuter, p. 412.

Parashergoldopsis levis Hinz-Schallreuter, 1993 (type
locality and horizon: Middle Cambrian, Late
Templetonian, Rogers Ridge, Queensland, Australia)
1993a *Parashergoldopsis levis* n. sp. – Hinz-
Schallreuter, p. 310, fig. 2.1a, b.
1993c *Parashergoldopsis levis* Hinz-Schallreuter, 1993
– Hinz-Schallreuter, fig. 20.2.

Pejonesia Hinz, 1992 (type species: *Mononotella sestina*
Fleming, 1973)
1992c *Pejonesia* gen. nov. – Hinz, p. 5.
1996 *Pejonesia* Hinz, 1993 – Hinz-Schallreuter &
Koppka, pp. 27, 36.
1997 *Pejonesia* – Hinz-Schallreuter, p. 13.
1998 *Pejonesia* Hinz, 1992 – Williams & Siveter,
p. 35.
1999 *Pejonesia* – McKenzie *et al.*, p. 462.

Pejonesia sestina (Fleming, 1973) Hinz, 1991 (type locality and horizon: Middle Cambrian, Georgina Basin,
Queensland, Australia)
1973 *Mononotella sestina* n. sp. – Fleming, p. 7.
1980 *Mononotella*? *sestina* Fleming, 1973 – Jones &
McKenzie, p. 220.
1986a *Mononotella sestina* Fleming, 1973 – Kempf,
p. 512.
1986b *Mononotella sestina* Fleming, 1973 – Kempf,
p. 538.
1987 *Mononotella sestina* Fleming, 1973 – Kempf,
p. 583.
1987 *Mononotella sestina* Fleming – Zhang, p. 1.
1990a *Mononotella sestina* Fleming, 1973 – Shu, p. 70.
1991a *Pejonesia sestina* (Jones & McKenzie, 1980) –
Hinz, p. 233.
1992c *Pejonesia sestina* (Fleming, 1973) – Hinz,
pp. 5, 7.
1999 *Pejonesia sestina* (Fleming) – McKenzie *et al.*,
fig. 33.4.

Phaseolella Zhang, 1987 (type species: *Phaseolella dimorpha* Zhang, 1987 by original designation)
1987 *Phaseolella* gen. nov. – Zhang, pp. 1, 2, 8–10,
16, 18.
1990 *Phaseolella* Zhang, 1987 – Abushik *et al.*, p. 46
(synonymised with *Dabashanella*).
1990 *Phaseolella* – Melnikova & Mambetov, pp. 57,
58, 60 (synonymised with *Dabashanella*).
1991 *Phaseolella* Zhang – Huo *et al.*, p. 180.
1991 *Phaseolella* – Melnikova & Mambetov, pp. 56,
57, 59 (synonymised with *Dabashanella*).
1996 *Phaseolella* Zhang, 1987 – Hinz-Schallreuter &
Koppka, pp. 28, 29, 38, 40, 41 (synonymised
with *Comleyopsis*).

Phaseolella curvata Zhang, 1987 (type locality and horizon: Lower Cambrian, Xichuan, Henan, China)
1987 *Phaseolella curvata* sp. nov. – Zhang, pp. 18,
figs. 2, 13A–E.
1989b *Phaseolella curvata* Zhang – Zhao, p. 412, table
1; pl. 1, fig. 13 (referred to *Dabashanella*).
1989b *Phaseolella curvata* Zhang – Zhao, p. 412, table
1; pl. 1, fig. 19 (referred to *Hesslandona*).
1990 *Phaseolella curvata* – Melnikova & Mambetov,
pp. 57, 61.
1991 *Phaseolella curvata* Zhang – Huo *et al.*, p. 181,
pl. 37, figs. 4–8.
1991 *Phaseolella curvata* – Melnikova & Mambetov,
pp. 56, 60.
1996 *Phaseolella curvata* Zhang, 1987 – Hinz-
Schallreuter & Koppka, p. 38.

Phaseolella dimorpha Zhang, 1987 (type locality and
horizon: Lower Cambrian, Xichuan, Henan, China)
1987 *Phaseolella dimorpha* sp. nov. – Zhang, pp. 1–3,
5, 6, 10, 16, 18, figs. 2–5, 12A–W.
1989b *Phaseolella dimorpha* Zhang – Zhao, p. 411
(synonymised with *Dabashanella hemicycla*).
1990 *Phaseolella dimorpha* – Melnikova &
Mambetov, pp. 57, 61 (synonymised with
Dabashanella retroswinga).
1991 *Dabashanella dimorpha* Zhang – Huo *et al.*,
p. 179, pl. 39, figs. 1–21; text-figs. 2–12, 13.
1991 *Phaseolella dimorpha* – Melnikova &
Mambetov, pp. 56, 60 (synonymised with
Dabashanella retroswinga).
1993b *Dabashanella dimorpha* (Zhang, 1987) – Hinz-
Schallreuter, pp. 329, 346 (on pp. 343, 345 as
Phaseolella dimorpha).
1993 *Phaseolella dimorpha* (Zhang, 1987) – Zhang
& Pratt, p. 94.
1996 *Phaseolella dimorpha* – Hinz-Schallreuter &
Koppka, pp. 38, 40.
1998 *Phaseolella dimorpha* – Hinz-Schallreuter,
p. 107.

Phosphatocopina Müller, 1964
1964a Phosphatocopina n. – Müller, pp. 1, 2, 3, 8,
18, 19–21, 41.
1965 Phosphatocopina Müller – Adamczak,
pp. 27–29, 32, 33.
1968 Phosphatocopida – Öpik, p. 9.
1969 Phosphatocopina Müller, 1964 – Andres,
pp. 169, 174, 179.
1970 ("phosphatocopine cladoceran-like form" –
McKenzie, p. 111 (referred to Adamczak
1965).)
1972 Phosphatocopina – Taylor & Rushton, p. 3.
1973 Phosphatocopina Müller, 1964 – Müller, p. 50.
1974 Phosphatocopina Müller – Kozur, pp. 823,
824, 827, 829.

1975   Phosphatocopina Müller, 1964 – Müller, pp. 177, 178.

1978   ("phosphatocopins" – Pokorny, p. 128.)

1979a Phosphatocopina – Müller, pp. 1–26.

1980   Phosphatocopina Müller, 1964 – Jones & McKenzie, pp. 203, 210, 219.

1980   Phosphatocopina Müller, 1964 – Landing, p. 757.

1982a Phosphatocopida nov. (ex suborder Phosphatocopina) – Müller, p. 278.

1982c Phosphatocopina Müller, 1982 – Müller, p. 251.

1982   ("phosphatocopine ostracods" – Schram, p. 111.)

1982   Phosphatocopina – Maddocks, pp. 225, 226.

1983   Phosphatocopina – Briggs, pp. 8–10, figs. 3, 4; table 10.

1983   Phosphatocopina – Huo & Shu, p. 84.

1983   Phosphatocopida Müller, 1982 – McKenzie *et al.*, pp. 35, 37.

1983   Phosphatocopina – Müller, p. 94.

1984   ("phosphatocopines" – Newman & Knight, p. 686.)

1984   Phosphatocopa – Schallreuter, p. 2.

1985   Phosphatocopina Müller, 1964 – Huo & Shu, p. 184.

1985   Phosphatocopina – Jiang & Xiao, pp. (179), 184.

1985a ("phosphatocopid ostracods" – Müller & Walossek, pp. 161, 171.)

1986   Phosphatocopina – Huo *et al.*, p. 23.

1986a Phosphatocopida – Müller & Walossek, p. 73.

1986   Phosphatocopida K.J. Müller, 1964 – Schram, pp. (415–)417, 544.

1987   Phosphatocopida Müller 1964 – Hinz, p. 59.

1987   Phosphatocopida Müller, 1982 – Tong, pp. 433, 436.

1987   Phosphatocopida – Zhang, pp. 1, 2, (8), 10.

1989   Phosphatocopina Müller, 1964 – Huo & Cui, p. 79.

1989b Phosphatocopina Müller, 1964 – Zhao, pp. (411), 415

1990   Phosphatocopina Müller, 1973 – Bengtson *in* Bengtson *et al.*, pp. 322, 323 (sic!).

1990   Phosphatocopina Müller, 1964 – Melnikova & Mambetov, p. 57.

1990a Phosphatocopida Müller, 1982 – Shu, pp. 40, 65, 79.

1990b Phosphatocopida Müller 1982 – Shu, pp. 321, 323.

1991a Phosphatocopida Müller, 1964 – Hinz, p. 232.

1991b ("phosphatocopid ostracods" – Hinz, p. 71.)

1991   Phosphatocopina Müller, 1964 – Huo *et al.*, pp. 177, 205.

1991   Phosphatocopina Müller, 1964 – Melnikova & Mambetov, p. 56.

1991a ("Phosphatocopinen" – Müller & Walossek, p. 283.)

1992   Phosphatocopida – Boxshall, p. 336.

1992b Phosphatocopina Müller, 1964 – Hinz, p. 123.

1992c ("phosphatocopinid ... ostracods" – Hinz, p. 7.)

1992   Phosphatocopina – Hinz & Jones, p. 9.

1993   Phosphatocopina Müller, 1964 – Hinz, pp. 3–5, 11.

1993a Phosphatocopida Müller, 1982 – Hinz-Schallreuter, pp. (306), 307.

1993b Phosphatocopina – Hinz-Schallreuter, pp. 329–331, 346–348.

1993c Phosphatocopina Müller, 1964 – Hinz-Schallreuter, pp. 385, 386, 391, 395.

1993   Phosphatocopina – Walossek, pp. 4, 112.

1993   Phosphatocopida – Zhang & Pratt, p. 94.

1994   Phosphatocopa – Hinz-Schallreuter, p. 13.

1995   Phosphatocopina Müller – Hinz-Schallreuter, p. 415.

1995   Phosphatocopina – Müller *et al.*, p. 112.

1996   (phosphatocopids – Fryer, p. 16.)

1996   Phosphatocopida – Hou *et al.*, pp. 1131, 1141.

1996   Phosphatocopina Müller – Hou *et al.*, p. 1131.

1996   Phosphatocopa Müller, 1964 – Hinz-Schallreuter & Koppka, p. 29.

1996   Phosphatocopina – Hinz-Schallreuter & Koppka, p. 38 (footnote).

1997   Phosphatocopa – Hinz-Schallreuter, p. 13.

1997   Phosphatocopida Müller, 1964 – Hou & Bergström, p. 112.

1997   Phosphatocopina Müller, 1964 – Melnikova *et al.*, p. 179.

1997   Phosphatocopida Müller – Siveter & Williams, p. 59.

1998   Phosphatocopida – Boxshall, p. 162.

1998   Phosphatocopina – Cohen *et al.*, pp. 251, 254.

1998   ("Phosphatocopinen" – Geyer, p. 18.)

1998   Phosphatocopa Müller, 1964 emended Schallreuter, 1984 – Hinz-Schallreuter, pp. (103–108), 109.

1998   Phosphatocopina – Schram & Hof, p. 244.

1998   ("phosphatocopids" – Vannier & Walossek, pp. 97, 98).

1998a ("phosphatocopines" – Walossek & Müller, p. 140.)

1998b Phosphatocopina – Walossek & Müller, p. 207.

1998   Phosphatocopida Müller, 1964 – Williams & Siveter, p. 27.

1998   Phosphatocopina – Ziegler, p. 223.

1999   Phosphatocopina – McKenzie *et al.*, pp. 459, 462, 507.

1999 Phosphatocopida – Shu *et al.*, pp. (281), 295, fig. 11.

1999 Phosphatocopida Müller, 1964 – Whatley *et al.*, p. 344.

2001 ("phosphatocopines" – Boxshall *in* Martin & Davis, p. 105.)

2001 Phosphatocopida – Chen *et al.*, p. 2184, fig. 4.

2001a Phosphatocopina – Maas & Waloszek, p. 99.

2001b Phosphatocopina – Maas & Waloszek, p. 152.

2001 ("phosphatocopid arthropods", "phosphato-copids" – Martin & Davis, pp. 10, (30).)

2001 Phosphatocopina – Schmidt-Rhaesa & Bartolomaeus, p. 129, fig. 9A.

2001 Phosphatocopida – Siveter *et al.*, pp. (479, 480), 481.

2001 ("phosphatocopids" – Vannier *et al.*, p. 75.)

2001 Phosphatocopida – Waloszek & Maas, p. 74.

Pseudindiana Zhao, 1989 (type species: *Indiana sipa* Fleming, 1973; assignment uncertain)
1989a *Pseudindiana* gen. nov. – Zhao, pp. 470, 472.
1989b *Pseudindiana* Zhao, 1989 – Zhao, pp. 412, 415.

Pseudindiana decliviovata (Zhao & Xiao *in* Xiao & Zhao, 1986) (type locality and horizon: Lower Cambrian, China)
1986 *Naviformella decliviovata* Zhao & Xiao – Zhao & Xiao *in* Xiao & Zhao, p. 81.
1989a *Pseudindiana decliviovata* (Zhao & Xiao, 1986) – Zhao, fig. 4.
1989b *Mononotella decliviovata* (Zhao & Xiao) – Zhao, p. 412, table 1, pl. 1, fig. 15 (referred to *Pseudindiana*).

Pseudindiana sipa (Fleming, 1973) Zhao, 1989 (type locality and horizon: Middle Cambrian, Georgina Basin, Queensland, Australia)
1973 *Indiana sipa* n. sp. – Fleming, p. 5.
1986a *Indiana sipa* Fleming, 1973 – Kempf, p. 413.
1986b *Indiana sipa* Fleming, 1973 – Kempf, p. 549.
1987 *Indiana sipa* Fleming, 1973 – Kempf, p. 583.
1989a *Pseudindiana sipa* Fleming, 1973 – Zhao, p. 472.
1989b *Indiana? sipa* Fleming – Zhao, p. 412, table 1 (synonymised with *Dabashanella hemicyclica*).
1989b *Indiana sipa* Fleming – Zhao, p. 415 (referred to *Pseudindiana*).
1990a *Indiana sipa* Fleming – Shu, p. 321 (synonymised with *Dabashanella hemicyclica*).

Pseudodabashanella Shu, 1990 (type species: *Pseudodabashanella striata* Shu, 1990 by original designation; to Dabashanella according to Shu 1990a, b)
1990a *Pseudodabashanella* gen. nov. – Shu, p. 69.
1990b *Pseudodabashanella* – Shu, p. 323.

Pseudodabashanella striata Shu, 1990 (type locality and horizon: Lower Cambrian, China)

1990a *Pseudodabashanella striata* sp. nov. – Shu, pp. 69, 91, pl. 10, figs. 7–10; text-fig. 44.
1990 *Pseudodabashanella striata* – Malz, table 1 (referred to Shu 1990b).

Pseudodahebaella Shu, 1990 (type species: *Pseudodahebaella striata* Shu, 1990; assignment uncertain)
1990a *Pseudodahebaella* gen. nov. – Shu, p. 62.

Pseudodahebaella striata Shu, 1990 (type locality and horizon: Upper Cambrian, China)
1990a *Pseudedahebaella striata* sp. nov. – Shu, pp. 62, 92, pl. 12, fig. 3.

Reticulocambria Müller, 1964 (type species: *Reticulocambria lobata* Müller, 1964 by original designation)
1964a *Reticulocambria* n. g. – Müller, p. 35.
1974 *Reticulocambria* Müller, 1964 – Kozur, p. 826 (referred to Hipponicharionidae Sylvester-Bradley, 1961).
1986a *Reticulocambria* Mueller, 1964 – Kempf, p. 660.
1986b *Reticulocambria* Mueller, 1964 – Kempf, p. 598.
1987 *Reticulocambria* Mueller, 1964 – Kempf, p. 136.
1993c *Reticulocambria* Müller, 1964 – Hinz-Schallreuter, p. 395 (footnote).
1998 *Reticulocambria* – Hinz-Schallreuter, p. 1.

Reticulocambria lobata Müller, 1964 (type locality and horizon: Upper Cambrian, Zone 5d–f, Trolmen, Kinnekulle, Sweden)
1964a *Reticulocambria lobata* n. sp. – Müller, p. 36, pl. 4, figs. 13–15.
1974 *Reticulocambria lobata* – Martinsson, pp. 208, 212.
1986a *Reticulocambria lobata* Mueller, 1964 – Kempf, p. 660.
1986b *Reticulocambria lobata* Mueller, 1964 – Kempf, p. 337.
1987 *Reticulocambria lobata* Mueller, 1964 – Kempf, p. 436.

Remarks. – This species is considered to be a nomen dubium because it proved to be merely a preservational state (Hinz-Schallreuter 1993c, 1998).

Schallreuterina Hinz-Schallreuter, 1993 (type species: *Schallreuterina campanae* Hinz-Schallreuter, 1993 by original designation)
1993a *Schallreuterina* n. g. – Hinz-Schallreuter, p. 310.
1993b *Schallreuterina* Hinz, 1993 – Hinz-Schallreuter, p. 343 (sic!; author's name wrong).
1993c *Schallreuterina* Hinz-Schallreuter, 1993 – Hinz-Schallreuter, pp. 410, 442.
1998 *Schallreuterina* Hinz-Schallreuter, 1993 – Hinz-Schallreuter, p. 115.

1998 *Schallreuterina* Hinz-Schallreuter, 1993 – Williams & Siveter, p. 30.

Schallreuterina campanae Hinz-Schallreuter, 1993 (type locality and horizon: Middle Cambrian, Late Templetonian, Mount Murray, Queensland, Australia)

1993a *Schallreuterina campanae* n. sp. – Hinz-Schallreuter, p. 310, fig. 2.3.

Schallreuterinidae Hinz-Schallreuter, 1993

1993c Schallreuterinidae n. fam. – Hinz-Schallreuter, pp. 386, 396, 410.

1996 Schallreuterinidae Hinz-Schallreuter, 1993 – Hinz-Schallreuter & Koppka, p. 29.

1998 Schallreuterinidae – Williams & Siveter, p. 30.

Semillia Hinz, 1992 (type species: *Semillia pauper* Hinz, 1992 by original designation)

1992a *Semillia* n. g. – Hinz, pp. 13–16.

1993b *Semillia* Hinz, 1992 – Hinz-Schallreuter, pp. 331, 341.

1993c *Semillia* Hinz, 1992 – Hinz-Schallreuter, p. 412.

1998 *Semillia* – Hinz-Schallreuter, p. 110.

Semillia pauper Hinz, 1992 (type locality and horizon: Middle Cambrian, Late Templetonian, Rogers Ridge, Queensland, Australia)

1992a *Semillia pauper* n. sp. – Hinz, pp. 13–16, pl. 19, figs. 1–4.

1992b *Semillia pauper* Hinz – Hinz, p. 125.

Shergoldopsis Hinz-Schallreuter, 1993 (type species: *Shergoldopsis marginoplana* Hinz-Schallreuter, 1993)

1993a *Shergoldopsis* n. g. – Hinz-Schallreuter, p. 312.

1993c *Shergoldopsis* Hinz-Schallreuter, 1993 – Hinz-Schallreuter, p. 412.

Shergoldopsis marginoplana Hinz-Schallreuter, 1993 (type locality and horizon: Middle Cambrian, Late Templetonian, Rogers Ridge, Queensland, Australia)

1993a *Shergoldopsis marginoplana* n. sp. – Hinz-Schallreuter, p. 312, fig. 2.2.

Spinella Shu, 1990 (type species: *Spinella amargina* Shu, 1990 by original designation)

1990a *Spinella* gen. nov. – Shu, pp. 61, 63.

Spinella amargina Shu, 1990 (type locality and horizon: Lower Cambrian, China)

1990a *Spinella amargina* sp. nov. – Shu, pp. 61, 62, 89, pl. 3, fig. 3, text-fig. 40.

Trapezilites Hinz-Schallreuter, 1993 (type species: *Aristozoe ? minima* Kummerow, 1931 by original designation, see p. 106)

1993c *Trapezilites* n. g. – Hinz-Schallreuter, pp. 399, 402.

1998 *Trapezilites* Hinz-Schallreuter – Williams & Siveter, p. 29.

Trapezilites minimus (Kummerow, 1931) Hinz-Schallreuter, 1993 (see p. 106; type locality and horizon: Upper Cambrian, Zone 1, Degerhamn, Isle of Öland, Sweden)

1924 *Aristozoë primordialis* Linnss. sp. – Kummerow, pp. 406, 445, 446.

. 1928 *Aristozoe* ? cf. *primordialis* Linnss. sp. – Kummerow, pp. 42, 59, pl. 2, fig. 19.

. 1931 *Aristozoe* ? *minima* n. sp. – Kummerow, pp. 254–255, text-fig. 18.

. 1934 "*Aristozoe*" *minima* (Kummerow), 1931 – van Straelen & Schmitz, pp. 176, 209, 228, 236, 238.

v. 1964a *Falites* (?) *minima* (Kummerow) – Müller, p. 29, pl. 4, figs. 8–12, 16.

. 1965 *Falites minima* (?) (Kummerow) – Adamczak, p. 28, pl. 1, fig. 3a, b; text-fig. 2.

. 1972 *Falites? minimus* (Kummerow) – Taylor & Rushton, p. 13, pl. 4 (borehole record).

. 1974 *Falites minimus* – Martinsson, p. 208.

p. 1978 *Falites? minimus* (Kummerow, 1931) – Rushton, p. 277 [*partim*: specimens BGS BDA 1167/1168, BGS BDA 1276/1277, BDA 1452/1453 (pl. 26, figs. 9, 10; text-fig. 2), *non* specimen BGS BDA 1771/1774 (= *Waldoria rotundata*)].

v. 1979a *Hesslandona* n. sp. – Müller, fig. 8.

v. 1979a *Falites? minima* (Kummerow) – Müller, p. 11.

. 1981 *Falites? minima* (Kummerow) – Gründel, p. 63, pl. III, figs. 7, 8.

. 1986a *Aristozoe? minima* Kummerow, 1931 – Kempf, p. 65.

. 1986a *Falites? minimus* (Kummerow, 1931) Mueller, 1964 – Kempf, p. 355.

. 1986b *Aristozoe? minima* Kummerow, 1931 – Kempf, p. 369.

. 1986b *Falites? minimus* (Kummerow) Mueller, 1964 – Kempf, p. 370.

. 1987 *Aristozoe? minima* Kummerow, 1931 – Kempf, p. 167.

. 1987 *Falites? minimus* (Kummerow, 1931) Mueller, 1964 – Kempf, p. 436.

. 1993c *Aristozoe? minima* Kummerow, 1931 – Hinz-Schallreuter, pp. 388, 402 (referred to *Trapezilites* n. g.).

. 1994 *Trapezilites minimus* – Hinz & Jones, p. 368.

. 1996a *Trapezilites minimus* (Kummerow) – Hinz-Schallreuter, pp. 85–88.

. 1998 *Aristozoe? minima* Kummerow, 1931 – Hinz-Schallreuter, p. 103 (historical review).

. 1998 *Trapezilites minimus* (Kummerow, 1931) – Williams & Siveter, pp. 29, 30, pl. 5, figs. 5, 6.

Tubupestis Hinz & Jones, 1992 (type species: *Tubupestis tuber* Hinz & Jones, 1992 by original designation)

1992 *Tubupestis* n. gen. – Hinz & Jones, pp. 9–12.

1993b *Tubupestis* Hinz & Jones, 1992 – Hinz-Schallreuter, pp. 331, 333, 341, 343.

1993c *Tubupestis* Hinz & Jones, 1992 – Hinz-Schallreuter, p. 412.

1998 *Tubupestis* – Hinz-Schallreuter, p. 110.

Tubupestis tuber Hinz & Jones, 1992 (type locality and horizon: Middle Cambrian, Late Templetonian, Mount Murray, Queensland, Australia)

1980 New pustulose genus – Jones & McKenzie, p. 205.

1992 *Tubupestis tuber* gen. et sp. nov. – Hinz & Jones, pp. 9–12, pl. 19.

1999 *Tubupestis tuber* Hinz & Jones – McKenzie et al., fig. 33.5B.

Ulopsidae Hinz-Schallreuter, 1993

1993c Ulopsidae n. fam. – Hinz-Schallreuter, pp. 386, 396, 410, fig. 5.

1993c Ulopsinae n. sf. – Hinz-Schallreuter, p. 412 (*Ulopsis, Shergoldopsis, Parashergoldopsis*).

1993c Tubupestidae – Hinz-Schallreuter, p. 396 (erroneously?).

1993c Tubupestinae n. sf. – Hinz-Schallreuter, p. 412 (only *Tubupestes*).

1998 Ulopsidae – Hinz-Schallreuter, p. 110.

Ulopsis Hinz, 1991 (type species: *Ulopsis ulula* Hinz, 1991 by original designation)

1991b *Ulopsis* n. gen. – Hinz, p. 69.

1992 *Ulopsis* Hinz, 1991 – Hinz & Jones, pp. 9–12.

1993 *Ulopsis* – Hinz, fig. 2b2.

1993b *Ulopsis* Hinz, 1991 – Hinz-Schallreuter, pp. 331, 333, 343.

1993c *Ulopsis* Hinz, 1993 – Hinz-Schallreuter, pp. 395, 412.

1998 *Ulopsis* – Hinz-Schallreuter, p. 110.

1999 *Ulopsis* – McKenzie et al., p. 461.

Ulopsis ulula Hinz, 1991 (type locality and horizon: Middle Cambrian, Zone B1, Rogers Ridge, Queensland, Australia)

1991b *Ulopsis ulula* sp. nov. – Hinz, pp. 69, 71, pl. 18 (p. 70), figs. 1–3, pl. 18 (p. 71), figs. 1–3.

1993 *Ulopsis ulula* Hinz, 1991 – Hinz, p. 7, fig. 4G, H.

1996 *Ulopsis ulula* Hinz, 1991 – Hinz-Schallreuter, p. 85.

1996 *Ulopsis ulula* – Hinz-Schallreuter & Koppka, p. 29.

Veldotron Gründel *in* Gründel & Buchholz, 1981 (type species: *Veldotron kutscheri* Gründel *in* Gründel & Buchholz, 1981 (= *Vestrogothia bratteforsa* Müller, 1964) by original designation, see p. 123)

1981 *Veldotron* n. g. – Gründel *in* Gründel & Buchholz, p. 66.

1986a *Veldotron* Gruendel, 1981 – Kempf, p. 7

1986b *Veldotron* Gruendel, 1981 – Kempf, p. 7

1987 *Veldotron* Gruendel, 1981 – Kempf, p. 7

1993b *Veldotron* Gründel & Buchholz, 1981 – Hinz-Schallreuter, p. 334.

1993c *Veldotron* Gründel *in* Gründel & Buchholz, 1981 – Hinz-Schallreuter, pp. 395, 402, 405, 408.

1998 *Veldotron* Gründel *in* Gründel & Buchholz, 1981 – Hinz-Schallreuter, p. 118.

1998 *Veldotron* Gründel – Williams & Siveter, 34.

Veldotron bratteforsa (Müller, 1964) Hinz-Schallreuter, 1993 (see p. 123; type locality and horizon: Upper Cambrian, Zone 2, Brattefors, Kinnekulle, Sweden)

v* 1964a *Vestrogothia bratteforsa* n. sp. – Müller, 34, pl. 3, figs. 1, 2.

. 1965 *Vestrogothia bratteforsa* Müller – Adamczak, p. 29.

. 1972 *Vestrogothia bratteforsa* Mueller – Taylor & Rushton, p. 18.

. 1974 *F. bratteforsa* – Martinsson, p. 208 (sic

. 1981 *Veldotron kutscheri* n. sp. – Gründel *in* Gründel & Buchholz, p. 66, pl. III, f. 11, 12, 15.

. 1986 *Vestrogothia bratteforsa* – Hua et al., 3-1 [cop. Müller (1964a, pl. 3, fig. 2b)].

. 1986a *Veldotron kutscheri* Gruendel, 1981 – Kempf, p. 745.

. 1986a *Vestrogothia bratteforsa* Mueller, 1964 – Kempf, p. 747.

. 1986b *Vestrogothia bratteforsa* Mueller, 1964 – Kempf, p. 101.

. 1986b *Veldotron kutscheri* Gruendel, 1981 – Kempf, p. 316.

. 1987 *Vestrogothia bratteforsa* Mueller, 1964 – Kempf, p. 436.

. 1987 *Veldotron kutscheri* Gruendel, 1981 – Kempf, p. 710.

. 1987 *Vestrogothia bratteforsa* Müller – Jiang, p. 433.

. 1993b *Veldotron kutscheri* Gründel & Buchholz, 1982 – Hinz-Schallreuter, p. 334, fig. 1

. 1993c *Vestrogothia bratteforsa* Müller, 1964 – Hinz-Schallreuter, p. 405 (synonymised).

. 1998 *Veldotron bratteforsa* (Müller, 1964) – Hinz-Schallreuter, p. 116.

. 1998 *V. [Veldotron] bratteforsa* – Williams & Siveter, pp. 34, 35.

. 1993c *Veldotron bratteforsa* (Müller, 1964) – Hinz-Schallreuter, pp. 393, 405, figs. 9.1, 9.2

Veldotron rushtoni Williams & Siveter, 1993 (type locality and horizon: Upper Cambrian, Zone 2, Merevale, Warwickshire, England)

1972   *Vestrogothia cf. bratteforsa* Müller – Taylor & Rushton, p. 18.

1998   *Veldotron rushtoni* sp. nov. – Williams & Siveter, pp. 34, 35, pl. 5, figs. 7–11.

Vestrogothia Müller, 1964 (type species: *Vestrogothia spinata* Müller, 1964 by original designation, see p. 139)

1964a   *Vestrogothia* n. g.   Müller, p. 30.

1972   *Vestrogothia* Mueller – Taylor & Rushton, p. 13.

non 1972   *Vestrogothia* Mueller – Taylor & Rushton, p. 18 (*Hesslandona*).

1974   *Vestrogothia* Müller, 1964 – Kozur, p. 827.

1980   *Vestrogothia* – Jones & McKenzie, p. 218.

1981a   *Hesslandona* – Müller, p. 147.

1982a   *Vestrogothia* – Müller, p. 287.

1983   *Vestrogothia* – Briggs, pp. 9, 10.

1983   *Vestrogothia* – Müller, p. 94.

1983   *Vestrogothia* – Reyment, p. 5.

1986   *Vestrogothia* – Huo *et al.*, p. 23.

1986   *Vestrogothia* – Schram, p. 415.

1986a   *Vestrogothia* Mueller, 1964 – Kempf, p. 747.

1986b   *Vestrogothia* Mueller, 1964 – Kempf, p. 707.

1987   *Vestrogothia* Mueller, 1964 – Kempf, p. 436.

1987   *Vestrogothia* – Zhang, pp. 5, 9.

1990   *Vestrogothia* – Bengtson *in* Bengtson *et al.*, p. 323.

1990a   *Vestrogothia* – Shu, pp. 61, 62, 77.

1993b   *Vestrogothia* Müller, 1964 – Hinz-Schallreuter, pp. 334, 342, 344, 347.

1993c   *Vestrogothia* Müller, 1964 – Hinz-Schallreuter, pp. 386, 395, 399, 402, 403, 409.

1995   *Vestrogothia* – Siveter *et al.*, p. 416.

1996b   *Vestrogothia* Müller, 1964 – Hinz-Schallreuter, p. 89.

1998   *Vestrogothia* Müller, 1964 – Hinz-Schallreuter, pp. 104, 107, 116, 118, 126, 132, text-figs. 1, 3.

1998   *Vestrogothia* – Ziegler, p. 223.

Vestrogothia bratteforsa Müller, 1964 (see *Veldotron bratteforsa*)

Vestrogothia granulata Müller, 1964 (type locality and horizon: Upper Cambrian, Zone 2, Brattefors, Kinnekulle, Sweden)

1964a   *Vestrogothia granulata* n. sp. – Müller, p. 33, pl. 1, figs. 13, 14; pl. 2, figs. 1, 3.

1974   *Vestrogothia granulata* Müller, 1964 – Kozur, p. 826 (referred to *Indiana* Matthew, 1902).

1974   *Vestrogothia granulata* – Martinsson, p. 208.

1979b   *Vestrogothia granulata* Müller, 1964 – Müller, p. 92, fig. 2.

1980   *V.* [*Vestrogothia*] *granulata* Müller, 1964 – Jones & McKenzie, p. 218.

1986a   *Indiana granulata* (Mueller, 1964) Kozur, 1974 – Kempf, p. 413.

1986a   *Vestrogothia granulata* Mueller, 1964 – Kempf, p. 747.

1986b   *Indiana granulata* (Mueller, 1964) Kozur, 1974 – Kempf, p. 250.

1986b   *Vestrogothia granulata* Mueller, 1964 – Kempf, p. 250.

1987   *Vestrogothia granulata* Mueller, 1964 – Kempf, p. 436.

1987   *Indiana granulata* (Mueller, 1964) Kozur, 1974 – Kempf, p. 605.

1987   *Vestrogothia granulata* Müller – Tong, p. 433.

1989   *Vestrogothia granulata* Müller – Zhao & Tong, p. 15.

1993c   ?*Vestrogothia granulata* Müller, 1964 – Hinz-Schallreuter, p. 412.

Vestrogothia hastata Müller, 1964 (type locality and horizon: Upper Cambrian, Zone 5c, Stenåsen, Falbygden, Sweden)

1964a   *Vestrogothia hastata* n. sp. – Müller, p. 32, pl. 2, fig. 9.

1986a   *Vestrogothia hastata* Mueller, 1964 – Kempf, p. 747.

1986b   *Vestrogothia hastata* Mueller, 1964 – Kempf, p. 262.

1987   *Vestrogothia hastata* Mueller, 1964 – Kempf, p. 436.

1993c   *Vestrogothia hastata* Müller, 1964 – Hinz-Schallreuter, p. 403.

Vestrogothia herrigi Hinz-Schallreuter, 1998 (type locality and horizon: Middle Cambrian, Isle of Bornholm)

1998   *Vestrogothia herrigi* n. sp. – Hinz-Schallreuter, pp. 103, 126, pl. 4, fig. 1a–c; pl. 8, figs. 6–8; table 8.

Vestrogothia longispinosa Kozur, 1974 (type locality and horizon: Middle Cambrian, erratic boulders from the Isle of Rügen, Germany)

1974   *Vestrogothia longispinosa* n. sp. – Kozur, p. 827, figs. 1, 2.

1986a   *Vestrogothia longispinosa* Kozur, 1974 – Kempf, p. 747.

1986b   *Vestrogothia longispinosa* Kozur, 1974 – Kempf, p. 341.

1987   *Vestrogothia longispinosa* Kozur, 1974 – Kempf, p. 605.

1993c   *Vestrogothia longispinosa* Kozur, 1974 – Hinz-Schallreuter, pp. 392, 396, 403, figs. 8.1–8.3.

1994a   *Vestrogothia longispinosa* Kozur, 1974 – Williams *et al.*, pp. 21, 23, 25; text-figs. 1, 2.

1998   *Vestrogothia longispinosa* Kozur, 1974 – Hinz-

Schallreuter, pp. 103–105, 107, 126, 130, 132, pl. 1, figs. 1–3; pl. 3, figs. 1–9; pl. 4, figs. 2–9; pl. 5, figs. 1–6; pl. 8, fig. 4; pl. 9, figs. 1, 2, 4; text-fig. 8; table 9.

1999   *Vestrogothia longispinosa* Kozur – McKenzie *et al.*, p. 464.

1999   *Vestrogothia longispinosa* Kozur, 1974 – Whatley *et al.*, p. 344.

Vestrogothia ? n. sp. Müller, 1964 (locality and horizon: Upper Cambrian, Zone 5d–f, Trolmen, Kinnekulle, Sweden)

1964a   *Vestrogothia* ? n. sp. – Müller, p. 35, pl. 2, fig. 2.

Vestrogothia minilaterospinata Hinz-Schallreuter, 1998 (type locality and horizon: Middle Cambrian, Isle of Bornholm)

1998   *Vestrogothia minilaterospinata* n. sp. – Hinz-Schallreuter, pp. 103, 132, pl. 8, fig. 5.

Vestrogothia? sp. Melnikova, Siveter & Williams, 1997 (locality and horizon: Middle Cambrian, area of the former Soviet Union)

1997   *Vestrogothia*? sp. – Melnikova *et al.*, p. 185, pl. 3, fig. 10, text-fig. 3.

Vestrogothia spinata Müller, 1964 (see p. 139; type locality and horizon: Upper Cambrian, Zone 5c, Stenåsen, Falbygden, Sweden)

v* 1964a   *Vestrogothia spinata* n. sp. – Müller, p. 30, pl. 2, figs. 4–8, 10, 11; pl. 5, figs. 1, 7–9.

.   1965   *Vestrogothica spinata* Müller – Adamczak, pp. 29, 32 (*V. spinata* on p. 29, sic! on p. 32).

.   1974   *Vestrogothia spinata* Müller, 1964 – Kozur, pp. 827, 828.

.   1979a   *Vestrogothia spinata* Müller – Müller, pp. 23, 24, figs. 3, 4, 13, 14, 16, 18, 29, 30, 31, 34A, D.

.   1979b   *Vestrogothia spinata* Müller, 1964 – Müller, p. 92, fig. 1.

.   1982b   *Vestrogothia spinata* Müller – Müller, fig. 1.

.   1982c   *Vestrogothia spinata* – Müller, fig. 4.

.   1982   *Vestrogothia spinata* – Schram, p. 111.

.   1983   *Vestrogothia spinata* – Briggs, p. 9.

.   1983   *Vestrogothia spinata* Müller, 1964 – McKenzie *et al.*, fig. 5.

.   1983   *Vestrogothia spinata* Müller, 1964 – Reyment, fig. 3.

.   1985a   *Vestrogothia spinata* Müller, 1964 – Müller & Walossek, fig. 2f.

.   1986   *Vestrogothia spinata* – Schram, fig. 33-10B–H.

.   1986a   *Vestrogothia spinata* Mueller, 1964 – Kempf, p. 747.

.   1986b   *Vestrogothia spinata* Mueller, 1964 – Kempf, p. 555.

.   1987   *Vestrogothia spinata* Mueller, 1964 – Kempf, p. 436.

.   1987   *Vestrogothia spinata* Müller – Tong, p. 3.

.   1989b   *Vestrogothia spinata* – Zhao, p. 471.

.   1989   *Vestrogothia spinata* Müller, 1979 – Z & Tong, p. 15 [referred to Müller (1979a, fig. 1)].

.   1993b   *Vestrogothia spinata* Müller, 1964 – inz-Schallreuter, pp. 330, 334, 344, 345, fi B.

.   1993c   *Vestrogothia spinata* Müller, 1964 – inz-Schallreuter, pp. 396, 403.

.   1996b   *Vestrogothia spinata* Müller, 1964 – inz-Schallreuter, p. 89.

.   1996   *Vestrogothia spinata* Müller, 1964 – inz-Schallreuter & Koppka, p. 38 (footnot

.   1998   *Vestrogothia spinata* Müller, 1964 – inz-Schallreuter, p. 126.

.   1998   *Vestrogothia spinata* Müller, 1964 – W ams & Siveter, p. 31.

.   1999   *Vestrogothia minuta* Müller, 1979 – Mc nzie *et al.*, p. 460, text-fig. 33.1.

.   1999   *Vestrogothia spinata* Müller – McKenzi *al.*, p. 463.

.   1999   *Vestrogothia spinata* Müller, 1964 – W tley *et al.*, p. 344.

Vestrogothia steffenschneideri Hinz-Schallreuter, 993 (type locality and horizon: Upper Cambrian, Z e 2, Bralitz, Oderberg, Brandenburg, Germany)

1993c   *Vestrogothia steffenschneideri* n. sp. – inz-Schallreuter, pp. 393, 403, fig. 6.4.

1994a   *Vestrogothia steffenschneideri* inz-Schallreuter, 1993 – Williams *et al.*, p.

Vestrogothiidae Kozur, 1974

1974   Vestrogothidae nov. – Kozur, p. 827.

1980   Vestrogothidae – Jones & McKenzie, p. 9.

1982a   Vestrogothiidae Müller, 1964 – Müller, 285 (emended).

1986   Vestrogothiidae K.J. Müller, 1964 – S am, p. 417.

1990   Vestrogothiidae Kozur, 1974 – Abushik al., p. 45.

1997   Vestrogothiidae Kozur – Siveter & Wi ms, p. 59.

1993b   Vestrogothiidae Kozur, 1974 – inz-Schallreuter, p. 347.

1993c   Vestrogothiidae Kozur, 1974 – inz-Schallreuter, pp. 391, 396, 399, 402.

1996b   Vestrogothiidae Kozur, 1974 – inz-Schallreuter, p. 89.

1997   Vestrogothiidae (=? Hesslandonidae) – inz-Schallreuter, p. 13.

1998   Vestrogothiinae Kozur, 1974 – inz-Schallreuter, pp. 106, 110, 116, 117 (subf nily of Hesslandonidae).

1998   Vestrogothiidae Kozur, 1974 – Willia s & Siveter, pp. 27, 31.

1999   Vestrogothiidae – McKenzie *et al.*, pp. 463, 464.

1999   Vestrogothiidae Kozur, 1974 – Whatley *et al.*, p. 344.

Vestrogothiina Müller, 1982

1982a Vestrogothiina nov. – Müller, p. 285.

1983   Vestrogothicopina Müller, 1982 – McKenzie *et al.*, pp. 35, 37.

1986   Vestrogothicopina K.J. Müller, 1982 – Schram, p. 417.

1987   Vestrogothicopina Müller, 1982 – Hinz, p. 59.

1987   Vestrogothiina – Zhang, p. 10.

1990   Vestrogothiacea Kozur, 1974 (nom. transl.) – Abushik *et al.*, p. 45.

1990a Vestrogothiina Müller, 1982 – Shu, pp. 40, 66, on p. 79 sic! as "vestrogothina".

1990b Vestrogothiina Müller 1982 – Shu, p. 323.

1993b Vestrogothiina [Müller, 1982] – Hinz-Schallreuter, pp. 346, 347.

1993c Vestrogothiina – Hinz-Schallreuter, p. 399 (on p. 395 as synonymous with Phosphatocopina).

1994   Vestrogothicopina – Hinz-Schallreuter, p. 13.

1995   Vestrogothicopina – Hinz-Schallreuter, p. 415.

1996   Vestrogothicopina = Vestrogothiina – Hinz-Schallreuter & Koppka, p. 38 (footnote).

1998   Vestrogothiina – Williams & Siveter, p. 27.

1999   Vestrogothiina Müller, 1982 – Whatley *et al.*, p. 344.

2001   Vestrogothicopina – Chen *et al.*, fig. 4.

Waldoria Gründel *in* Gründel & Buchholz, 1981 (type species: *Waldoria buchholzi* Gründel, 1981, see p. 113)

1981   *Waldoria* n. g. – Gründel *in* Gründel & Buchholz, p. 60 (referred to Bradoriidae Matthew, 1902).

1986a *Waldoria* Gruendel, 1981 – Kempf, p. 749.

1986b *Waldoria* Gruendel, 1981 – Kempf, p. 707.

1987   *Waldoria* Gruendel, 1981 – Kempf, p. 710.

1998   *Waldoria* Gründel *in* Gründel & Buchholz, 1981 – Hinz-Schallreuter, p. 118.

1998   *Waldoria* Gründel – Williams & Siveter, p. 31.

Waldoria buchholzi Gründel *in* Gründel & Buchholz, 1981 (type locality and horizon: Upper Cambrian, Zone 3, erratic boulders from the Isle of Rügen, Germany)

1981   *Waldoria* n. g. – Gründel *in* Gründel & Buchholz, p. 60.

1986a *Waldoria* Gruendel, 1981 – Kempf, p. 749.

1986b *Waldoria* Gruendel, 1981 – Kempf, p. 707.

1987   *Waldoria* Gruendel, 1981 – Kempf, p. 710.

1998   *Waldoria* Gründel *in* Gründel & Buchholz, 1981 – Hinz-Schallreuter, p. 118.

1998   *Waldoria* Gründel – Williams & Siveter, p. 31.

Waldoria rotundata Gründel *in* Gründel & Buchholz,

1981 (see p. 115; type locality and horizon: ?Upper Cambrian, erratic boulders from the Isle of Rügen, Germany)

1978   *Bradoria* sp. – Rushton, p. 275, pl. 26, figs. 13, 14 (see Williams & Siveter 1998).

p. 1978 *Falites fala* Müller, 1964 – Rushton, p. 276 (BGS BDA 1824 only), *non* BGS BDA 1820 (pl. 26, fig. 12), BDA 1844, BDA 1855, BDA 1863 (= *Falites fala*) (see Williams & Siveter 1998).

p. 1978 *Falites? minimus* (Kummerow, 1931) – Rushton, p. 277 (*partim*: BGS BDA 1771/1774 part and counterpart only), *non* BGS BDA 1167/1168, BDA 1452/1453 (= *Trapezilites minimus*) (see Williams & Siveter 1998).

1978   *Walcottella* sp. – Rushton, p. 276, pl. 26, figs. 6, 7.

* 1981 *Waldoria rotundata* n. sp. – Gründel *in* Gründel & Buchholz, p. 61, pl. II, fig. 8; text-fig. 4.

. 1981 *Waldoria* n. sp. 1 – Gründel *in* Gründel & Buchholz, p. 62, pl. II, figs. 9, 10; text-fig. 5.

v. 1985a *Waldoria* sp. – Müller & Walossek, figs. 4c, 6a, b (UB 770).

. 1986a *Waldoria rotundata* Gruendel, 1981 – Kempf, p. 749.

. 1986b *Waldoria rotundata* Gruendel, 1981 – Kempf, p. 514.

. 1987 *Waldoria rotundata* Gruendel, 1981 – Kempf, p. 710.

1998   *Waldoria* cf. *rotundata* Gründel, 1981 – Williams & Siveter, p. 31, pl. 6, figs. 7, 8.

. 1998 *W. rotundata* – Williams & Siveter, p. 31.

Xiangzheella Shu, 1990 (type species: *Xiangzheella alta* Shu, 1990 by original designation; assignment uncertain)

1990a *Xiangzheella* gen. nov. – Shu, p. 60.

Xiangzheella alta Shu, 1990 (type locality and horizon: Lower Cambrian, China)

1990a *Xiangzheella alta* sp. nov. – Shu, pp. 60, 61, 92, pl. 12, fig. 6.

Xiangzheella taoyuanensis Shu, 1990 (type locality and horizon: Lower Cambrian, China)

1990a *Xiangzheella taoyuanensis* sp. nov. – Shu, pp. 61, 92, pl. 12, fig. 7.

"*Xiaoyangbaella*" *nudata* Zhao & Xiao *in* Xiao & Zhao, 1986 (to *Dabashanella* according to D. J. Siveter, pers. comm. 2002; type locality and horizon: Lower Cambrian, China)

1986   *Xiaoyangbaella nudata* Zhao & Xiao (sp. nov.) – Zhao & Xiao *in* Xiao & Zhao, p. 82, pl. 2, figs. 5–10.

1989b *Hesslandona nudata* Zhao & Xiao – Zhao, p. 412, pl. 1, fig. 18.

Xichuanella Zhao, 1989 (see *Monasterium*).
Zepaera Fleming, 1973 (type species: *Zepaera rete* Fleming, 1973 by original designation)
1973　*Zepaera* – Fleming, p. 7.
1980　*Zepaera* Fleming, 1973 – Jones & McKenzie, p. 213.
1986　*Zepaera* Fleming – Huo *et al.*, p. 22.
1986a　*Zepaera* Fleming, 1973 – Kempf, p. 759.
1986b　*Zepaera* Fleming, 1973 – Kempf, p. 708.
1987　*Zepaera* Fleming, 1973 – Kempf, p. 583.
1987　*Zepaera* Fleming – Zhang, pp. 8, 10, 12.
1989b　*Zepaera* – Zhao, pp. 410, 411, 413.
1990　*Zepaera* Fleming, 1973 – Abushik *et al.*, p. 43.
1993　*Zepaera* – Hinz, p. 5.
1993c　*Zepaera* – Hinz-Schallreuter, pp. 395, 440.

Zepaera brevidorsa Zhou *in* Huo & Shu, 1985 (type locality and horizon: Lower Cambrian, Yutaishan, Huoqiu, Anhui, China)
1985　*Zepaera brevidorsa* n. sp. – Zhou *in* Huo & Shu, p. 197, pl. 36, fig. 1a–d.
1986　*Zepaera brevidorsa* Zhou – Huo *et al.*, pl. 1, fig. 13; text-fig. 1-13.
1989b　*Zepaera brevidorsa* Zhou – Zhao, p. 410, pl. 1, fig. 2.

Zepaera primitiva Shu, 1990 (type locality and horizon: Lower Cambrian, Shaanxi, China)
1990a　*Zepaera primitiva* sp. nov. – Shu, pp. 49, 75, 91, pl. 8, figs. 1–8, text-fig. 26.
1990b　*Zepaera primitiva* – Shu, pl. 2, fig. 20a, b.
1990　*Zepaera primitiva* sp. nov. – Malz, table 1 (referred to Shu 1990b).
1994　*Zepaera primitiva* – Shu & Chen, fig. 5j.

Zepaera rete Fleming, 1973 (type locality and horizon: Middle Cambrian, Georgina Basin, Queensland, Australia)
1973　*Zepaera rete* – Fleming, p. 8, pl. 1, figs. 18, 19; pl. 2, figs. 1–4, text-figs. A8, A9.
1980　*Zepaera rete* Fleming, 1973 – Jones & McKenzie, p. 213.
1985　*Zepaera rete* – Zhou *in* Huo & Shu, p. 197.
1986　*Zepaera rete* Fleming – Huo *et al.*, pl. 1, fig. 14; text-fig. 1-12.
1986a　*Zepaera rete* Fleming, 1973 – Kempf, p. 759.
1986b　*Zepaera rete* Fleming, 1973 – Kempf, p. 498.
1987　*Zepaera rete* Fleming, 1973 – Kempf, p. 583.
1987　*Zepaera rete* Fleming – Zhang, pp. 1, 9.
1989b　*Zepaera rete* Fleming – Zhao, p. 410.
1990　*Zepaera rete* Fleming, 1973 – Bengtson *in* Bengtson *et al.*, p. 323.
1993　*Zepaera rete* Fleming, 1973 – Hinz, fig. 4A, B.

1996　*Zepaera rete* Fleming, 1973 – Whatley al., p. 344.
1999　*Zepaera rete* Fleming – McKenzie *et al.*, 161.

Zepaera sinensis (Zhou, 1985) Zhao, 1989 (type locality and horizon: Lower Cambrian, China)
1985　*Ophoisema sinensis* n. sp. – Zhou, p. 98, l. 1, fig. 2.
1989b　*Zepaera sinensis* Zhou – Zhao, p. 410.

Zepaera xichuanensis (Zhang, 1987) Zhao, 1989 (type locality and horizon: Lower Cambrian, China)
1987　*Houlongdongella xichuanensis* sp. nov. – ing, p. 12, figs. 6A–C, 10A–O.
1989b　*Zepaera xichuanensis* (Zhang) – Zhao, l. 1, fig. 4.

Possible Phosphatocopines, described in open nomenclature

"*Bradoria*" sp. B Melnikova, 1990 (type locality and horizon: Lower/Middle Cambrian, area of the former Soviet Union)
1990　*Bradoria* sp. B – Melnikova, p. 73.
1997　*Bradoria* sp. B of Melnikova, 1990 (= *Liangshanella*?) – Melnikova *et al.*, p. 187, l. 4, fig. 7, text-fig. 3.

Bradoriidae species E Hinz, 1987 (locality and horizon: Lower Cambrian, Comley, England)
1987　Species E – Hinz, p. 60, pl. 3, fig. 13; ta 1.

Gen. et sp. A Hinz, 1987 (locality and horizon: Lower Cambrian, Comley, England)
1987　*Vestrogothicopina* gen. et sp. A – Hinz, 59, pl. 2, fig. 12; pl. 3, figs. 1–3, 5, 10; table

Gen. et sp. indet. E Williams & Siveter, 1998 (type locality and horizon: Upper Cambrian, Merevale, Warwickshire, England)
1972　Elongate *Cyclotron*? – Taylor & Rushton, 25, pl. 2 (borehole record).
1998　Gen. et sp. indet. E – Williams & Siveter, 35, pl. 5, fig. 12.

Gen. et. sp. indet. F Williams & Siveter, 1998 (type locality and horizon: Lower Cambrian, Comley Quarry, Shropshire, England)
1998　Gen. et sp. indet. F – Williams & Siveter, 35, pl. 6, figs. 9, 10.

"Muschelkrebs mit vollständig erhaltenen Gliedmaßen" Müller 1981 (locality and horizon: Upper Cambrian, Zone 5, Sweden)
1981b　"Muschelkrebs mit vollständig erhaltenen Gliedmaßen" – Müller, p. 8 (text-fig.).

Phosphatocopida sp. Siveter, Williams & Waloszek, 2001

(type locality and horizon: Lower Cambrian, Comley, Shropshire, England)

2001   Phosphatocopida sp. – Siveter *et al.*, p. 479, figs. 1, 2.

Phosphatocopina n. gen. n. sp. Müller, 1973 (type locality and horizon: Upper Cambrian, Derenjal Mountains, Iran)

1973   Phosphatocopina n. gen. n. sp. – Müller, p. 50, pl. 4, fig. 9a, b.

Phosphatocopines from the Middle Cambrian of Siberia Müller, Walossek & Zakharov, 1995 (type locality and horizon: Middle Cambrian, Lena mouth area, Siberia, Russia)

1995   Phosphatocopines from the Middle Cambrian of Siberia – Müller *et al.*, fig. 4A–E.

2001   Undetected Phosphatocopina – Chen *et al.*, fig. 4 (referred to Müller *et al.* 1995).

Species C Hinz, 1987 (locality and horizon: Lower Cambrian, Comley, England)

1987   Species C – Hinz, p. 59, pl. 3, figs. 6, 9; table 1.

Species D Hinz, 1987 (locality and horizon: Lower Cambrian, Comley, England)

1987   Species D – Hinz, p. 60, pl. 3, figs. 11, 12, 14; table 1.

Species F Hinz, 1987 (locality and horizon: Lower Cambrian, Dairy Hill, England)

1987   Species F – Hinz, p. 60, pl. 3, figs. 4, 7; table 1.